STUDIES IN
NUMERICAL ANALYSIS

STUDIES IN NUMERICAL ANALYSIS

Papers in Honour of Cornelius Lanczos

Edited by

B. K. P. SCAIFE
School of Engineering,
Trinity College,
Dublin, Ireland.

Published for

THE ROYAL IRISH ACADEMY
by ACADEMIC PRESS-LONDON AND
NEW YORK

ACADEMIC PRESS INC. (LONDON) LTD
24–28 Oval Road
London NW1 7DX

U.S. Edition published by
ACADEMIC PRESS INC.
111 Fifth Avenue,
New York, New York 10003

Library of Congress Catalog Card Number: 72–12280

ISBN: 0-12 621150–7

Film set by the European Printing Corporation
Limited, Dublin and printed
by Unwin Brothers Limited
The Gresham Press Old Woking Surrey
A member of the Staples Printing Group

Contributors

ACTON, FORMAN S. *School of Engineering, Princeton University, 08540, U.S.A.*

BIRELINE, JANICE, *231 Bowery, New York, N.Y. 10002, U.S.A.*

BLANCH, GERTRUDE *3625 1st Avenue, San Diego, California 92103, U.S.A.*

DAVIS, WILLIAM R. *Department of Physics, North Carolina State University, Raleigh, North Carolina 27607, U.S.A.*

GALLAGHER, A. P. *Department of Engineering Mathematics, The Queen's University of Belfast, Belfast, BT9 5AH Northern Ireland.*

CLENSHAW, C. W. *Department of Mathematics, Cartmel College, Bailrigg, Lancaster, England.*

HAMMING, R. W. *Bell Telephone Laboratories, 600 Mountain Avenue, Murray Hill, New Jersey 07974, U.S.A.*

HOUSEHOLDER, ALTON S. *Department of Mathematics, Ayres Hall, The University of Tennessee, Knoxville 37916, U.S.A.*

JÁNOSSY, L. *Magyar Tudományos Akadémia, Központi Fizikai Kutató Intézet, Budapest XII, Konholy Thege Ut, Hungary.*

LORD, K. *Department of Mathematics, Cartmel College, Baillrigg, Lancaster, England.*

MAGNUS, WILHELM *Courant Institute for Mathematical Sciences, 251 Mercer Street, New York 10012, U.S.A.*

MASKELL, S. J. *Department of Mathematics, University of Exeter, Exeter, Devon, England.*

METROPOLIS, N. *Los Alamos Science Laboratory C-DO, Los Alamos, New Mexico 87554, U.S.A.*

MILLER, JOHN J. H. *School of Mathematics, Trinity College, Dublin 2, Ireland.*

McMahon, J. J. *St. Patrick's College, Maynooth, Co. Kildare, Ireland.*

Ortiz, Eduardo L. *Department of Mathematics, Imperial College of Science and Technology, Exhibition Road, London SW7 2RH, England.*

Peters, G. *Division of Numerical Analysis, National Physical Laboratory, Teddington, Middlesex TW11 OLW, England.*

Quinlan, P. M. *Department of Mathematical Physics, University College, Cork, Ireland.*

Rhodes, Ida *National Bureau of Standards, Washington, D.C. 20234, U.S.A.*

Rota, Gian-Carlo *Los Alamos Science Laboratory C-DO, Los Alamos, New Mexico 87554, U.S.A.*

Sack R. A. *University of Salford, Salford, England.*

Schoenberg, I. J. *1111 Oak Way, Madison, Wisconsin 53705, U.S.A.*

Synge, J. L. *School of Theoretical Physics, Dublin Institute for Advanced Studies, 10 Burlington Road, Dublin 4, Ireland.*

Wilkinson, J. H. *Division of Numerical Analysis, National Physical Laboratory, Teddington, Middlesex TW11 OLW, England.*

Editor's Foreword

In editing these papers in honour of Cornelius Lanczos on his 80th birthday I have had invaluable assistance from Miss E. Wills and I express to her my warm thanks.

Without the willing help of Prof. J. R. McConnell, Dublin Institute for Advanced Studies, Dr. W. G. S. Scaife, Trinity College, Dublin, Academic Press Inc. (London) Limited, and of Mr. E. W. Webb, European Printing Corporation Limited, publication of this Festschrift would have been impossible.

Financial assistance provided by the Governing Body of the School of Theoretical Physics of the Dublin Institute for Advanced Studies is gratefully acknowledged.

The valuable help in proof reading and in the preparation of the subject index from Mr. W. T. Coffey is greatly appreciated.

B. K. P. Scaife.

Dublin
2 February, 1974.

Cornelius Lanczos — A Biographical Note

Cornelius Lanczos was born in Szekesfehervar, Hungary, on February 2nd, 1893, the son of Carolus Loewy and Adele Hahn. The name Lanczos is a Hungarianization of Loewy.

In 1927 he married Maria Rump and their son, Elmar, was born in 1933. His wife died in 1939 and in 1955 he married Ilse Hildebrand.

From 1911 until his graduation in 1916 Cornelius Lanczos attended the University and the Polytechnicum in Budapest where he had as teachers Fejér and Eötvös. He then studied under Ortvay, a student of Sommerfeld, at the University of Szeged. He obtained his Ph.D. degree for a thesis entitled "Function Theoretical Relations of Maxwell's Vacuum Equations". The next three years were spent at the University of Freiburg i.B., Germany. In 1924 he moved to Frankfurt where he became an assistant to E. Madelung. In 1926 he published a paper which put forward an interpretation of quantum mechanics on a continuum basis in terms of integral equations. This preceded the publication by Schrödinger of his partial differential equation. During his stay in Frankfurt Lanczos came in contact with a group of very able mathematicians such as Dehn, Hellinger, Siegel and Szasz. During the academic year 1928–29, at the personal invitation of Einstein, Lanczos went to Berlin and collaborated with him. It was during this year that he met such people as J. Franck, Kallmann, Laue, Nernst, Neumann, Pringsheim, Rubens, Schrödinger and Wigner at seminars held in the Physikalische Technische Reichsanstalt organised by Laue and Pringsheim.

1931 saw the publication of his first paper on the quadratic action principle in relativity. This is a subject in which he has maintained an interest ever since. In the same year he accepted an invitation to go to the United States where he was appointed to the Chair of Mathematical

Physics at the University of Purdue, Indiana, a post which he held until 1946. As will be seen from the bibliography at the end of this note, up to the War years all of Lanczos's work was concerned with relativity theory and quantum theory. However, before he left Germany, and afterwards, he took an ever-increasing interest in areas of mathematics which would now come under the heading of numerical analysis. His first publication in this field appeared in 1938 and described what is now known as the Lanczos Tau-method. This work was followed in 1942 by two papers in conjunction with G. C. Danielson on practical techniques of Fourier analysis. These papers anticipated the Fast-Fourier Transform.

During the second world war he spent the year 1943–44 on the staff of the National Bureau of Standards working on the Mathematical Tables Project. He left Purdue in 1946 to take up an appointment as a senior research engineer with the Boeing Airplane Company in Seattle. In 1947 he was appointed Walker-Ames lecturer at the University of Washington, Seattle and gave a series of lectures on "The Fourier Series and its Applications". In 1949 he moved to the National Bureau of Standards Institute for Numerical Analysis at the University of California at Los Angeles where he remained until 1953. He spent the year 1953–54 with North American Aviation as a specialist in computing. In 1951 he published the first exact method for obtaining all the eigenvectors and eigenvalues of an arbitrary matrix. His theorem on the decomposition of an arbitrary matrix was published in 1958 and it was for this work that he was awarded the Chauvenet Prize in 1960 by the Mathematical Association of America.

The precision approximation to the Gamma function appeared in 1964 and the first account of his Sigma factors for smoothing Gibbs' oscillations in Fourier Series was given in his book *Discourse on Fourier Series*, published in 1966.

Shortly after moving to Los Angeles his first text-book was published in 1949 by the University of Toronto Press entitled *Variation Principles of Mechanics*, a book which is now in its 4th edition.

In 1954, at the invitation of Mr. Eamon de Valera, who was at that time Prime Minister of the Republic of Ireland, Cornelius Lanczos accepted the post of Senior Professor in the School of Theoretical Physics of the Dublin Institute for Advanced Studies. Apart from occasional visits abroad he has remained in Dublin since that time. He retired from his post at the Dublin Institute for Advanced Studies in 1968.

His period in Dublin has been one of remarkable productivity in which, apart from the publication of many papers, he has had published six books, namely, *Applied Analysis*, 1956; *Linear Differential Operators*, 1961; *Albert Einstein and the Cosmic World Order*, 1965; *Discourse on*

Fourier Series, 1966; *Numbers Without End*, 1968; and *Space through the Ages*, 1970. His latest book, *Einstein Decade: 1905–1915*, is due to be published shortly.

Lanczos has travelled widely since he came to Dublin and his activities have included a visiting Professorship at Oregon State College, 1957–58, visits to the United States Army Mathematical Research Centre, Wisconsin, 1959–60; the Computer Centre, Chapel Hill, North Carolina, 1960; University of Michigan, 1962; Science Laboratory, Ford Motor Company, 1964–65; North Carolina State University, 1965 and 1968. He is a Fellow of the American Association for the Advancement of Science, of the American Physical Society, he is a member of the American Mathematical Society, of the Mathematical Association of America and of the Society for Industrial and Applied Mathematics. He is a member of Sigma Xi and an honorary member of Sigma Pi Sigma.

With such a distinguished career it is not surprising that he has received many honours and these include Membership of the Royal Irish Academy, 1957, the award of the Sc.D. (h.c.) by Trinity College, Dublin in 1962, the degree of D.Sc. (h.c.) by the National University of Ireland in 1970, an honorary degree of Dr.Nat.Phil. from the Johann Wolfgang Goethe University, Frankfurt, in 1972 and the degree of D.Sc. (h.c.) from the University of Lancaster in 1972.

List of publications by Cornelius Lanczos

(Publications marked thus * appeared in German; the titles are here translated into English. † indicates the more important items).

*Simplifying coordinate system for the Einsteinian equations of gravity. *Phys. Z.* **23** (1922), 537–9.

*Remark on the de Sitter universe. *Phys. Z.* **23** (1922), 539–43.

†*On the theory of the Einsteinian equations of gravity. *Z. Phys.* **13** (1923), 7–16.

*On the rotation problem of general relativity. *Z. Phys.* **14** (1923), 204–19.

*On the red-shift in the de Sitter universe. *Z. Phys.* **17** (1923), 168–89.

*Remark on the theory of perihelion-precession of Mercury. *Naturwissenschaften* **11** (1923), 910–11.

*On a stationary cosmology based on Einstein's equations of gravity. *Z. Phys.* **21** (1924), 73–110.

*Surface distribution of matter in Einstein's theory of gravity. *Annln Phys.* **74** (1924), 518–40.

*On the problem of infinitely weak fields in Einstein's theory of gravitation. *Z. Phys.* **31** (1925), 112–32.

*On a universe periodic in time and a new approach to the problem of radiation. *Z. Phys.* **32** (1925), 56–80.

*On the problem of radiation in a universe closed in space. *Z. Phys.* **32** (1925), 135–49.

*On the action principle of general relativity. *Z. Phys.* **32** (1925), 163–72.

*On tensorial integral equations. *Math Annln.* **95** (1925), 143–53.

*On the application of a variational principle in general relativity. *Acta Litt. Acad. Scient. R. Univ. hung. Francisco-Josephina.* **2** (1925), 182–92.

*On a field theoretical representation of the new quantum mechanics. *Z. Phys.* **35** (1926), 812–30.

*Variational principle and quantum condition in the new quantum mechanics. *Z. Phys.* **36** (1926), 401–9.

*On the complex nature of the matrices of quantum mechanics. *Z. Phys.* **37** (1926), 405–13.

*On the dynamics of general relativity. *Z. Phys.* **44** (1927), 773–92.

*On the law of motion in general relativity. *Phys. Z.* **28** (1927), 723–6.

†*The tensor-analytical relations of Dirac's equation. *Z. Phys.* **57** (1929) 447–73.

*Covariant formulation of Dirac's equation. *Z. Phys.* **57** (1929), 474–83.

*The conservation laws in the field theoretical representation of Dirac's equation. *Z. Phys.* **7** (1929), 484–93.

*Invariant characterization of the conservation laws of general relativity. *Math. naturw. Ber. Ung.* **46** (1929), 554–75.

†*Invariant formulation of the conservation laws of general relativity *Z. Phys.* **59** (1930), 514–39.

*Dirac's wave-mechanical theory of the electron and its field theoretical interpretation. *Phys. Z.* **31** (1930), 120–30.

*On the anomalous intensities of the Stark effect lines in very strong fields. *Naturwissenschaften* **18** (1930), 329–30.

†*On the theory of the Stark effect in strong fields. *Z. Phys.* **62** (1930), 518–44.

†*Displacement of the hydrogen terms in strong electric fields. *Z. Phys.* **65** (1930), 431–55.

†*On the weakening of spectral line-intensities in strong electric fields. *Z. Phys.* **68** (1931), 204–32.

*Einstein's new field theory. *Ergebn. d. exakt. Naturw.* **10** (1931), 97–132.

*Electromagnetism as a natural property of Riemannian geometry. *Z. Phys.* **73** (1931), 147–68.

Electricity as a natural element of Riemannian geometry. *Phys. Rev.* **39** (1932), 188.

†*Electricity as a natural property of Riemannian geometry. *Phys. Rev.* **39** (1932), 716–36.

*The theory of relativity in comparison with other theories of physics. *Naturwissenschaften* **20** (1932), 113–6.

*On the appearance of the vector potential in Riemannian geometry. *Z. Phys.* **75** (1932), 63–77.

*On the problem of the regular solutions of Einstein's equations of gravity. *Annln Phys.* **13** (1932), 621–35.

*Wave mechanics as a Hamiltonian dynamics of the function-space. A new derivation of Dirac's equation. *Z. Phys.* **81** (1933), 703–32.

*On the Hamiltonian dynamics of the function-space. *Z. Phys.* **85** (1933), 107–27.

A fundamental connection between Hamiltonian dynamics and wave mechanics. *Phys. Rev.* **44** (1933), 318.

†*A new transformation theory of linear canonical equations. *Annln Phys.* **20** (1934), 653–88.

A new transformation theory of linear canonical equations. *Phys. Rev.* **45** (1934), 560.

*A new formulation of world-geometry. *Z. Phys.* **96** (1935), 76–106.

†Trigonometric interpolation of empirical and analytical functions. *J. Math. Phys.* **17** (1938), 123–99.

†A remarkable property of the Riemann-Christoffel Tensor in four dimensions, *Ann. Math.* **39** (1938), 842–850.

Total mass of a particle in general relativity. *Phys. Rev.* **59** (1941), 708–16.

Dynamics of a particle in general relativity. *Phys. Rev.* **59** (1941), 813–9.

†Some improvements in practical Fourier analysis and their application to X-ray scattering from liquids. (with G. C. Danielson). *J. Franklin Inst.* **233** (1942), 365–80 and 432–52.

Matter waves and electricity. *Phys. Rev.* **61** (1942), 713–20.

The variational principles of mechanics. Mathematical Expositions, No. 4, Toronto: University of Toronto Press, 1949.

Lagrangian multiplier and Riemannian spaces. *Rev. mod. Phys.* **21** (1949), 497–502.

†An iteration method for the solution of the eigenvalue problem of linear differential and integral operators. *J. Res. nat. Bur. Stand.* **45** (1950), 255–81.

The separation of close eigenvalues of a real symmetric matrix. (with Rosser, Hestenas and Karush). *J. Res. nat. Bur. Stand.* **47** (1951), 291–7.

Solution systems of linear equations by minimized iterations. *J. Res. nat. Bur. Stand.* **49** (1952), 33–53.

†Analytical and practical curve fitting of equidistant data. *Nat. Bur. Stand. Reports No. 1591* (1952), 1–102.

Chebyshev polynomials in the solution of large-scale linear equations. Toronto Symposium on Computing Techniques (1952), 124–33.

†"Introduction" to the "Tables of Chebyshev polynomials $S_n(x)$ and $C_n(x)$". *Nat. Bur. Stand., Appl. Math. Ser. No. 9* (1952).

Albert Einstein and the theory of relativity. *Nuovo Cim., Suppl.*, **2** (1955), 193–220.

Science and Society. *Icarus* (Trinity College, Dublin) **5** (1955), No. 16, 31–7.

Spectroscopic eigenvalue analysis. *J. Wash. Acad. Sci.* **45** (1955), 315–23.

Applied analysis. Englewood Cliffs, N. J.: Prentice-Hall, (1956).

Electricity and general relativity. *Rev. mod. Phys.* **29** (1957), 337–50.

†Iterative solution of large-scale linear systems. *SIAM J.* **6** (1958), 91–109.

Electricité relativité générale. *Cah. Phys.* **95** (1958), 247–55.

†Linear systems in self-adjoint form. *Am. math. Mon.* **65** (1948), 665–79.

Tensor analysis. Handbook of Physics, New York: McGraw-Hill, 1958. Part I, ch. 10, 111–22.

Albert Einstein and the role of theory in contemporary physics. *Am. Scien.* **47** (1959), 41–59.

Extended boundary value problems. *Proc. Internat. Congr. Math., Edinburgh 1958*, 154–181. Cambridge Univ. Press, 1960.

Solution of ordinary differential equations by trigonometric interpolation. *Sympos. numer. treatment ord. diff. eqns, integ. & integro-diff. eqns, Rome*, 1960, 22–32. Basle: Birkhauser, 1960.

Linear differential operators. London: van Nostrand, 1961.

The splitting of the Riemann tensor. *Rev. mod. Phys.* **34** (1962), 379–89.

Some properties of the Riemann–Christoffel curvature tensor. Recent Developments in General Relativity (Volume dedicated to Prof. Infeld), pp. 313–21. Warsaw: Pergamon/PWN, 1962.

An integral approach to the calculus of variations. Studies in math. anal. & related topics, ed. Gilbarg *et al.*, 191–8. Stanford Univ. Press, 1962.

Undulatory Riemannian spaces. *J. math. Phys.* **4** (1963), 951–59.

Intégration globale. *Ann. Fac. Sci. Univ. Clermont, Actes Coll. Math Tricent. mort B. Pascal*, II, 97–107, 1962.

Modern physics in perspective. *Studies* **52** (1963), 283–93.

Signal propagation in a positive definite Riemannian space. *Phys. Rev.* **134B** (1964), 476–80.

†Evaluation of noisy data. *J. SIAM Numer. Anal. B* **1** (1964), 76–85.

†A precision approximation of the Gamma function. *SIAM J. Numer. Anal. B* **1** (1964), 86–96.

Méthodes locales et globales pour l'intégration des problèmes de trajectoires. *Coll. anal. num, Mons, 1961*. 37–49. Centre Belge Rech. Math. (Louvain), 1961.

The inspired guess in the history of physics. *Studies* **53** (1964), 398–412.

Albert Einstein and the cosmic world order. New York: Wiley, 1965.

Discourse on Fourier series. Edinburgh: Oliver & Boyd, 1966.

Metrical lattice and the problem of electricity. *J. math. Phys.* **7** (1966), 316–324.

†Boundary value problems and orthogonal expansions. *SIAM J. Appl. Math.* **14** (1966), 831–63.

William Rowan Hamilton, an appreciation. *Am. Scient.* **55** (1967), 129–43.

Einstein equations and electromagnetism. *J. math. Phys.* **8** (1967), 829–36.

Rationalism and the physical world. *Proc. Boston Coll. Phil. Sci. 1964/66*, Vol. 3, ed. R. S. Cohen & M. W. Wartofsky (In memory of N. R. Hanson), 181–198. Dordrecht-Holland: Reidel, 1967.

Why mathematics? *Irish Math. Teachers Assoc. Newslett.* **9** (1967), 7–11.

Entstehung, Entwicklung und Perspektiven der Einsteinschen Gravitations-theorie. Einstein-Symposium, Deutsch Akad. Wiss. Berlin, 1965. Akad-Verlag, 1966, 38–56.

Numbers without end. Edinburgh: Oliver & Boyd, 1968. (Contemporary Science Paperbacks).

Boundary value problems and the algebraic method. Coll. Inter. Centre Nat. Rech. Sci., No. 165: Programmation en math. num., Besancon, 1966, 205–15. Eds. Centre Nat. Rech. Sci., Paris, 1968.

Variational principles. Scott, Univs. Summer School 1967: Math. methods in solid state and superfluid theory, pp. 1–45. Eds. Clark & Derrick, Edinburgh: Oliver & Boyd, 1969.

†Quadratic action principle of relativity. *J. math. Phys.* **10** (1969), 1057–65.

Judaism and science. 11th Selig Brodetsky Mem. Lect. Leeds Univ. Press, 1970.

Space through the ages. London: Academic Press, 1970.

L'Univers des Nombres, a la Decouverte des Mathematiques. (Translation of "Numbers without end") Paris: Dunod, (1971).

Einstein's path from special to general relativity. General relativity: Papers in honour of J. L. Synge. Ed. L. O'Raifeartaigh. Oxford: Clarendon Press, 1972, 1–19.

Vector potential and quadratic action. *Foundations in Physics*, **2** (1972) 271–285.

Poisson Bracket. Aspects of Quantum theory: Papers in honour of P. A. M. Dirac. Ed. A. Salam and E. Wigner. Cambridge: University Press, 1972, 169–178.

Emmy Noether and Calculus of variations. *Bull. Inst. Math. and Appl.* **9** (1973), 253–8.

Legendre versus Chebyshev polynomials. Topics in Numerical Analysis. Ed. J. J. H. Miller. London: Academic Press, 1974, 191–201.

Vector Potential and Riemannian Space. *Found. Phys.* **4** (1974), 137–147.

Einstein Decade: 1905–1915. London: Paul Elek (To be published).

Contents

Table-making at the National Bureau of Standards

GERTRUDE BLANCH
San Diego, California, U.S.A.

AND

IDA RHODES
National Bureau of Standards, Washington, D.C., U.S.A.

In the pre-electronic era, the production of mathematical tables was a much more difficult task than it is today, when a high-speed computer can spew out a complete table within minutes—a task that would have required many man-years of labor as late as 1938. Moreover, the need for tables, as working tools, is far less important nowadays than it was three decades ago, since the computer can generate solutions to many problems directly, without resort to tables. However, it is perhaps worth stating that even nowadays, some tables of functions that have not yet been sufficiently explored are comforting to have; if only for key values in checking computer codes.

The numerical evaluation of a mathematical function, whether for a table or for other uses, often presents challenging problems in numerical analysis. Such tasks have interested mathematicians of past generations and ours—from Euler and Gauss to von Neumann and Lanczos. Because Professor Lanczos has contributed so abundantly in many areas, it is perhaps not generally known that he had a hand in table-making, too. It is therefore fitting that we devote this paper to sketching the history (in some ways, a romantic one) of an organization in the United States that started with table-making.

The Mathematical Tables Project (MTP) was organized under the auspices of the National Bureau of Standards and supported with funds of the Depression's Works Projects Administration. The aim was to utilize unemployed high school graduates for computing tables of the higher mathematical functions. To accomplish this, Dr. Lyman J. Briggs—then Director of the National Bureau of Standards—appointed Dr. Arnold N. Lowan to head the project. Several members of the learned societies helped in the initial organization and planning, among them the

1

National Research Council and the U.S. Geological Survey. The first meeting, which invited suggestions for a work-plan, included the following:

A. A. Bennett, Brown University
R. C. Archibald, Brown University
H. T. Davis, Northwestern University
D. H. Lehmer, then at Lehigh University
C. E. van Orstrand, U.S. Geological Survey
E. C. Crittendon, National Bureau of Standards
L. B. Tuckerman, National Bureau of Standards.

Wisely, the group selected for the first task the computation of the exponential function, to be tabulated to fifteen decimal places.

In 1938, the MTP opened in New York, with a staff of some 120 raw recruits and seven mathematicians to supervise the work. Over ninety per cent of the workers were supplied with no more sophisticated implements than pencil and paper. Among the group there were about twenty better qualified; these were supplied, in addition, with primitive desk calculators, for the all-important task of generating key values of the function to be tabulated. To add to the initial difficulties, it soon became apparent that the vast majority of the workers had lost whatever acquaintance they once had with the law of signs. Thus, one of the primary tasks was that of education. In addition, the mathematicians had to design and test the work sheets to be used and to decide on the formulas that would be most suitable. The work sheets had to incorporate enough iron-clad checks along the way, so that an earnest worker could find his own mistakes and end up with a correct result.

Two of the initial technical staff, F. G. King and W. Kaufman, left during the first year for more lucrative positions. They contributed considerably to the first two tables published by the project. Dr. Milton Abramowitz, the first recruit of MTP, remained, and was destined for increasing responsibilities until his untimely death in 1958. Within two years there were added to the staff a group of younger mathematicians, most of whom remained for many years, and then contributed notably in other institutions. Among them were Drs. Jack Laderman, Herbert E. Salzer, Abraham Hillman, Eugene Isaacson, Meyer Karlin; Mr. W. Horenstein and the Misses Irene Stegun and Ruth Zucker. The authors of the present chronicle were also in this group.

What talents did the technical staff bring initially? We had training in a variety of mathematical fields, but not in numerical analysis. It did not take us long to discover the lacunae in our knowledge of this field. With the enthusiasm of youth for socially desirable accomplishments—the more so because of the mathematics involved—we concentrated on the acquisition of the required skills. In this task, we had the most generous cooperation of that remarkable group of table-makers, the British Association for the Advancement of Science. Among the men who gave freely of their time were Bickley, Comrie, J. C. P. Miller, Milne-Thomson, and Sadler, to name a few. We aspired to publish tables as accurate mathematically as those of the British school; we knew that we could not equal the aesthetic appearance of their tables. Some men in our own country also gave generous help and encouragement—among them Professors D. H. Lehmer, R. C. Archibald, and P. M. Morse.

As we look back upon the early history of MTP, it is perhaps worth realizing that the production of scores of error-free volumes was not the only triumph achieved by its members. There is the human element; at the risk of growing a bit sentimental, some of these facts will be recalled. The majority of the working staff had come to the MTP from the relief rolls. A considerable number were physically handicapped—some with polio disabilities, epilepsy, and arrested tuberculosis. The beneficial effects of a steady job and (we like to believe) the sympathetic appreciation of the supervisory personnel, gradually became evident. There was a heartwarming *camaraderie* among our "tablémakers" (the accent along with the designation is due to our own witty Dr. Salzer). In the ensuing three decades, the humble MTP was to be transformed into the opulent Computation Laboratory with its million-dollar machines; but this first genuine *esprit de corps*, the just pride in achievement, which animated our $67-a month amateurs in computing, has never been equalled. The method of relying on the earnestness of the computing staff, as well as the conviction of the young men and women that their work was of scientific importance, were prime factors in generating enthusiasm and cooperation for the tasks in hand. By 1939, the Table of the Exponential Function was in the hands of the printer. Then came a succession of volumes which required considerably more mathematical sophistication.

Recognition of the worth of our efforts did not come easily. To the lay public of that period, the imprint "Works Projects Administration" suggested mediocrity and "boondoggling", and the scientific community remained skeptical for what seemed too long. When our country entered World War II in 1941, it was clear that radical changes were in order. The high school graduates could now find more lucrative employment, and the staff dwindled. It seemed that the work of the MTP would have to terminate, and the somewhat ambitious plans for new mathematical tables would have to be dropped. Unexpectedly, important support came from recognized authorities not only here but also abroad, to keep the MTP in operation. The National Defense Research Council took over the support of our Project. The greater part of our efforts went into special computations that were called for in the defense efforts; but some time was still devoted to mathematical research relating to tables.

By 1942 our group consisted of some sixty individuals. The National Bureau of Standards now absorbed on its own payroll the small group of mathematicians, who by now had considerable experience in numerical analysis, and in addition, a few very able computer specialists. The era of electronic computing was not yet at hand, but we now acquired tabulators, relay multipliers, adding machines of a special type, and the best desk calculators that were available.

The year 1943 was one of special significance for the MTP. The successive tan-covered volumes of tables that came from the printer now numbered about a dozen. Many gratifying letters of appreciation came from eminent authorities. Among those was one from Cornelius Lanczos, then at Purdue University, praising the scope of our tables. A few months later, we were pleasantly surprised to learn that Professor Lanczos consented to join our staff for a year's stay. We were then located in an older building in lower Manhattan. Our offices were on the 19th floor, with a magnificent view of the Hudson and East rivers. We hoped this might offer some aesthetic compensation to Professor Lanczos, in lieu of a private office, appropriate to his professional eminence.

The year of his stay was indeed a stimulating one, in ways which far exceeded our expectations. His lectures attracted a wide audience, not only from the Project, but from mathematicians at the local universities. His listeners were especially impressed by the elegance of his expositions, and the clarity and humor which rendered advanced concepts meaningful not only to the experts, but even to some who did not have a college education. All of us were charmed by his complete lack of awareness of his own importance, and his genuine appreciation of all workers, regardless of their rank. The computations for the Tables of Chebyshev Polynomials had just been completed, and Professor Lanczos was requested to write the Introduction for it. Now, a generation later, the continued popularity of this volume can perhaps be ascribed more to his scholarly Introduction than to the tables proper.

If one were to seek reasons why the Mathematical Tables Project was able to accomplish as much as it did, the writers of this record would ascribe the achievements to the "non-competitiveness" of the technical staff. No one tried to take credit for what some one else had done, and no one edged his way to administrative prominence at the expense of a fellow worker. A breakthrough by one mathematician received the enthusiastic applause of his peers, and there was no hint of jealousy. Such a spirit is rare, even in scientific organizations. Perhaps we all sensed the greatness of spirit in Professor Lanczos. He not only lent importance to the organization through his presence, but he rendered substantial help in computing matters. The writers recall one particularly troublesome initial-value problem which the computers were struggling with—unsuccessfully. A suggestion of Professor Lanczos to combine computing with some appropriate mathematical approximation in the difficult region solved the problem! When he left, the staff presented him with a scroll, as a token of its affection for him. A material gift, we felt, was not fitting.

The basic tasks of the Mathematical Tables Project underwent a radical change in 1948, when the organization was transferred physically to its main office at the National Bureau of Standards in Washington, D.C. Dr. A. N. Lowan resigned his position, to continue his connections with Yeshiva University as professor of physics. The Director of the National Bureau of Standards was now Dr. E. W. Condon and the chief of the applied mathematics division in charge of our group was Dr. J. H. Curtiss, who envisaged a different role for the computing branch, in line with the explosive developments in computer technology. The Eniac Computer, which only three years before had startled the mathematical community, was now obsolete. Engineers and mathematicians were building new machines, and the National Bureau of Standards itself was engaged in completing one and planning another. At that time, one of the present chroniclers was devoting her main efforts in developing the software for the Seac Computer, while the other went to the Los Angeles branch, operating out of the University of California, where the Institute for Numerical Analysis was started. Professor Lanczos joined the staff at Los Angeles in 1949. The computing laboratories, both in Washington and California, were now essentially service organizations, whose aim was to help with many types of problem in which computing was involved, and to prepare computations for high-speed computer handling. A few tables still in process were to be completed, but no new ones were undertaken.

The many important papers which Professor Lanczos wrote in this period on the solutions of large systems of linear equations, are now a matter of history. In addition, many of the other papers on the same subject, which were contributed by a distinguished group of visiting mathematicians, can be said to have been directly stimulated by the work of Professor Lanczos. His room (he now had a private office!) was the center for those who wanted mathematical advice on diverse subjects, and he gave of his time generously. In addition, his help was solicited in a variety of day-to-day problems which needed special attention. More than any one else, he could be relied upon to find a successful attack on what at first sight seemed an impossible task.

By 1952, the new Institute for Numerical Analysis in Los Angeles had earned a measure of just renown, for the stimulus it provided to mathematics as a whole, as the pleasant gathering place for mathematicians from various parts of the world. Yet this boon to international science was not destined to last long. The political upheavals of the period which fostered suspicions and hate began to take their toll, even in the quiet corners of non-political scientists. Dr. Curtiss resigned, and the Institute was disbanded. Dr. E. W. Cannon took charge of a much reduced staff in Washington. Professor Lanczos left for Ireland, to join the Dublin Institute for Advanced Studies; and Paul Erdos, who was a visitor at the time, embarked upon his frequent trips to distant countries.

Many times, after sanity had returned again to the political scene, our loss to Ireland of Professor Lanczos has been cited as one of the sad consequences of political barbarism. Yet in some ways this is an unreal assessment. A mathematician of the stature of Professor Lanczos belongs to no one nation—the whole world benefits from his contributions. A number of private organizations in this country have succeeded in enticing him for short periods of stay, to enrich our mathematics and our culture. His visits are immensely appreciated, even by the younger "abstract" generation of mathematicians. May he continue his important work for many, many years to come.

Vignette of a Cultural Episode

WILHELM MAGNUS

*Courant Institute of Mathematical Sciences,
New York University, New York, U.S.A.*

In his notes and comments on Persian and Arabic literature, Goethe describes the time of the Barmecides, a noble family whose members administered the empire of the caliphs before and during the time of Harun al-Rashid. Goethe says:

"Proverbially it was a time when, in a particular locality, all human endeavors interacted in such a fortunate way that the recurrence of a similar period could be expected only after many years and in very different places under exceptionally favorable circumstances".

It is the purpose of this article to describe such a period. It nearly coincides with the years 1922–1931 when Cornelius Lanczos lived and taught in Frankfurt am Main in Germany. It is not the whole of the city and not even the whole of the university about which I wish to report, but only a segment of the academic life which I know from first hand experience as a student of mathematics and physics.

What I shall have to say must not be mistaken for the sentimental reminiscences of an old man. I am still in contact with many of the surviving members of the small group of students and professors who met at that time, and if any two of these people meet again after decades, they recall their common experience as something of lasting value and greet each other as old friends. An additional proof of the exceptional nature of this Barmecidian period is the essay by Carl Ludwig Siegel on the history of the Mathematical Seminar in Frankfurt which was reprinted in the third volume of his collected papers. I shall try to describe the same period, but from the point of view of a student rather than that of a professor.

The University of Frankfurt was founded in 1914 with donations from wealthy members of the community. It was the only private university

7

which ever existed in Germany and, as far as I know, the only one in continental Europe. The German inflation after the First World War put an end to the private character of the University. However, the creation of a private institution of higher learning was only the crowning (and, unfortunately, the last) event in a long sequence of remarkable manifestations of the public spirit in the city of Frankfurt. Bridges and hospitals, museums and libraries were among the earlier contributions from private citizens. Of these, the Senckenberg Museum of Natural History and the Rothschild Library may deserve special mention. Although Frankfurt could not compete with Berlin in attracting celebrities, its theatre, opera, concerts and art galleries were of a high quality and contributed to a stimulating cultural atmosphere. So did the many remnants of a great past. The city is first mentioned in 793 by Einhard, the biographer of Charlemagne. Apart from brief interruptions, it was an independent free city until 1866, when it was incorporated into Prussia. Since 1152, the Holy Roman Emperors were elected there. The spirit of moderation which is frequently a characteristic of trade centers seems to have been well developed. There never was a pogrom in the free city. During the Fettmilch riots in the early seventeenth century the city council tried to protect the Jewish community whose members were eventually compensated for their losses. That the first German Parliament convened in Frankfurt after the revolutionary upheavals of 1848 was an acknowledgement of the honorable tradition of the city which had upheld a municipal constitution since the early thirteenth century.

It is hard to say how much the past had contributed to the cultural atmosphere of the city in the nineteen-twenties. But a very simple physical characteristic certainly was of importance. The city was not very large in population (about half a million) and in size. It had a definite center, within the boundaries of the old city, it had a beautiful (and, at that time, only moderately polluted) river and it was easy to get out of it, into forests and mountains which even on a hot summer Sunday offered lonely places to those who knew where to go. At the same time the city was large enough to protect privacy from the intrusive type of gossip which is generated by boredom and to allow the development of special relations which are based on choice rather than on mere proximity.

The university was young, without traditions and the fame based on past achievements. In a surprising and very successful manner it turned this situation to its advantage. The social sciences which were considered as upstarts in many older universities were strongly represented in Frankfurt from the very beginning. The Faculty of Natural Sciences (which included the mathematicians) could not have attracted established celebrities. It managed to appoint a surprising number of future celebrities, in particular the physicists Otto Stern and Max Born who later were awarded the Nobel Prize. The university was not able to hold them for very long and they had already left when I went there in 1926. The mathematicians were more successful. In 1922, the faculty appointed Carl Ludwig Siegel as successor of the highly distinguished mathematician Arthur Schoenflies to the position of a full professor ("Ordinarius"). Siegel was then twenty six, he stayed in Frankfurt for fourteen years. It would be unbecoming for a former student to eulogize his teacher publicly in general terms; for any reader who knows any of the fields in which Siegel has worked it is also completely superfluous. However, it should be pointed out that it was a very unusual procedure to fill a position of highest rank with a scholar so young, and it speaks highly for the good judgement of the mathematicians who were responsible for this action that they resisted the objections of more conservative faculty members which undoubtedly must have been raised against such an untraditional step. The names of these mathematicians were Max Dehn and Ernst Hellinger. Together with the mathematicians Paul Epstein and Otto Szasz, they formed a group of close friends which included also Cornelius Lanczos whose field of research and teaching at that time was exclusively theoretical physics.

The first encounter of a student with the professors of a university takes place in the classroom, and I was incredibly fortunate in meeting teachers whose styles were very much their own, each one being excellent in a particular way. I never took a course with Epstein. His function— which he fulfilled admirably — was to introduce those students into basic parts of mathematics for whom this field was not of primary interest but who had to acquire some mathematical knowledge for an understanding of their major field. The fact that mathematics is, to some degree, the handmaiden of many other sciences is frequently forgotten by Mathematicians, but this was not the case in Frankfurt.

My own experiences started with a course by Szasz and another one by Siegel.

Szasz spoke slowly and carefully. What made his course so attractive was his ability to bring out the specific merit of each idea and to show why it worked. Euler had the same ability, and I believe that this is the reason why, among all of the great mathematicians, he appears to be the most lovable one.

Siegel was demanding. He had very high standards, in the first place for himself, but he also expected his students to work thoroughly and hard. His lectures expressed his complete sovereignty over the material. They were incredibly well prepared — Siegel never used a manuscript and even wrote down the greatest prime number known at that time from memory. He proceeded at a fast pace, but he never glossed over details or brushed computations aside with a deprecating manner. He carried them out, briefly but lucidly. On rare occasions he made remarks of a general nature which expressed his attitude towards mathematics. The mathematical universe, he said, is inhabited not only by important species but also by interesting individuals. As an example, he mentioned the elliptic functions. The insight that the particular as well as the general are needed to make the world of mathematics complete is not always remembered by mathematicians.

Instinctively, everyone in the class knew that none of us would ever be as powerful a mathematician as Siegel. Contrary to all the talk from psychologists and educators who warn against oppressing the developing student, this need not be a depressing experience at all. The opposite is true: it is beneficial to know early what high standards really mean. And Siegel was encouraging when he felt that this was justified. And his word then carried weight.

Hellinger was probably the most widely appreciated teacher among the mathematicians. He, too, was very well prepared. His lectures were highly polished, but he never forgot to mention the motivation for a theorem and he always pointed out connections between different parts of mathematics. His presentation was less austere than Siegel's. He liked to make entertaining remarks and to suspend the need for concentration for a moment, giving the audience a brief respite. He was an outstanding psychologist in the best sense: He always knew exactly how far a student could go, and his advice was given in a tactful manner which, at the same time, left no room for doubt.

When I attended a course taught by Lanczos for the first time, I had already changed my original plan to become a physicist, realizing that I was more at home and at ease in mathematics. Perhaps, this was fortunate, because Lanczos might have made me stay in physics if I had met him earlier. To work in theoretical physics requires an uncanny combination of talents: a specific type of intuitive understanding of the realities of physics and a well developed ability to handle the necessary mathematical tools with complete ease. What made Lanczos such a fascinating teacher for a mathematician was his ability to inject some of the intuition of the physicist into mathematics. Even the supreme clarity of Lanczos' lectures would not have sufficed to produce this effect. What one really could learn from him was the over-riding importance of motivation for the development of a theory.

Max Dehn was my Ph.D. adviser and I have been influenced deeply by him. I took courses taught by him only in my last year at the university, and they had a lasting effect on me in spite of the fact that they were not as polished and smooth as those which I had attended before. Dehn communicated ideas. One had to be ready for this. In fact, one had to be able to enter into a dialogue with him. Even if one had only a tiny contribution to make, and even if one expressed it in a confused way, this was enough. Dehn always understood. He had the ability which Socrates claimed for himself: to act as a midwife at the birth of an idea. This ability went far beyond mathematics. Dehn had an extensive knowledge of philosophy and of history, and he used it to gain the proper perspective for any particular fact or occurrence. He was very undogmatic and did not belong to any philosophical school, but he always tried to see the significance of ideas and facts within the general framework of human experience.

I have tried to describe the characteristic qualities of my teachers in some detail because of a current fashion prevailing at least in the United States to set up standards for good teaching. The questionaires issued in many universities to students for the evaluation of their teachers show that this is considered to be an important problem. I believe that it may be possible to define bad teachers. But good teaching has too many aspects to be evaluated on a quantitative scale.

I have not nearly mentioned all of the outstanding scholars and teachers to whom I owe my education. There was K. W. Meissner, a brilliant experimental physicist whose lectures were models of perfection in planning and execution. And there were many other remarkable scientists, some of whom also contributed to a program which, at that time, may have been unique in Germany. It consisted of lectures for high school students which were given on a reasonably high level and attracted a large audience. And there was the Faculty of Philosophy which could take pride in counting many celebrities among its members. Adhemar Gelb and Max Wertheimer, two of the founders of *gestalttheorie*, were also personal friends of the mathematicians. Their lectures have enabled me to see the limitations of a positivist philosophy which many scientists still consider as the only one which is compatible with their professional work.

In his essay on the history of the Mathematical Seminar in Frankfurt, Siegel has described the close cooperation between the professors and the personal relations between faculty and students. There would be no point in repeating his presentation here, but I should like to supplement it with a few remarks. The studies of the history of mathematics which were conducted by four professors and a few students over many years and were based on the original works of mathematicians from Euclid to Newton have had aftereffects which were not visible at that time. One of them is the fact that many years later the university created a chair for the history of science and appointed to it Willy Hartner, then an internationally known authority in this field and a close friend of Dehn, Hellinger, Siegel and Lanczos who had joined in the historical studies for a long time. Another fact is that A. Prag, the most active of the students who participated in these studies, is now collaborating on a definitive edition of Newton's scientific work.

Although all of the better students took their studies very seriously, the absence of any visible outside pressure on their work was a remarkable and beneficial phenomenon. Even more remarkable was the fact that the students did not compete with each other for attention or distinction. We would have thought it indecent if anyone had tried to do so. I believe that this was a reflection of the harmonious relations between the professors which were of course obvious to the students. But it may be worthwhile to note that a high level of achievement is possible without the stimulus of competition.

We never lacked encouragement or advice, and the general atmosphere was one of informality. This did not diminish our respect for the professors whose authority was based on personal and intellectual qualities which were fully recognized by the students.

The intellectual world in the twenties was very much smaller than it is now. The number of scientists engaged in research was a tiny fraction of today's number, and the amount of learning needed to reach the frontiers of knowledge could be acquired in a much shorter time than today. This had two effects: a greater universality was possible and even required. A knowledge of physics and, to a much lesser degree, of psychology or philosophy, was an indispensable part of the education of a mathematician. The other effect was that the time of studying was much shorter. It was possible and not unusual to obtain a Ph.D. in mathematics four years after leaving high school. The legal minimum was three years, and even that occurred on some rare occasions. It was possible to give much leeway to the student who planned his program. There were no bad marks for dropping out of a course. There were no intermediate examinations which forced the pace of progress for the student. And although the teacher-student ratio was at times rather small, the absolute numbers involved were also small, and it was still possible for a professor to know every student who worked in his field.

All of these things contributed to the existence of a truly academic atmosphere, the word being understood in the original sense of Plato's Academy. However, it would have been impossible to plan, or organize, the composition of the mathematical faculty at Frankfurt in the twenties. Harmonious relations cannot be created by bureaucratic coordination, even if the coordinator is a trained psychologist. No objective test—as, for instance, the number of papers published in a given time — can replace the good judgment of the senior faculty members when the question of tenure or promotion arises.

It should be added here that the economic conditions in Germany at that time were not very good for the academic professions and became increasingly worse with the arrival of the great depression. The chances for entering an academic career were infinitesimal for a young mathematician. The only other profession open at that time, that of a high school teacher, had a very limited number of openings. Lanczos was fortunate to be offered a position at Purdue University in the United States of America; thus it was that Frankfurt lost an established and brilliant scholar and teacher.

It is important to note that life in general was not at all carefree in the Germany of the twenties. That those associated with the mathematicians at Frankfurt remember these years as rich and fruitful ones has to be explained with an insight which was found long ago and which has largely been discredited or forgotten in recent times. The common pursuit of intellectual goals, the search for truth and knowledge, create bonds between us which are stronger and more lasting than those based on common origin and nationality.

The time of the Barmecides came to an end when a distrustful caliph destroyed the illustrious family. The end of the episode I have tried to describe in this essay was a tiny part of a much more horrible and much more encompassing catastrophe. Siegel, in his essay quoted above, has described the destruction of mathematical life in Frankfurt wrought by the Nazi government. There remains a question which I cannot answer. The disaster came very suddenly. No dikes had been built against the pestilential flood which swept away nearly everything. How could the dikes have been built and by whom? The life of the sciences, like that of the arts, is easily destroyed. Adversaries much less monstrous than those who arose in Germany can be deadly. But if some of our works are fragile, this does not mean that we can do without them.

BIBLIOGRAPHY

Siegel, C. L., *Zur Geschichte des Frankfurter Mathematischen Seminars. Collected papers*, Vol. 3, No. 81, pp. 462–474, Springer-Verlag, 1966.

The Physicist as Poet

JANICE BIRELINE

Raleigh, North Carolina, U.S.A.

This is a poem, a "poem of the mind in the act of finding what will suffice" — an ode to Einstein and to Lanczos whose travels in this universe were under the guise of physicist, mathematician, and philosopher. Nonetheless, they have never been able to conceal their true roles as metaphysicians and poets. May the multitudes give thanks.

"If we go back to the collection of solid, static objects extended in space, . . . and if we say that the space is blank space, nowhere, without color, and that the objects, though solid, have no shadows and, though static, exert a mournful power, and without elaborating this complete poverty, if suddenly we hear a different . . . description of the place: . . . if we have this experience we know how physicists help people to live their lives. This illustration must serve for all the rest. There is, in fact, a world of physics indistinguishable from the world in which we live, or, I ought to say, no doubt, from the world in which we shall come to live, since what makes the physicist the potent figure that he is, or was, or ought to be, is that he creates the world to which we turn incessantly and without knowing it and that he gives to life the supreme fictions without which we are unable to conceive of it."†

In 1905, Einstein described the world of physics with respect to some judiciously chosen reference frame much in the way that many other activities of the human mind are described. Frames of reference are

---

† All quotations, unless otherwise noted are from Wallace Stevens' essay "Art as Establisher of Value" from *The Modern Tradition* by Ellmann and Feidelson, Oxford University Press, New York, 1965 and from *The Collected Poems of Wallace Stevens*, Alfred A. Knopf, New York, 1967. With apologies to Mr. Stevens, all italicized quotes have been paraphrased such that the words *poet, philosopher, physicist, poetry, philosophy,* and *physics* have been interchanged.

established in literature according to a period such as classical, or a style, such as realistic or romantic. Systems of religion are anchored in Judaism, Pantheism, Christianity. Philosophies find their footing in positivism, Kantianism, idealism — Economic systems are based on socialism, Fascism, capitalism. Since no one of these reference systems can be chosen as *the* correct frame for any of these disciplines, in physics also there cannot be an absolute or preferred frame of reference. On this tenet is founded the celebrated principle of special relativity of Einstein which demands the following: It is possible to formulate the laws of the physical universe in such fashion that they hold in all possible frames of reference which are in uniform relative motion with respect to each other. That is to say that all properly formulated equations of physics have the property of remaining invariant with respect to arbitrary coordinate transformations made within these systems.

This concept, though magnificent, was also troubling to Einstein, for its reformulation in the geometrical terms of Minkowski in 1908 was not in harmony with Newton's mysterious "gravitational force" which demanded an absolute space and an absolute time. Minkowski's geometry was a four-dimensional one in which space and time became interpretable as coordinates of equal significance and, as in the geometry of Euclid, these coordinates described a flat space. Einstein noticed, however, that all forces apparent in reference systems that are different from Newton's absolute system have in common the same property as Newton's "gravitational force" in that they are always strictly proportional to the mass of the moving body. It was at this time that Einstein was introduced to the geometry of Riemann which had been pioneered by Gauss. Einstein immediately recognized that a four-dimensional Riemannian structure for spacetime could be fundamental to physics. This resulted in his startling discovery that the peculiar phenomenon of universal attraction was an inevitable consequence of this geometry. From this investigation he obtained quite naturally the time-honored gravitational equations which give a full account of all the phenomena of gravitation on the basis of pure geometric structure, including that of Newton, as the linear statical approximation of his theory.

"We do not prove the existence of the poem.
It is something seen and known in lesser poems.
It is the huge, high harmony that sounds
A little and a little, suddenly,
By means of a separate sense. It is and it
Is not and, therefore, is. In the instant of speech,
The breath of an accelerando moves,
Captives the being, widens — and was there."

Poincaré, the French mathematician, first enunciated the principle of relativity in 1900, a fact which was unknown to Einstein. But, no matter, for while Poincaré outlined a program which should be followed, he gave no solution. Einstein quite independently came up with a principle of relativity which included a solution. A most striking feature of his theory was the first postulate which expresses the equal admissibility of all reference systems which move with constant velocity relative to each other. More astonishing was the second fundamental postulate which states that light travels in every direction with the same constant velocity in every internal reference system independent of the speed of the source. Einstein's first postulate was quite a natural explanation of the so-called negative result of the Michelson-Morley experiment of 1887. The self-same experiment verified the validity of Einstein's second postulate.

"One poem proves another and the whole,
For the clairvoyant men that need no proof:
The lover, the believer and the poet.
Their words are chosen out of their desire,
The joy of language, when it is themselves.
With these they celebrate the central poem,
The fulfillment of fulfillments, in opulent,
Last terms, the largest, bulging still with more,"

And so at a time in history when mechanistic thinkers believed that the results of the Michelson-Morley experiment spelled the doom of all further investigations of the physical universe, Einstein's insights together with his interpretations of the work of Minkowski, Lorentz, Gauss, Riemann, Ricci, and Levi-Civita brought forth "The fulfillment of fulfillments, in opulent, Last terms, the largest bulging still with more,".

The space time coordinate system of Minkowski and the transformations of Lorentz were helpful to Einstein in reformulating the principle of equal admissibility of Lorentz inertial reference systems in uniform motion relative to each other (special relativity) in a useful mathematical language. From these investigations also evolved his famous principle of the equivalence of mass and energy. With the help of his principle of equivalence in conjunction with the earlier work of Gauss and Riemann, Einstein saw that gravitational phenomena were fundamentally embraced by the geometric structure of spacetime. The absolute differential calculus (tensor calculus) of others of his predecessors, the Italian mathematicians Ricci and Levi-Civita, led him to make great strides in expanding his principle of equal admissibility of arbitrary reference systems (general covariance).

These monumental discoveries which historically, were the essence of what Einstein referred to as the theory of general relativity, marked a turning point in the history of physics. Whereas the operation of physical laws had theretofore been recognized, now came the astounding cognizance of the operation of interconnecting principles. The position of logical positivism in the development of physics had been assumed by poetic imagination.

"*In physics we attempt to approach truth through reason. Obviously this is a statement of convenience. If we say that in poetry we attempt to approach truth through the imagination, this, too, is a statement of convenience. We must conceive of physics as at least the equal of poetry . . . . It seems to be elementary from this point of view, that the physicist in order to fulfill himself, must accomplish a physics that satisfies both the reason and the imagination.*"

In Einstein's own words, "The emotional state which leads to such achievements resembles that of the worshipper or the lover; the daily striving does not arise from a purpose or a programme, but from an immediate need." The magnitude of Einstein's achievements, including his contribution to quantum and statistical mechanics, has stood out above and beyond all the myriad of discoveries since that time, yet he died unfulfilled. For while it indeed seemed reasonable to expect the electromagnetic and quantum field theories to fit in quite naturally with his gravitational theory, resulting in the fundamental unity of the universal scheme, these expectations never came to pass. He retired into semi-isolation in 1925 and spent most of the remainder of his life in voluntary exile while searching for this "insane" truth.

"We have been a little insane about the truth. We have had an obsession. In its ultimate extension, the truth about which we have been insane will lead us to look beyond the truth to something in which the imagination will be the dominant complement. It is not only that the imagination adheres to reality, but also that reality adheres to the imagination and that the interdependence is essential . . . . *But just as the nature of truth changes, perhaps for no more significant reason than that philosophers live and die, so the nature of physics changes, perhaps for no more significant reason than that physicists come and go.*"

About the time that Einstein went into semi-seclusion, he was assisted in his work on the equations of motion by J. Grommer, a Russian mathematician. Because of ill health, Grommer returned to his native land never to return, for he vanished during the Stalin purge. When Grommer left, Einstein turned to his friend and colleague, Leo Szilard, to aid him in procuring a new assistant. Szilard recommended the young Hungarian mathematician, Cornelius Lanczos. Lanczos had received his Ph.D. in Hungary in 1921, whereupon he came to Freiburg to continue his work. His first publication was on isometric coordinates, showing how the contracted Riemann tensor, R_{ik}, is made simpler in these coordinates which are generalizations of the well known normalization of the reference system in Einstein's infinitesimal equations for arbitrary strong fields. As a consequence, Lanczos produced a subsequent paper which demonstrates a method of successive approximations which solve Einstein's equation to any degree of approximation. Starting with $R_{ik} = 0$, the first approximation is obtained. Higher approximations require the solution of the wave equation with a given right side. While the method is simple in principle, in actuality, it is somewhat complex. The solutions of the gravitational equation are obtained by expanding into powers of the parameter κ which is an exceedingly small constant even for gravitational fields as strong as that of the sun or other very large bodies. This is the same constant that appears in the Einstein equation $R_{ik} - \frac{1}{2} R g_{ik} = \kappa T_{ik}$. To expand the field equations in powers of κ is a very natural procedure, but complications arise due to the fact that the equations must satisfy certain identities. This requirement can be eliminated, however, since the solution of the wave equation is a sufficient condition in a normalized system when isometric coordinates are used. This work was published in 1923. Later in 1927, Lanczos published two other papers, one on the dynamics of general relativity and the other on the principle of motion in general relativity. So it is seen that Lanczos had already developed an interest in the work of Einstein, and when the opportunity presented itself in 1928 to become the successor of Grommer, he immediately accepted. His first task was to investigate the equations of motion of general relativity.

By way of background, general relativity is based on two independent roots, the first of which is that the equations which determine the field of the sun or the planets have spherically symmetric solutions. But this is only part of the picture—the second consideration is the assumption that the paths of the planets describe a geodesic of the external field, or that is to say, the shortest Riemannian line, which directly implies that the law of inertia is in effect. This implication negates the need for a special force of gravity, and this is the great achievement of Einstein's theory. That the gravitational force of the internal field can be ignored is not self-evident. If we look at the gravitational field of the earth, it is infinitely stronger near the earth's surface than the gravitational field of the sun which possesses a tremendously greater mass; yet in calculating the physical effects of these fields on the path of the earth, the earth's field is entirely negligible. That is to say that the self field and the self energy of a body have no effect on the equation of motion. These results were published in a paper by Einstein and Grommer in 1927 where they concluded that these "independent" roots of the theory are not independent at all, since the field equations already contain and, therefore, establish the motion law. They had not, however, provided a complete solution of the problem and this was to be the first task of Lanczos in collaboration with Einstein. Lanczos was asked to investigate the interrelation of the field equations with the equations of motion—that is, to what extent can one obtain from the field equations the equation of motion, that means, the motion along a geodesic line. He was able to show that for a first approximation, i.e., for weak fields, the law of inertia can be demonstrated if the self field is spherically symmetric and essentially static.

The approach of Lanczos was first to ask what conclusions can be drawn from the Einstein equation $R_{ik} = 0$? Since these equations are non-linear, one is therefore able to obtain interactions between fields and, consequently, also a motion law. The primary difficulty with this type of equation, however, is that it is not possible to operate with solutions that satisfy everywhere and are free of singularities, in other words, points where the field quantities approach infinity. In order to avoid these singularities, Lanczos used the following scheme. In the Schwarzschild line element there is a singularity of the type $1/R$ where, of course, $R = 0$ would give a singular point where the $1/R$ goes to infinity. The method of Lanczos was to round off the $1/R$ near to the singularity. That is, instead of following the hyperbola to infinity, the curve is peaked at a high point and brought down again. This process gives a continuity of both function and derivatives so that there is a regular solution of the equation, but it is no longer a solution of the potential equation. Something must now be put on the right side at the point where the rounding off takes place since $\nabla^2\phi$ is no longer equal to zero. Instead of having a solution of a homogeneous equation with singularities, there is now an inhomogeneous equation, whose solution has no singularities, which describes a portion of space that can be made quite small. This allows the use of the Einstein equation $R_{ik} = 0$ almost everywhere except in the immediate neighborhood of the singularities. Then one can operate with a function which is regular, continuous and differentiable up to the first order. Although at the peak the function is discontinuous in the second derivative because $\nabla^2\phi \neq 0$, this poses little difficulty because the right hand side can be chosen quite arbitrarily in the vicinity of the rounding off.

It is interesting here to examine how a motion law for weak fields can be obtained using the Maxwell equations which are linear. This requires first of all an interaction between fields. Since the equations are linear, a superposition principle can be used for the static fields of two spherically symmetric charges. The Maxwell tensor, an electromagnetic energy-momentum tensor, is of second rank and is symmetric. The divergence of this tensor is everywhere zero where the homogeneous Maxwell equations are satisfied. From this fact, assuming certain properties of the field, the Lorentz force acting on one charge in the field of the other can be obtained. This is a peculiar paradox, because under the proper conditions a motion law can be derived for two particles existing under static conditions.

With the Einstein equations, it is different on account of the non-linearity, but the basic difficulty is still the same in that there is the presence of singularities. If there were a hypothetical field theory which included more than just the gravitational equations for which singularity-free solutions, automatically highly non-linear, (the non-linearity implies interaction between fields from which a motion law can be derived) could be obtained, then a motion law would quite naturally result, having been determined by the field equations. Suppose now that the matter tensor $T_{ik} \neq 0$ is considered at the round-off points. The choice of the right hand side is no longer arbitrary because the divergence of T_{ik} must be zero at these points. This appeared to be a weak condition because the matter tensor has ten components, and here only four conditions must be satisfied. These consist of four differential equations of the first order (the divergence is a vector equation), and, therefore, the evolution of the world line of the particle can be described. It appeared that the system was so underdetermined that any kind of motion law could be derived, since the evolution of the world line can still be described in a very arbitrary way. As it turned out, this is not so, because on the basis that the T_{ik} is symmetric and that the $\partial T_{ik}/\partial x_k = 0$, the tensor has not 16 but only 10 independent components from which 10 conservation laws are obtained. The first four represent the three equations of the conservation of momentum, Newton's second law of motion, and the one equation of conservation of energy which is at the same time conservation of mass (these would hold even if the tensor were not symmetric). The symmetry of the tensor adds six more conservation laws, three of which represent the conservation of angular momentum, embodied in Euler's equations for rotating bodies. These three are of particular importance because they lead to a definite motion law for the center of mass-energy of the particle. By taking the first approximation of the T_{ik} for an infinitely weak field, the law of inertia is obtained purely on the basis of the divergence and the symmetry of the T_{ik} and the center of mass is found to move in a straight line in a Minkowskian four-dimensional world. The next approximation obtains the Newtonian gravitational law of motion along a geodesic. In order to do this, the assumptions must be made that the observer is in a comoving frame, leading to practically stationary conditions, and that the system is at least quasisymmetrical. Later Papapetrou showed that the motion of a spinning planet would not be along a geodesic in an external field. Strong dipole effects could also influence the motion law, but both of these cases are beyond the scope of Lanczos' work at that time. The remaining three conservation laws are of very great importance. They are defined as laws of momentum, or of the density of momentum, which in the field theoretical sense has no kinematic significance in itself. However, on account of the symmetry of the matter tensor, these three laws correspond directly to the total momentum of the field or, in other terms, to the total mass of the particles times the velocity of their center of mass — which is precisely the Newtonian definition of the total momentum for a solid body. Herein lies the major significance of these last three conservation laws, for as a consequence of the fact that the time rate of change of momentum is equal to zero and the momentum is defined in the Newtonian sense, the Newtonian equation of motion for a free particle obtains. As long as the demand is made that $\partial T_{ik}/\partial x_k = 0$, in Minkowskian space the motion of the center of mass of the particle is in a straight line in accordance with the law of inertia. The results of this investigation were published in 1930. Now if the Euclidean geometry is modified making it a weak Riemannian geometry, the divergence equations become modified by further terms, which leads in this approximation to the path of the centre of mass of a particle becoming a geodesic of this space.

Other work done by Lanczos in collaboration with Einstein during 1928–29 was in connection with Einstein's concept of distance parallelism in Riemannian geometry which he had hoped would lead to a natural interconnection between gravitation and electromagnetism. Lanczos had little confidence in this hypothesis since it meant foisting parallelism into curved space where it does not naturally occur. To begin with an arbitrary choice of field equations will result in overdetermined or underdetermined solutions. In the first instance, solutions will be trivial, and in the second instance, they will be incomplete. Using a carefully chosen Lagrangian, (Einstein's papers label this a "Hamiltonian", which is, of course, in contemporary literature a totally different concept), Lanczos was able to derive, through the use of a variational principle, a generally covariant set of equations. The advantage of using a Lagrangian is that the field equations derived through a variational principle are neither over- nor underdetermined. While there is a possibility of a minimum or a stationary solution, these are not trivial. This bit of magic provided a beautiful mathematical solution, but Einstein became discouraged when he concluded that the resulting field equations which he had hoped would reveal a unified theory not only did not produce any evidence for electromagnetism, but the gravitational case seemed to imply the presence of two gravitational centers whose fields did not interact. Physically, the existence of a static gravitational field with two centers is not possible. This investigation was not published by Lanczos though Einstein made reference to it in his later works. While in Berlin, Lanczos completed an important paper on the covariant integral formulation of the field conservation laws of general relativity.

"Phoebus is dead, ephebe. But Phoebus was
A name for something that never could be named.
There was a project for the sun and is.

There is a project for the sun. The sun
Must bear no name, gold fluorisher, but he
In the difficulty of what it so to be."

(Arguments raised by the mathematician E. Trefftz on the basis of gravitational equations developed by H. Weyl appear to contradict Einstein's idea that field equations and motion laws are related to each other. However, from Trefftz' calculation, Einstein was able to clear up the mystery of the presumably non-interacting gravitation particles as well as to corroborate his guess that the motion of the particles is interrelated to the field equations. But that is another story –).

Lanczos returned to Frankfurt in 1929 where he continued his work in relativity. After publishing the work that he had begun with Einstein, he started to work on his own investigations using the quadratic action principle. The primary difficulty with the theory as far as the metric tensor, g_{ik}, goes, is that one gets ten fourth-order differential equations. Now if one uses the matter tensor, T_{ik}, from which to obtain the field equations by the quadratic action principle, one gets twenty second-order differential equations, and so the difficulty is not so great as it at first seems. But, if one uses the Einstein method of starting with the Minkowskian background and assuming for weak fields a small deviation from the Minkowskian metric 1, 1, 1, -1, then little results beyond the Einstein equation. The Einstein equation, $R_{ik} = 0$, as well as the cosmological equation, $R_{ik} + \lambda g_{ik} = 0$, are *exact* solutions of the quadratic action principle, even though trivial. Though the theory appeared promising because of the appearance of a quantity which might be interpreted as a vector potential, it turned out that it was not. While dipoles could be obtained, a free charge would cause the corresponding g_{ik} to increase to infinity. A reasonable metric tensor which would satisfy the conditions of the vector potential could not be found.

In later years, Einstein believed that one must go to more general geometries in order to go beyond general relativity; while Lanczos believed that the four-dimensional Riemannian geometry was fundamental.

Although the first attempt was unsuccessful, the idea was not dropped, and in 1942, Lanczos concluded that the assumption that the g_{ik} = constant is mistaken and that perhaps instead of a flat Minkowski universe the background is a dynamic highly oscillatory field which causes enormous curvatures. The deviation from Minkowskian values could be exceedingly small, but on account of high frequency oscillations, the usually neglected second order terms of the Riemannian tensor bring out the squares of the frequencies which can be extremely large quantities. This dynamical approach to general relativity was published. Now when Lanczos discussed this with Einstein, Einstein said that he himself had considered such a possibility, but had discarded the notion because he could not reconcile it with the Lorentz transformation. Einstein felt that certain phenomena would be predicted which would contradict the physical facts. Lanczos countered that since a high frequency background does occur in the vicinity of g_{ik} = constant, perhaps the resulting physical constants were indeed valid. But Einstein rejected this suggestion on empirical grounds. He believed that the oscillatory background would create certain statistical fluctuations which would not agree with physical facts.

Undaunted, Lanczos continued to contemplate his oscillatory field theory. In 1960, he began to experiment with the idea of periodic boundary conditions. For Lanczos, the field equations which follow from the quadratic action principle were a natural choice from which to look for solutions. Recall that these equations number ten of fourth-order for g_{ik} and twenty of second-order for T_{ik}, so it is necessary to specify the type of desired solution since the possibilities are many. Einstein had looked for spherically symmetric and static solutions. Discarding the notion that the background is static, then, a logical choice in a dynamic space-time would be a solution of four-fold periodicity. The basic period could be submicroscopic, and in four dimensions could be thought of as forming a lattice structure. But wait—might this assumption preclude a preferred set of axes? Not necessarily, because of its infinitesimal dimensions, the crystalline substructure could be considered to be isotropic, having three mutually orthogonal principal axes with identical eigenvalues. This turns out to be a very interesting metaphysical hypothesis which leads to certain possibilities which go well beyond anything which one may look for in connection with the Einstein's gravitational equations, or even the unified field theory which includes the Maxwell tensor. Suppose this submicroscopic lattice is the basic metrical plateau on which the physical fields are constructed. This plateau can be interpreted as the aether which is empty from the physical standpoint. It is the imperfections in the monotonous lattice which are the material particles observed as superposition effects. The superposed solutions which represent the material particles are free and of very low frequency compared with the lattice, else they would disappear into the background. The boundary conditions for these particles are free; the particles are static or quasi-static, and do not have submicroscopic vibrations, so that the basic lattice would only show up in terms of its average values. Now a Lagrangian is definitely determined (except for a numerical factor which is so far free) from which everything else must follow, e.g., $\mathcal{L} = R_{ik}R^{ik} - \kappa R^2$.

To obtain solutions from so complicated a set of field equations under the assumption of periodicity is a monstrous assignment, nevertheless it is taken for granted that such solutions do exist. Conclusions are drawn only in terms of superposition effects because physically, only macro-scopic effects are of interest, since the basic lattice constant is so small that it is directly unobservable, except in terms of its quadratic average values.

Much work remains to be done on this theory, but there are many ex-citing ideas left to pursue, for example, what is the role of the vector potential in general relativity? The Einstein theory leaves a free vector which is completely undetermined, but this vector field which puzzled Einstein can be explained under the conditions of the lattice theory. The wave equation for it is obtained and can be identified with the vector potential. The conservation of charge would automatically follow by setting the divergence of the vector potential equal to zero (the Lorentz condition).

Another more intriguing pursuit was the splitting of the Riemann tensor, resulting in, in addition to the electromagnetic vector potential, the metric tensor and the spin tensor, a third-order tensor which is re-garded as a new element of four-dimensional Riemannian geometry. With the choice of a quadratic Lagrangian, the matter tensor is expres-sible in terms of this new element. These results, in the opinion of Lanczos, "seem to provide all the building blocks for a rational explana-tion of electricity and quantum phenomena."

"Begin, ephebe, by perceiving the idea
Of this invention, this invented world,
The inconceivable idea of the sun.

You must become an ignorant man again
And see the sun again with an ignorant eye
And see it clearly in the idea of it.

Never suppose an inventing mind as source
Of this idea nor for that mind compose
A voluminous master folded in his fire.

How clean the sun when seen in its idea,
Washed in the remotest cleanliness of a heaven
That has expelled us and our images . . .

They will get it straight one day at the Sorbonne.
We shall return at twilight from the lecture
Pleased that the irrational is rational,"

The proliferation of the work of Lanczos is nothing short of astounding, for that which is mentioned here is but a fragment of his endeavor in general relativity. There has been a conscientious attempt to pick up the threads of his early interest in the relativistic universe and maintain the continuity in relating his expanding passion for a magnificient order. He is a perfect mathematician who, in the words of Weierstrass, must also be "somewhat of a poet". The excitement and the wonder that he exudes in his grand plan for universal order leave us no doubt as to his poetic abilities. He tells the story about Newton, but one cannot help but compare Lanczos with the small boy playing on the sea-shore who occasionally finds an exceptionally pretty shell, knowing all the while that these are simply clues to the presence of the great ocean of truth far beyond.

Lanczos is clearly deeply influenced by Einstein, the man, but his independence of thought is unquestionable. He believes, as did Einstein, that the positivistic viewpoint voids the possibility of deeper meaning. The real truth, or what is behind the phenomena, can only be seen clearly by the mind, since experimental facts do not give us the essence of things.

Positivism met its demise with the advent of atomism. Atomism cannot exist without metaphysical concepts, since the atom is a hypothetical construct for which there is no tangible evidence. Boltzmann introduced atomism as a construct for explaining thermodynamics. He showed that upon the application of sufficient heat, oriented macroscopic motion changes over into disoriented molecular motion, that orderly motion changed to disorderly motion and could not change back. This was his famous concept of entropy, a tremendous achievement.

A. "A violent order is disorder; and
B. A great disorder is an order. These
   Two things are one. . . ."

Boltzmann was hounded to suicide in 1906, just one year after the publication of Einstein's famous paper on Brownian motion which made the existence of molecules almost palpable, because the idea of a hypothetical construct was rejected by the positivists as having no place in science. The atom today is still a construct, but it is accepted as a fact which rather well illustrates that positivism is an empty shell. It is now recognized universally that only by making proper constructs can progress be made in physics. Lanczos believes that the possibilities of Riemannian geometry are not yet exhausted, and will provide a key to discovering new universes. He has the uncanny ability to go beyond the visible to come to the true universe, but the last truth is still very far away.

In the words of the Psalmist,

"Marvellous are thy works: and that my
Soul knoweth right well.
How precious are thy thoughts unto me, O God!
How great is the sum of them.
If I should count them, they are more in
Number than the sand: when I am
Awake, I am still with thee."

Conservation Laws in Einstein's General Theory of Relativity*

WILLIAM R. DAVIS

Department of Physics, North Carolina State University,
North Carolina, U.S.A.

I. INTRODUCTION

Any attempt at a comprehensive discussion of conservation laws in Einstein's general theory of relativity will necessarily proceed initially along two different lines: (1) the conservation expressions that can be formulated in generally covariant field theories independent of any particular geometric considerations, and (2) the particle and field conservation expressions following in consequence of the symmetry properties that may be admitted by particular Riemannian spacetimes V_4. Later, it will be seen that these two different lines of discussion are related to several types of investigations in these areas, which have been or are being made, that actually may often complement each other. A number of the historically important contributions will be mentioned in the appropriate sections of this paper.† For the present suffice it to say that after a long, difficult, and confused history for the conservation laws of Einstein's theory it is now possible to state certain definite results and conclusions as

*The work reported on in Sections VI and X was supported in part by a Guggenheim Foundation Fellowship.
†In view of the topic and the occasion of this volume celebrating the 80th birthday of Professor Cornelius Lanczos it is appropriate to point out that he contributed a very significant early paper [C. Lanczos, "Über eine invariante Formulierung der Erhaltungssätze in der allgemeinen Relativitätstheorie", Z. Physik, **59**, 514 (1930)] giving a covariant integral formulation of the conservation laws when certain boundary conditions hold for physically closed systems (see Sections II and VII). Unfortunately, the results of this important paper have been largely overlooked. Most of this work on conservation laws was carried out in Berlin in 1929 during a one year period while Lanczos was closely associated with Einstein.

29

well as to shed more light on a number of the remaining problems and questions.

For over half a century considerable effort and attention have been given to the question of conservation laws of Einstein's general theory of relativity and other types of generally covariant field theories. This is not a surprising fact in view of the continuing great importance of conservation laws in all the other particle and field theories (which are not generally covariant or formulated in general Riemannian spacetimes) forming the present-day foundations of physics. In particular, interest has centered on the problem of finding satisfactory formulations of the conservation laws in general relativity that are analogous to those of energy, momentum, and angular momentum familiar in all Lorentz-covariant field theories. For those outside this field it might be rather surprising to find that even by the end of the 1950's there was no general agreement regarding the form that these field conservation laws should take in general relativity. By 1960 the problems with all attempts to obtain a satisfactory formulation for the differential form of the field conservation laws[1] of general relativity could be summarized under one or more of the following difficulties discussed at that time[2]. (1) The conservation expressions proposed were not of tensor form[3]. (2) The "superpotentials" of the "strong" conservation expressions that follow in consequence of general coordinate covariance make possible the construction of an infinity of distinct vanishing divergence expressions[4]. (3) A generally covariant theory in general Riemannian spacetime does not admit symmetries [e.g., groups of motions[5] (isometries)] that could lead to conservation laws analogous to those of Lorentz-covariant field theories[6]. Without the developments and considerations given in the later sections of this paper we can only comment that most aspects of the above difficulties have been resolved by more recent formulations or by a deeper understanding of these apparent earlier difficulties.

The most important early work that directly bears on many aspects of the present considerations is that of Noether[7] (1918) who showed that the invariance groups of physical theories based on variational principles always lead to conservation expressions and, conversely, that the existence of these conservation expressions requires that the variational principle of the given theory be invariant under a certain group of transformations[8] (namely, the invariance group admitted by the variational formulations of the given physical theories). In the case of particle mechanics, the consequences of Noether's theorem are most simply realized at the level of the linear first integrals of the equations of motion. The maximum number of these which may be realized in a given mechanical problem formulated in an n-dimensional flat space is $n(n+1)/2$, corresponding to the n-dimensional space admitting an $r = n(n+1)/2$ group of motions G_r (isometries)[5]. Of course, in special relativity this ten parameter group of motions G_{10} corresponds to the inhomogenous Lorentz group and the first integrals are the familiar expressions for energy, momentum, and angular momentum. The same considerations follow essentially parallel lines for any Lorentz covariant field theory. On the other hand, the invariance of general relativity under the group of general coordinate transformations $G_{\infty 4}$ appears to present one with the puzzling problem of an infinity of conservation expressions as mentioned above. It is the purpose of this paper to indicate the present status of the problem of conservation laws in general relativity and to discuss briefly most of the more important results that have been obtained. Also, a few new developments will be considered. The topics to be considered in the various sections of this paper can be broadly grouped under the following: (1) particle conservation laws—first integrals of m-th order, (2) uses of the identities of generally covariant field theory, (3) field conservation laws in general relativity—conservation law generators and symmetry properties, (4) field conservation laws admitted for particular curve congruences with special kinematical properties and symmetry properties, (5) the role of the tetrad formulation of general relativity in the formulation and application of conservation laws. Unfortunately, in the space available, we cannot even attempt to mention all the papers that have contributed to the above topics. Thus, it will be necessary to limit consideration to those works that we regard as the most important, representative, and/or definitive.

II. HISTORICAL COMMENTS—FIELD CONSERVATION LAWS PROPOSED FOR GENERAL RELATIVITY 1916–1959

Historically, the conservation laws for the general theory of relativity have been a subject of controversy since Einstein first proposed a formulation[3] involving his gravitational stress-energy pseudotensor in 1916. The early controversy arose primarily because Einstein's formulation of the conservation laws was nontensorial. In particular, Einstein's formulation was found to yield physically tenable results for the total energy of physically closed systems (e.g., Schwarzschild metric) only when they are expressed in quasi-Galilean coordinates[9]. While several alternative expressions of the conservation laws were proposed, Einstein's formulation was the generally accepted formulation for at least 30 years[10]. In this period it was also generally believed that no acceptable tensor formulation of the conservation laws had been discovered. However, as we have already mentioned, in 1930 Lanczos published an important paper giving a covariant (tensor) integral formulation of the conservation laws for Riemannian spacetimes representing physically closed systems admitting asymptotic Killing vectors[11]. This important contribution and its relation to presently accepted results appears to have been largely overlooked even through the 1960's. In Section VII we will consider the relation of Lanczos' results to the other formulations of current interest.

Around 1950 a new level of controversy began developing with the proposal of a new symmetric version of the conservation laws for general relativity by Landau and Lifshitz[12]. The new pseudotensor proposed had the advantage of being symmetric in addition to containing only up to first derivatives of the metric tensor as in the case of Einstein's pseudotensor. In 1958 an already disturbing situation was converted into extreme confusion and controversy by the demonstrations of Goldberg[13] and Bergmann[14] that a multiple infinity of distinct pseudotensors could be formulated in general relativity. Each of these distinct complexes[15] played the usual part in the differential form of a conservation law. Also, other specific formulations of the conservation laws for general relativity were proposed and applied by Møller[16], Komar[17], and Dirac[18].

We now proceed to indicate the formal relationships that exist between these various formulations. In this connection it is important to introduce explicitly the notion of a so-called superpotential; namely, any quantity of the general form $S^{[\alpha\beta]}$ which leads immediately to a strong conservation law[4] $(S^{[\alpha\beta]}_{,\beta})_{,\alpha} \equiv 0$. The superpotential underlying Einstein's original form of the conservation laws was first given by Freud[19] in the form $U^{[\alpha\beta]}_{\gamma}$. This expression is a third order complex, which is well defined in all coordinate frames giving $U^{[\alpha\beta]}_{\gamma,\beta} \equiv T^{\alpha}_{\gamma} + t^{\alpha}_{\gamma}$ where T^{α}_{γ} is the matter tensor density and t^{α}_{γ} is the Einstein pseudotensor. More generally, it is worth stressing that for any generally covariant field theory based on a variational principle the superpotentials, as well as the related strong conservation laws, always follow as a direct consequence of the differential identities. In turn, the differential identities are the most general set of identities following in consequence or demand of general covariance (see Section III). Here we might also call attention to the fact that on account of the properties of the superpotential it is always possible to reexpress integral conservation expressions involving hypersurface integrals in terms of two-dimensional surface integrations[20].

With the help of the superpotential, following Goldberg[13], we can implicity define $[(-g)^{m/2}g^{\delta\gamma}U^{[\alpha\beta]}_{\delta}]_{,\gamma}$ (where $\{[(-g)^{m/2}g^{\delta\gamma}U^{[\alpha\beta]}_{\delta}]_{,\alpha,\beta}\} \equiv 0$ which gives a distinct gravitational pseudotensor for every different choice of the weighing factor m. For example, for the choice $m = 1$ the

Landau–Lifshitz complex is obtained. Furthermore, Bergmann[14] observed that one also obtains other distinct gravitational complexes

$$[U_\gamma^{[\alpha\beta]}\phi^\gamma]_{,\beta} \qquad (\text{where } \{[U_\gamma^{[\alpha\beta]}\phi^\gamma]_{,\beta}\}_{,\alpha} \equiv 0) \qquad (2.1)$$

for every distinct choice of the general vector (or vector density of weight *m*) field ϕ^γ. Beyond these elements of arbitrariness one can also easily see that these expressions are arbitrary to the extent of the addition of terms of the form $W_{\gamma,\delta}^{[\alpha\beta\delta]}$ or $\bar{W}_{,\delta}^{[\alpha\beta\delta]}$. Of course, one is perfectly free to modify any one of the above superpotential expressions by additions of the form ($V_{,\beta}^{[\alpha\beta]} = 0$). For example, in 1958 Møller[16] modified the Freud superpotential by a term of the form $W_\gamma^{[\alpha\beta]} = -\delta_\gamma^\alpha U_\lambda^{\beta\lambda} + \delta_\gamma^\beta U_\lambda^{\alpha\lambda}$ which was found to be the term necessary to add to any given $U_\gamma^{[\alpha\beta]}$ in order to give the combined expression tensor properties with respect to spatial coordinate transformations. Thus, with Møller's development it is no longer necessary to restrict applications of the calculations of the total energy of closed systems to quasi-Galilean coordinates. However, this complex has other serious defects[21] that make it impossible to view it in any general sense as a tenable formulation of the conservation laws. In 1959 Komar[17] combined the essential aspects of the above mentioned results of Bergmann and Møller to construct the simplest covariant conservation expression requiring that they reduce to Møller's result for $\phi^\alpha = k^0\delta_0{}^\alpha$, a timelike displacement. Komar found the following identically vanishing expression:

$$P_{;\alpha}^\alpha \equiv 0 \quad \text{where} \quad P^\alpha \equiv [\sqrt{-g}(\phi^{\alpha;\beta} - \phi^{\beta;\alpha})]_{;\beta}. \qquad (2.2)$$

Clearly, there are as many distinct "conservation laws" following from this expression as there are distinct vector (or vector densities of weight *m*) fields ϕ^α.

Another independent development in 1959 due to Pirani[22] starts with a physical argument to define the gravitational energy in a region in terms of the curvature of a normal curve congruence which leads to an expression of the same form as that found by Komar. The significance of this result will be considered in Sections IX and X where we consider other conservation expressions that are related to the properties of curve congruence and tetrads.

Most of the above mentioned developments will be further correlated and discussed in the following sections. The reader is reminded that the general problems associated with the above mentioned works on the conservation laws of general relativity have already been summarized in the introduction.

III. PROPERTIES OF GENERAL COVARIANT FIELD THEORIES

In accord with the previously mentioned work of Noether[7], it will be seen that the invariance of a theory (based on a variational principle in n-dimensional space) with respect to the group $G_{\infty n}$ results in a number of differential identities[23] which can be combined to obtain n principal identities which are linear and homogeneous in the field equations. In addition, these differential identities always lead to an expression for the superpotential of the theory[24]. The principal identities are often referred to as the generalized Bianchi identities because in the case of Einstein's general theory of relativity they correspond to the four twice-contracted Bianchi identities which take the form $(R_\alpha^\beta - \frac{1}{2}\delta_\alpha^\beta R)_{;\beta} \equiv 0$ where R_α^β is the Ricci tensor. Thus, in accord with the concept of covariance with respect to general coordinate transformations in spacetime, the field variables g_{ij} (including their first and second partial derivatives) in Einstein's field equations $R_\beta^\alpha - \frac{1}{2}\delta_\beta^\alpha R = \kappa T_\beta^\alpha$ satisfy four identities. It is the object of this section to summarize briefly some of the properties that are central to the structure of generally covariant field theories and to show how it is possible that such theories lead to an infinite number of formally distinct, identically vanishing divergence expressions (strong "conservation laws"). In turn, this will help us to understand better the various works mentioned in the previous section. Also, the results of this section will serve in part to facilitate the discussion of symmetry properties which may be admitted by generally covariant field theories that will be taken up in the next section. The formal results to be obtained in this section will play a role in the discussion of Section VII dealing with covariant conservation law generators in the general theory of relativity.

We now proceed to indicate the form of the above mentioned results on the basis of the following simple field structure. The variational principle will be assumed to take the form [25]

$$\delta \int L(\Psi_A, \Psi_{A,\alpha}, \Phi_\Omega, \Phi_{\Omega,\alpha})d_n x = 0 \quad (d_n x = dx^1 dx^2 \cdots dx^n) \quad (3.1)$$

where the field quantities Ψ_A $(A = 1, \cdots, r)$ and Φ_Ω $(\Omega = 1, \cdots, s)$ are varied independently and L differs from a scalar density of weight one by no more than a pure divergence term. We then obtain the field equations

$$\delta L/\delta\Psi_A \equiv L^A \equiv \partial L/\partial\Psi_A - \partial_\alpha(\partial L/\partial\Psi_{A,\alpha}), \quad (3.2a)$$

$$\delta L/\delta\Phi_\Omega \equiv L^\Omega \equiv \partial L/\partial\Phi_\Omega - \partial_\alpha(\partial L/\partial\Phi_{\Omega,\alpha}). \quad (3.2b)$$

where $\partial_\alpha \equiv \partial/\partial x^\alpha$ and $\Psi_{A,\alpha} = \partial \Psi_A/\partial x^\alpha$. The infinitesimal transformation law for L that gurantees the invariance of (3.1) under the infinite group $G_{\infty n}$ is given by

$$\bar{\delta}L = -\epsilon(L\xi^\alpha)_{,\alpha}, \tag{3.3}$$

assuming L is a scalar density of weight one where $x^\alpha \rightarrow x^\alpha + \epsilon\xi^\alpha$. For simplicity, the corresponding transformation laws for the field variables are assumed to be of the following form:

$$\bar{\delta}\Psi_A = C_{A\alpha}\xi^\alpha + C^\beta_{A\alpha}\xi^\alpha_{,\beta}, \tag{3.4a}$$

$$\bar{\delta}\Phi_\Omega = C_{\Omega\alpha}\xi^\alpha + C^\beta_{\Omega\alpha}\xi^\alpha_{,\beta}. \tag{3.4b}$$

Here the $C_{A\alpha}$, $C^\beta_{A\alpha}$ and $C_{\Omega\alpha}$, $C^\beta_{\Omega\alpha}$ may be independent functions of the Ψ_A and Φ_Ω (including their derivatives to finite order) respectively. The C's are arbitrary except for the demand that the group property be satisfied[26]. If we now write

$$\bar{\delta}L = (\partial L/\partial\Psi_A)\bar{\delta}\Psi_A + (\partial L/\partial\Phi_\Omega)\bar{\delta}\Phi_\Omega + (\partial L/\partial\Psi_{A,\alpha})\bar{\delta}\Psi_{A,\alpha} + (\partial L/\partial\Phi_{\Omega,\alpha})\bar{\delta}\Phi_{\Omega,\alpha}$$

and then introduce (3.3), (3.4a), and (3.4b) one can bring the resulting expression in the form

$$^0I_\alpha\xi^\alpha + {}^1I^\beta_\alpha\xi^\alpha_{,\beta} + {}^2I^{(\beta\gamma)}_\alpha\xi^\alpha_{,\beta,\gamma} = 0.$$

Clearly, the coefficients $^0I_\alpha$, $^1I^\beta_\alpha$, and $^2I^{(\beta\gamma)}_\alpha$ of the ξ^α and their partial derivatives in (3.5) must vanish identically in accord with the general covariance demand requiring the ξ^α to be viewed as algebraically independent arbitrary functions. The general differential identities corresponding to the particular field theoretical structure defined above may be given the following explicit form

$$^0I_\alpha \equiv L^A C_{A\alpha} + L^\Omega C_{\Omega\alpha} + \left[\frac{\partial L}{\partial\Psi_{A,\beta}}C_{A\alpha} + \frac{\partial L}{\partial\Phi_{\Omega,\beta}}C_{\Omega\alpha} - L\delta^\beta_\alpha\right]_{,\beta} \equiv 0, \tag{3.6a}$$

$$^1I^\beta_\alpha \equiv L^A C^\beta_{A\alpha} + L^\Omega C^\beta_{\Omega\alpha} + \left[\frac{\partial L}{\partial\Psi_{A,\beta}}C_{A\alpha} + \frac{\partial L}{\partial\Phi_{\Omega,\beta}}C_{\Omega\alpha} - L\delta^\beta_\alpha\right]$$

$$+ \left[\frac{\partial L}{\partial\Psi_{A,\gamma}}C^\beta_{A\alpha} + \frac{\partial L}{\partial\Phi_{\Omega,\gamma}}C^\beta_{\Omega\alpha}\right]_{,\gamma} \equiv 0, \tag{3.6b}$$

$$^2I^{\beta\gamma}_\alpha \equiv \tfrac{1}{2}\left[\frac{\partial L}{\partial\Psi_{A,\gamma}}C^\beta_{A\alpha} + \frac{\partial L}{\partial\Psi_{A,\beta}}C^\gamma_{A\alpha}\right] + \tfrac{1}{2}\left[\frac{\partial L}{\partial\Phi_{\Omega,\gamma}}C^\beta_{\Omega\alpha} + \frac{\partial L}{\partial\Phi_{\Omega,\beta}}C^\gamma_{\Omega\alpha}\right] \equiv 0. \tag{3.6c}$$

If the identities (3.6a) and (3.6c) are now introduced into (3.6b) one can write the resulting expression in the form

$$\tau^\beta_\alpha \equiv L^A C^\beta_{A\alpha} + L^\Omega C^\beta_{\Omega\alpha} + t^\beta_\alpha \equiv U^{[\beta\gamma]}_{\alpha,\gamma} \tag{3.7}$$

where

$$t_\alpha{}^\beta \equiv \left[\left(\frac{\partial L}{\partial \Psi_{A,\beta}} C_{A\alpha} + \frac{\partial L}{\partial \Phi_{\Omega,\beta}} C_{\Omega\alpha} - \delta_\alpha^\beta L \right] \right.$$

and

$$U_\alpha^{[\beta\gamma]} \equiv \tfrac{1}{2} \left[\left(\frac{\partial L}{\partial \Psi_{A,\beta}} C^\gamma_{A\alpha} - \frac{\partial L}{\partial \Psi_{A,\gamma}} C^\beta_{A\alpha} \right) + \left(\frac{\partial L}{\partial \Phi_{\Omega,\beta}} C^\gamma_{\Omega\alpha} - \frac{\partial L}{\partial \Phi_{\Omega,\gamma}} C^\beta_{\Omega\alpha} \right) \right].$$

Here (3.7) gives us n identically vanishing divergence expressions (strong "conservation laws") because $U_{\alpha,\gamma,\beta}^{[\beta\gamma]} \equiv 0$ and implicitly defines the super-potential $U_\alpha^{[\beta\gamma]}$. Clearly, $U_\alpha^{[\beta\gamma]}$ and $\tau_\alpha{}^\beta$ are arbitrary to the extent of pure divergence terms of the following type:

$$U'_\alpha{}^{[\beta\gamma]} = U_\alpha^{[\beta\gamma]} + W_{\alpha,\delta}^{[\beta\gamma\delta]}, \qquad (3.8a)$$

$$\tau'_\alpha{}^\beta = \tau_\alpha{}^\beta + K_{\alpha,\gamma}^{[\beta\gamma]}. \qquad (3.8b)$$

The n principal identities may be conveniently obtained from the identities by eliminating t_α^β from (3.7) using (3.6a) and then taking the divergence with respect to the index β which gives

$$(L^A C^\beta_{A\alpha} + L^\Omega C^\beta_{\Omega\alpha})_{,\beta} - L^A C_{A\alpha} - L^\Omega C_{\Omega\alpha} \equiv 0. \qquad (3.9)$$

Finally, it might be noted that the general approach used here to obtain identities does not have to be limited to the Lagrangian of variational principle; however, it is certainly simpler to find the identities using a Lagrangian when one is known. Also, it might be mentioned that these identities can be used to identify the consequences of symmetry demands and to help determine suitable modifications of field structures admitting symmetries (see Davis and York[24]).

It is not difficult to show that if L is taken to be the usual non-invariant Lagrangian[12] $\mathcal{L} = \sqrt{-g}R - S^\alpha_\alpha$ for general relativity (3.7) leads to $\tau_\alpha{}^\beta$ and $U_\alpha^{\beta\gamma}$ being identified with the Einstein gravitational pseudotensor and the Freud superpotential[19] respectively [where $\delta\mathcal{L} = -(\epsilon\sqrt{-g}R\xi^\alpha + \delta S^\alpha)_{,\alpha}$]. Another general method for obtaining general identities for field theories called the commutator method is due to Heller[27]. This method involves the application of two variations performed in reverse order. Since this method cannot give any new results it will not be considered further. However, it should be pointed out that this method does, in general, produce the identities and conservation laws in a different form and it could have some formal advantages in working directly with the trans-formation properties of the field equations.

All of the methods for obtaining the differential identities suffer from the disadvantage that, in general, the identities and conservation expres-sions that follow from them have a high degree of arbitrariness and they are nontensorial. This makes it difficult to even consider directly identifying these expressions with a formulation of the conservation laws that would be physically acceptable. However, any one or all of these identities could be useful and important in various mathematical reformulations.

Again in accord with the observations of Goldberg[13] and Bergmann[14] (cf. Section II) using the superpotential in (3.7) one can write $[U_\alpha^{[\beta\gamma]}\phi^\alpha]_{,\beta,\gamma}$ $\equiv 0$ which defines as many distinct gravitational complexes and corresponding identically vanishing divergence expressions as there are distinct choices of independent vector fields of weight m.

Of course, any of these strong conservation expressions are actually arbitrary by a term of the form $K^{[\beta\gamma]}$. As an example of the use of such a term we mention that Komar[17] started with the Møller[16] (cf. Section II) superpotential and demanded that $K^{[\beta\gamma]}$ be so determined that a tensor expression would result. On the other hand Fletcher[2], using a method more directly related to the original Noether[7] work, obtained Komar's result $[\sqrt{-g}\,(\phi^{\alpha;\beta} - \phi^{\beta;\alpha})]_{;\beta,\alpha} = 0$ directly[28].

IV. REMARKS ON SYMMETRY PROPERTIES AND INVARIANCE PRINCIPLES IN PHYSICAL THEORIES

The subject of symmetry properties and invariance principles in physical theories is inherently difficult in its most general aspects and it is far too large a subject to do justice to in the short space available here[29]. However, by limiting ourselves to the aspects of this important subject that are of particular importance to understanding the various results relating to the works on conservation laws in general relativity and by use of examples drawn from more familiar areas we attempt to give a brief coherent discussion that will also serve to define our terminology. In spite of this limitation to certain aspects of general relativity, it is suggested that many of the relevant considerations given here are of importance and broadly applicable in many other areas. One domain of difficulty in this subject of symmetry properties and invariance principles is the differences in terminologies, notations, and mathematical methods[30]. Some of the simpler examples of symmetry properties are drawn from particle mechanics. In Sections V and VI several examples of this type are given that can help one better understand the more difficult case of fields.

Ordinarily, when we speak of a symmetry property (or symmetry transformation) in mechanics we refer to the fact that, mathematically, the given equations of motion are invariant under a certain transformation group (representing the symmetry property) and, physically, to the fact that the actual physical problem has a certain symmetry in the geometric space of the problem. Similarly, in field theory we speak of a symmetry property if the field equations of the theory are invariant in detailed functional form under a certain transformation group. In Lorentz-covariant theories one has symmetry properties [the G_{10} group of motions (isometries) admitted by the geometric spacetime in which the theories are formulated] that correspond to an invariance principle that is physically interpreted in terms of the equivalence of all inertial reference frames. Of course, these cases of invariance relating to symmetry properties are to be distinguished from general coordinate covariance where the use of the term invariance refers to the invariance of the general form or tensor form but not the invariance of the detailed functional form. Nonetheless, a given general covariant field structure (including geometry) could admit a symmetry property; for example, a G_1 isometry that would be a symmetry subgroup of the general covariance group. Similarly, in the case of any given general covariant theory one may find that the given theory admits symmetry subgroups (not necessarily isometries) of its general covariance group. Of course, as remarked earlier, it is well known that general Riemannian spaces do not admit isometries.

For the present purposes, a general definition of a physical symmetry property could be stated as follows: A symmetry property is expressed by the invariance of a given object of interest in a physical theory which may be mathematically represented as a mapping or transformation where the group property can be established[31]. Next, it is convenient to distinguish between the symmetry properties of specific geometric space(s) (e.g., isometries) in which a theory may be formulated and the invariance or symmetry property admitted by certain Riemannian spaces (often referred from those of the geometric space (e.g., gauge transformations in electromagnetic theory). Clearly, the given structure of a field theory may or may not share the symmetries admitted by the geometry in terms of which it is formulated.

In addition to the above considerations it is necessary to distinguish those symmetry properties that are representable as coordinate mappings or point deformations from those that are not. For example, the projective symmetry property admitted by certain Riemannian spaces (often referred to as geodesic correspondence[32]) cannot be represented as a coordinate mapping because here we are involved with a mapping of one Riemannian space onto another distinct Riemannian space. In the special case that a given Riemannian space admits a projective mapping of this type onto itself it can be represented as a symmetry point deformation called a projective collineation[33].

In this paper we shall give little consideration to the case of symmetry properties that are not representable by coordinate mappings because they are more difficult to deal with and less is known about them[34]. In the case of point deformations the symmetry properties admitted by the geometric space as a continuous group G_r, a number of examples have been discussed and may be conveniently defined in terms of Lie derivatives operating on various geometric objects of interest (see Figure 1)[35]. We shall limit ourselves, in the sections that follow, to considerations of symmetry properties that can be characterized as (Lie) continuous groups involving a finite number of parameters. Infinite groups ("gauge groups") involving arbitrary functions are not symmetry properties unless their existence depends on special conditions being fulfilled relating to the given geometry or field structure. We now proceed to some examples of the above considerations.

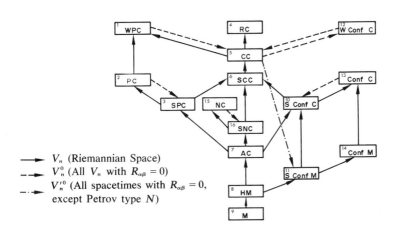

$\longrightarrow$ V_n (Riemannian Space)

$--\rightarrow$ V_n^0 (All V_n with $R_{\alpha\beta} = 0$)

$-\cdot\cdot\rightarrow$ $V_n^{\prime 0}$ (All spacetimes with $R_{\alpha\beta} = 0$, except Petrov type N)

1. *WPC*—Weyl Projective Collineation: $\mathcal{L}W^{\alpha}_{\beta\gamma\delta} = 0$, $(n > 2)$.
2. *PC*—Projective Collineation: $\mathcal{L}\Gamma^{\alpha}_{\beta\gamma} = \delta^{\alpha}_{\beta}\phi_{:\gamma} + \delta^{\alpha}_{\gamma}\phi_{:\beta}$.
3. *SPC*—Special Projective Collineation: $\mathcal{L}\Gamma^{\alpha}_{\beta\gamma} = \delta^{\alpha}_{\beta}\phi_{:\gamma} + \delta^{\alpha}_{\gamma}\phi_{:\beta}$, $\phi_{:\gamma} = 0$.
4. *RC*—Ricci Collineation: $\mathcal{L}R_{\alpha\beta} = 0$.
5. *CC*—Curvature Collineation: $\mathcal{L}R^{\alpha}_{\beta\gamma\delta} = 0$.
6. *SCC*—Special Curvature Collineation: $(\mathcal{L}\Gamma^{\alpha}_{\beta\gamma})_{;\delta} = 0$.
7. *AC*—Affine Collineation: $\mathcal{L}\Gamma^{\alpha}_{\beta\gamma} = 0$.
8. *HM*—Homothetic Motion: $\mathcal{L}g_{\alpha\beta} = 2\sigma g_{\alpha\beta}$, $\sigma = $ constant.
9. *M*—Motions: $\mathcal{L}g_{\alpha\beta} = 0$.
10. *S* Conf *C*—Special Conformal Collineation: $\mathcal{L}\Gamma^{\alpha}_{\beta\gamma} = \delta^{\alpha}_{\beta}\sigma_{:\gamma} + \delta^{\alpha}_{\gamma}\sigma_{:\beta} - g_{\beta\gamma}g^{\alpha\delta}\sigma_{:\delta}$, $\sigma_{:\beta:\gamma} = 0$.
11. *S* Conf *M*—Special Conformal Motions: $\mathcal{L}g_{\alpha\beta} = 2\sigma g_{\alpha\beta}$, $\sigma_{:\beta:\gamma} = 0$.
12. *W* Conf *C*—Weyl Conformal Collineation: $\mathcal{L}C^{\alpha}_{\beta\gamma\delta} = 0$, $(n > 3)$.
13. Conf *C*—Conformal Collineation: $\mathcal{L}\Gamma^{\alpha}_{\beta\gamma} = \delta^{\alpha}_{\beta}\sigma_{:\gamma} + \delta^{\alpha}_{\gamma}\sigma_{:\beta} - g_{\beta\gamma}g^{\alpha\delta}\sigma_{:\delta}$.
14. Conf *M*—Conformal Motion: $\mathcal{L}g_{\alpha\beta} = 2\sigma g_{\alpha\beta}$.
15. *NC*—Null Geodesic Collineation: $\mathcal{L}\Gamma^{\alpha}_{\beta\gamma} = g_{\beta\gamma}g^{\alpha\delta}\psi_{:\delta}$.
16. *SNC*—Special Null Geodesic Collineation: $\mathcal{L}\Gamma^{\alpha}_{\beta\gamma} = g_{\beta\gamma}g^{\alpha\delta}\psi_{:\delta}$, $\psi_{:\beta:\gamma} = 0$.

FIGURE 1. Relations between symmetries[35].

V. CONSERVATION LAWS OF PARTICLE
MECHANICS IN RIEMANNIAN SPACETIME

Here we briefly consider the particle conservation laws following in consequence of Riemannian spacetimes admitting symmetry properties in order that we may better appreciate the role of related considerations when we return to the case of field conservation laws. In this and the next section we will limit ourselves to the discussion of first integrals (FI) of the geodesic differential equations [i.e., $d^2x^\alpha/ds^2 + \left\{{\alpha \atop \beta\gamma}\right\} (dx^\beta/ds)$ $\times (dx^\gamma/dsx) = 0$] of mass pole test particles. It is well known that the necessary and sufficient condition for the existence of an m-th order first integral[36] of the form

$$A_{\alpha_1\alpha_2} \cdots {}_{\alpha_m} \frac{dx^{\alpha_1}}{ds} \frac{dx^{\alpha_2}}{ds} \cdots \frac{dx^{\alpha_m}}{ds} = \text{constant.} \tag{5.1}$$

is

$$P(A_{\alpha_1,\alpha_2, \cdots \alpha_m;\beta}) = 0 \tag{5.2}$$

where P indicates the sum of all $m + 1$ terms obtained by a cyclic permutation of the subscripts $\alpha_1, \alpha_2, \ldots, \alpha_m, \beta$.

We are particularly interested in demonstrating that various symmetry properties admitted by Riemannian spacetimes (see Fig. 1) can lead to (FI) as discussed earlier[37]. In view of the fact that a survey article[38] is available on this subject (in the area of symmetry properties that can be described by point deformations) only a few typical results will be given. A few examples of particle conservation laws relating to symmetry properties that cannot be characterized as point deformations will be briefly considered in Section VI.

For spacetimes admitting groups of motions (isometries) one can immediately write down the following conservation law generator (where $ds^2 = g_{\alpha\beta}dx^\alpha dx^\beta$)[39]

$$\frac{d}{ds}\left(g_{\alpha\beta}\frac{dx^\alpha}{ds}\xi^\beta\right) = \tfrac{1}{2}(\xi_{\alpha;\beta} + \xi_{\beta;\alpha})\frac{dx^\alpha}{ds}\frac{dx^\beta}{ds} = 0 \tag{5.3}$$

giving an independent linear first integral (LFI) for each independent vector ξ^α which is a solution of Killing equations $\xi_{\alpha;\beta} + \xi_{\beta;\alpha} = 0$. That is, in the given spacetime the ξ^α, in the infinitesimal transformation $\bar{x}^\alpha = x^\alpha + \epsilon\xi^\alpha$, represents a G_r group of motions (involving r independent parameters) and leads to r independent (LFI)[40]. Of course, Killing's equations are just a special case of (5.2) corresponding to (LFI).

Quadratic first integrals (QFI) can be obtained, for example, in the case of spacetimes admitting at least a G_1 group of nontrivial projective collineations [i.e., $(\xi_{\alpha;\beta} + \xi_{\beta;\alpha})_{;\gamma} = 2g_{\alpha\beta}\Psi_{,\gamma} + g_{\alpha\gamma}\Psi_{,\beta} + g_{\beta\gamma}\Psi_{,\alpha}$ with $\Psi_{,\gamma} = (\frac{1}{5}) g^{\alpha\beta}\xi_{\alpha\beta;\gamma}$].

In this case the QFI can be written in the form

$$(\xi_{\alpha;\beta} + \xi_{\beta;\alpha} - 4g_{\alpha\beta}\Psi) \frac{dx^\alpha}{ds} \frac{dx^\beta}{ds} = \text{constant}. \tag{5.4}$$

It can be shown that the conservation law generator for these QFI takes the form[37]

$$\frac{d}{ds} \left[\left(\frac{dx^\alpha}{ds}\frac{dx^\beta}{ds} - \frac{2}{5}g^{\alpha\beta} \right) (\xi_{\alpha;\beta} + \xi_{\beta;\alpha}) \right] = 0. \tag{5.5}$$

As a simple formal example of an m-th order (FI) consider the symmetry property (admitted by some special Riemannian spacetime) defined by $(\xi_{\alpha;\beta} + \xi_{\beta;\alpha})_{;\gamma_1;\gamma_2;\cdots;\gamma_{m-1}} = 0$ [where $(\xi_{\alpha;\beta} + \xi_{\beta;\alpha})_{;\gamma_1;\gamma_2;\cdots;\gamma_{m-2}} \neq 0$]. Clearly, in accord with (5.2) we have $(\xi_{\alpha;\beta} + \xi_{\beta;\alpha})_{;\gamma_1;\gamma_2;\cdots;\gamma_{m-2}}(dx^\alpha/ds)(ds^\beta/ds) \times (dx^{\gamma_1}/ds) \cdots (dx^{\gamma_{m-2}}/ds) = \text{constant}$. Of course, there is no guarantee that such higher order integrals will not be trivial products of lower order integrals following in consequence of simpler symmetry properties admitted by the given specetime[41]. In the case of any given spacetime admitting certain conservation laws [e.g., (FI) of geodesics and the corresponding symmetry properties] it is of physical importance to be able to interpret the resulting expressions physically in terms of the detailed properties of the spacetime[42]. Conversely, through such expressions (representing characteristic properties of the geodesics) one gains essential information about the properties of the given spacetime.

VI. PARTICLE CONSERVATION LAWS FOR SYMMETRY PROPERTIES NOT REPRESENTABLE BY POINT DEFORMATIONS

Our considerations in this section will be limited to one general type of mapping on the affine connection (i.e., the Christoffel symbols of the second kind in Riemannian spacetimes) that could lead to quadratic first integrals (QFI) of the geodesic differential equations. It will be seen that this mapping on the given spacetime has to satisfy certain special conditions for it to be a symmetry mapping leading to (QFI) and a symmetric Killing tensor[42] of second order. We will find that the general type of mapping considered here includes as special cases several types of more familiar mappings.

The mapping of the affine connection that we are interested in here is defined by[43]

$$\overline{\left\{ \begin{matrix} \gamma \\ \alpha\beta \end{matrix} \right\}}_A = \left\{ \begin{matrix} \gamma \\ \alpha\beta \end{matrix} \right\} + \tfrac{1}{2} A^{\gamma\delta} (A_{\alpha\delta;\beta} + A_{\beta\delta;\alpha} - A_{\alpha\beta;\delta}) \qquad (6.1)$$

where the barred and unbarred Christoffel symbols are formed in terms of the $A_{\alpha\beta}$ and the $g_{\alpha\beta}$ respectively with $|A_{\alpha\beta}| \neq 0$, $A_{\alpha\beta} A^{\beta\gamma} = \delta^\gamma_\alpha$, $A_{\alpha\beta} = A_{\beta\alpha}$. For $|A_{\alpha\beta}| \neq 0$, the above mapping $(A_{\alpha\beta} \rightarrow g_{\alpha\beta})$, onto any given Riemannian spacetime with metric tensor $g_{\alpha\beta}$, constitutes an identity relation[44] [i.e., given $|A_{\alpha\beta}| \neq 0$ the mapping relation (6.1) is an unrestricted mapping]. With this fact in mind, it is not difficult to show that a V_4 that admits a (QFI) always admits a restricted version of the above mapping and conversely. Of course, symmetry mappings of this type related to (QFI) would not be admitted by Riemannian spacetimes in general.

Given a Riemannian spacetime with metric tensor $g_{\alpha\beta}$ that admits a (QFI) and the related symmetric Killing tensor of second order:

$$A_{\alpha\beta} \frac{dx^\alpha}{ds} \frac{dx^\beta}{ds} = \text{constant}$$

Case I

$$A_{\alpha\beta;\gamma} + A_{\gamma\alpha;\beta} + A_{\beta\gamma;\alpha} = 0, \qquad (6.2)$$

$$f(x) A_{\alpha\beta} \frac{dx^\alpha}{ds} \frac{dx^\beta}{ds} = \text{constant}$$

Case II

$$A_{\alpha\beta;\gamma} + A_{\gamma\alpha;\beta} + A_{\beta\gamma;\alpha} = A_{\alpha\beta} (\ln f^{-1})_{,\gamma} + A_{\gamma\alpha} (\ln f^{-1})_{,\beta} + A_{\beta\gamma} (\ln f^{-1})_{,\alpha}, \qquad (6.3)$$

it follows with the help of (6.2) and (6.3) that the symmetric Killing tensors associated with the QFI define the following symmetry mapping
Case I

$$\overline{\left\{ \begin{matrix} \gamma \\ \alpha\beta \end{matrix} \right\}}_A = \left\{ \begin{matrix} \gamma \\ \alpha\beta \end{matrix} \right\} - A^{\gamma\delta} A_{\alpha\beta;\delta} \qquad (6.4)$$

or

$$\overline{\left\{ \begin{matrix} \gamma \\ \alpha\beta \end{matrix} \right\}}_A = \left\{ \begin{matrix} \gamma \\ \alpha\beta \end{matrix} \right\} + A^{\gamma\delta} (A_{\alpha\beta;\delta} + A_{\beta\delta;\alpha}),$$

Case II

$$\overline{\left\{ \begin{matrix} \gamma \\ \alpha\beta \end{matrix} \right\}}_A = \left\{ \begin{matrix} \gamma \\ \alpha\beta \end{matrix} \right\} + \tfrac{1}{2} A^{\gamma\delta} [A_{\alpha\delta} (\ln f^{-1})_{,\beta} + A_{\beta\delta} (\ln f^{-1})_{,\alpha}$$
$$+ A_{\alpha\beta} (\ln f^{-1})_{,\delta} - 2 A_{\alpha\beta;\delta}]. \qquad (6.5)$$

Conversely, starting with the mappings (6.4) and (6.5), when admitted, one can construct[45] the integrability conditions (6.2) and (6.3) and, thus, determine the $A_{\alpha\beta}$ in the (QFI) to within a multiplicative constant and an additive term consisting of a constant times the given $g_{\alpha\beta}$ (i.e., $A'_{\alpha\beta} = C_1 A_{\alpha\beta} + C_2 g_{\alpha\beta}$).

We note that if $\bar{g}_{\alpha\beta;\gamma} = 2\bar{g}_{\alpha\beta}\psi_{,\gamma} + \bar{g}_{\alpha\gamma}\psi_{,\beta} + \bar{g}_{\beta\gamma}\psi_{,\alpha}$ [alternative definition for geodesic correspondence[32] ($\bar{g}_{\alpha\beta} \rightarrow g_{\alpha\beta}$)] is substituted ($A_{\alpha\beta} = \bar{g}_{\alpha\beta}$) in (6.5) with $\ln f = -4\psi$ one obtains the usual defining equations

$$\overline{\left\{\begin{matrix}\gamma\\\alpha\beta\end{matrix}\right\}} = \left\{\begin{matrix}\gamma\\\alpha\beta\end{matrix}\right\} + \delta^\gamma_\alpha \psi_{,\beta} + \delta^\gamma_\beta \psi_{,\alpha}.$$

Thus the projective symmetry mapping usually referred to as geodesic correspondence[46] is a rather special case of the mapping (6.5) representing a symmetry property that is not describable by point deformations. The defining property of this mapping, if admitted, is that it maps one Riemannian space (metric $\bar{g}_{\alpha\beta}$) onto another distinct Riemannian space (metric $g_{\alpha\beta}$) such that the geodesic differential equations (in their parameter independent forms) of these spaces have the same detailed functional form. When geodesic correspondence is admitted by a given Riemannian spacetime it is easily seen that QFI immediately follow.[47]

VII. FIELD CONSERVATION LAW GENERATORS IN GENERAL RELATIVITY

The discussion and results of the previous sections strongly suggest that for the Riemannian spacetimes of Einstein's general theory of relativity we can only find conservation laws with at least some analogies to those of the familiar type in Lorentz-covariant theories in the special cases of spacetimes admitting groups of motions (isometries). Thus, in view of the fact that general Riemannian spacetimes do not admit isometries, it seems clear that these types of conservation laws cannot be as universally important in Einstein's general theory as in previous theories and, particularly, as in all Lorentz-covariant theories. Nonetheless, the Riemannian spacetimes admitting isometries are important because it is precisely in these special "simple" situations that the physical content of general relativity becomes more readily comprehensible. For more general Riemannian spacetime, not admitting isometries, there is always the possibility of other symmetry properties being admitted together with the related conservation laws that would have the disadvantage that, in general, they would be far more difficult to interpret. In any case, we wish to emphasize that the intimate relation of symmetry properties and the physical conservation laws known to be fundamental in other areas of physics is equally fundamental to the general theory of relativity[48]. When no symmetry properties can be recognized we still can use the identities and formal "conservation expressions" (see Section III) for mathematical reformulations and in conjunction with the problems of motion.[49] In any of the above cases it would appear that there is little hope of giving a definite local invariant meaning to a conservation expression unless it is suitably formulated in terms of (or referred to) special curve congruences or tetrad fields (i.e., special matter "streamlines" or local "observers").

In this section we are particularly interested in finding the simplest expressions that could play the role of generator of the physical conservation laws corresponding to symmetry properties being admitted by the given Riemannian spacetime. Here we limit ourselves to symmetry properties that can be characterized by continuous groups of point deformations ($x^\alpha \rightarrow x^\alpha + \epsilon \xi^\alpha$). We will be particularly interested in those explicit conservation expressions that are reasonably analogous to those of Lorentz-covariant theories. Presumably, the only really good analogies are for the integral expressions that follow for physically closed systems where the corresponding Riemannian spacetimes are asymptotically flat. Clearly, if suitable tensor expressions can be found, we will not have to be worried about our results being well defined in all coordinate frames.

In terms of the considerations of Section III it is not difficult to see that the simplest generator that can be given an explicit tensor formulation may be written as

$$(U_\gamma^{[\alpha\beta]}\xi^\gamma + W^{[\alpha\beta]})_{;\alpha,\beta} = 0. \tag{7.1}$$

Here, in accord with the above considerations, the ξ^α is to be taken as the mathematical representation of an actual symmetry property admitted by the given physical spacetime. Of course, the given $g_{\alpha\beta}$ determine the $U_\gamma^{[\alpha\beta]}$. In turn the $W^{[\alpha\beta]}$ (which is a function of the $g_{\alpha\beta}$ and the ξ^α and their derivatives) are fixed by demanding that the identity be of tensor form. The actual conservation laws are obtained from (7.1) after suitably eliminating the arbitrary parameters in the ξ^α representing the symmetry properties admitted. This method of obtaining the physical conservation laws (for Riemannian spacetimes admitting symmetry properties) that proceeds from a generator, we call the "symmetry method"[50]. It can be shown[50,51] that these demands lead to the following generator[52]:

$$[\sqrt{-g}\,(\xi^{\alpha;\beta} - \xi^{\beta;x})]_{;\alpha,\beta} = 0. \tag{7.2}$$

This expression is formally identical to the expression first obtained by Komar[17] but the interpretation and significance attributed to it here are quite different[50,51]. The above conservation law generator (7.2) can also be directly obtained by a procedure that only employs tensorial terms and operations and which is free of the addition of arbitrary elements[28].

Another expression, that meets the above demands when one restricts the symmetry properties to be only isometries, was first proposed by Trautman[53] in the context of symmetry consideration. Trautman's expression takes the form

$$[T^{\alpha\beta}\xi_\beta]_{,\alpha} = 0 \tag{7.3}$$

where $T^{\alpha\beta}$ is the Einstein tensor (weight one) and ξ^α has the same meaning as above. Actually, it was the integral form of the expression (7.3) that was actually proposed by Lanczos[11] as the invariant formulation of the conservation laws of Einstein's general theory of relativity. In particular, Lanczos treated the case of the total integral conservation laws of physically closed systems finding what he referred to as "boundary conditions" to be the solution of Killing equations in the asymptotically flat region of the given spacetime. Thus Lanczos was able to obtain properly all of the integral conservation laws of energy, momentum, and angular momentum if the corresponding symmetries were admitted by the given Riemannian spacetime.

It follows that the generators (7.2) and (7.3) cannot disagree by more than multiplicative constants and pure divergence terms in the case of motions[54]. The integral expressions that follow from them will fix these arbitrary elements to agree with each other and with dimensionally correct physically acceptable results. Here we are more interested in (7.2) because it is more general[55] than (7.3). In the next section these conservation generators will be further discussed in terms of specific applications. It will be particularly important to consider these comparatively simple applications before turning to still more general considerations.

VIII. COMMENTS ON APPLICATION OF CONSERVATION LAWS FOR PHYSICALLY CLOSED SYSTEMS

The most important early application in this area was to determine the total energy associated with the Schwarzschild metric using $(T^\alpha_\beta + t^\alpha_\beta)_{,\alpha} = 0$ where t^α_β is the Einstein complex[10,20]. In this section we first briefly consider such energy calculations before considering applications relating to momentum and angular momentum all based on (or related to) the conservation law generators discussed in Section VII. Here we will sketch the total Schwarzschild energy calculation using the more recent Landau–Lifschitz[12] complex $h^{\alpha\beta\gamma}$ evaluated in isotropic coordinates which gives the following result for the required components

$$h^{44\alpha} = (c^4/16\pi k)\,[(-g)g^{44}g^{ii}] = \frac{c^4 r_0 x^i}{8\pi k r^3} \qquad (i = 1,2,3). \tag{8.1}$$

Then we find (for more details see Ref. 12, pages 331, 334–346)

$$\mathscr{E} = \int [(-g)(T^{44}+\tau^{44})]dV = \int h^{44i}dS_i = mc^2$$

where $dS_i = (x_i/r)r^2 \sin\theta\,d\theta\,d\phi$ ($r \to \infty$) and $r_0 = 2mk/c^2$. Of course, the same result is obtained with the Einstein complex. Møller's[16] superpotential gives the same result in quasi-Galilean coordinates (for a "system at rest") and continues to give the same result for any purely spatial coordinate transformation $[J(x/\bar{x}) \neq 0]$.

It is not difficult to demonstrate that the conservation law generators (7.2) or (7.3) lead to the result (8.1) provided $\xi^\alpha = \delta_4{}^\alpha k$ (see footnote 52). In terms of symmetry considerations we would describe this situation by saying that a physically closed system possesses a total energy conservation law provided it admits a G_1 group of motions that can be physically identified as a displacement of the timelike coordinate. Of course, it is evident that such symmetry exists from the fact that the Schwarzschild metric admits static isometric coordinate representations. Thus, for physically closed systems, the integral conservation law for energy (aside from multiplicative constants) takes the form[56]

$$\int [\sqrt{-g}(\xi^{4;i} - \xi^{i;4})]dS_i \qquad (i = 1,2,3) \qquad (8.2)$$

where ξ^α is a solution of Killing's equations that can be physically identified as a time translation (just as in the case of Lorentz-covariant field theories). Clearly, after elimination of the arbitrary parameter in ξ^α (which is most simply and clearly accomplished while the generator is in differential form) one will obtain the same value for energy in all physically interesting coordinate frames if one continues to use the same appropriately transformed ξ^α. Thus, the problem of the nontensor character of the conservation laws of general relativity is solved in this instance by basing all calculations, in any given coordinate frame, on a tensorial conservation law generator.

In a similar way, it can easily be seen that integral conservation expressions for total momentum and total angular momentum can be found for physically closed systems admitting the required types of isometries. Again, it should be stressed that to make the identifications of "linear momentum" or "angular momentum" the corresponding groups of motions (admitted) have to be identified, in the context of the physical properties of the spacetime, as a spatial $r \leq 3$ parameter displacement group and a $p \leq 6$ parameter rotation group ($r+p+1 \leq 10$). As another example we might mention the total momentum of the Kerr metric has been calculated by independent methods using the generators (7.2) and (7.3) in conjunction with the appropriate symmetry considerations[57].

No problem is encountered in writing out the differential form of the conservation laws explicitly for symmetry properties more general than motions once the explicit ξ^α admitted by the given spacetime are known. Again, the number of conservation laws depends on the number of independent parameters characterizing the particular symmetry property admitted. The real problems here are involved with the discovery of the symmetry property admitted, finding the explicit form of the ξ^α (including some characterization of the group structure it represents), and then proceeding with the physical interpretation of the symmetry property. Of course, one must be particularly concerned with the question of useful applications for the conservation expressions that follow from the existence of the given symmetry property.

We can proceed with the specialization of the conservation law generator (7.2) under the assumption that certain symmetries are admitted. Of course, such expressions will only be valid for particular Riemannian spacetimes that admit the given symmetry property. To write out these specialized conservation law generators one simply makes use of the necessary and sufficient conditions (see, for example, defining equations under Fig. 1). In some cases it might be necessary to use their integrability conditions[30,32,33] together with some identities. As an example, it is easily seen that[37]

$$[\sqrt{-g}\,(\xi^{\alpha;\beta} - \tfrac{1}{4} g^{\alpha\beta} g_{\gamma\delta}\xi^{\gamma;\delta})]_{;\alpha,\beta} = 0 \qquad (8.3)$$

is a suitably specialized conservation law generator for spacetime admitting conformal motion represented by the ξ^α. Similarly, for spacetime admitting projective collineations[33,58] one finds[37]

$$[\sqrt{-g}\,(\xi^{\alpha;\beta} - (7/10)g^{\alpha\beta} g_{\gamma\delta}\xi^{\gamma;\delta})]_{;\alpha,\beta} = 0. \qquad (8.4)$$

More recently several examples of spacetimes admitting curvature collineations[35] have been discussed including examples representing gravitational pp waves[59].

IX. TENSORIAL TETRAD CONSERVATION LAW GENERATORS

In this section we wish to call attention to the fact that other types of interesting expressions can be constructed, with the help of tetrad fields[60] $h_A^\alpha(x)$ $(h_A^\alpha h_\beta^A = \delta_\beta^\alpha)$, that could justify the term conservation law generator[61]. For particular choices of the class of tetrad fields these expressions [tensorial tetrad conservation law generators (TTCLG)] include the results given earlier in Sections VII and VIII. For any type of essentially local application of conservation laws, assuming the relevant symmetry properties are admitted, it appears that a tetrad formulation is quite natural[61-62], if not essential.

A little reflection shows that identically vanishing (conservation-law generators) of the type (7.2) could be constructed from any antisymmetric tensor of second order that can be formed from the $g_{\alpha\beta}$, $g_{\alpha\beta,\gamma}$, $g_{\alpha\beta,\gamma,\delta}$ and the ξ^α and its derivatives. However, all the known expressions of this type, which differ essentially from the expression (7.2), are rather complicated and do not reduce to the established results given, for example, by the Møller superpotential for situations where the latter are known to lead to valid results. It will be seen that the situation is quite different with regard to formulating conservation-law generators in terms of tetrads and tensors that can only be constructed in terms of tetrads.

In attempting to formulate suitable TTCLG, it should be borne in mind that the expressions obtained are not definite or determined, even in the case of a given Riemannian spacetime, until a tetrad field $h_A^\alpha(x)$ is specified in some manner. Thus, in any given case, the reduction of a suitable TTCLG to familiar results can only be expected for a specific class of tetrads defined in terms of the components of the metric $g_{\alpha\beta}(x)$ at each point of spacetime and for choices of the ξ^α which represent symmetry properties of the given spacetime, if any are admitted. This type of specification is only suitable for applications over extended regions. However, the arbitrariness in specifying the tetrad field plays a more essential role, in general, in that it is precisely the freedom that is needed in specifying a frame of reference for local applications of conservation laws. In this connection, the intrinsic or invariant forms of TTCLG's are of particular interest.

The simplest tensor with a pair of antisymmetric indices that can be formed from the h_A^α and its first derivatives is

$$\gamma^{\alpha\beta\gamma} \equiv h_A^\alpha h^{A\beta;\gamma} = \tfrac{1}{2}(h_A^\alpha h^{A\beta;\gamma} - h_A^\beta h^{A\alpha;\gamma}). \tag{9.1}$$

The third-order tensor $\gamma^{\alpha\beta\gamma}$ is the contravariant form of the associate tensor of the Ricci coefficients of rotation $\gamma^{ABC} = h_{\alpha;\beta}^A h^{B\alpha} h^{C\beta}$. Using this tensor it is easily seen that the simplest tentative TTCLG that can be found[61] is given by

$$(\{U_\gamma^{[\alpha\beta]}\xi^\gamma\}_{,\beta})_{,\alpha} \equiv (\{|h| [a\gamma^{\alpha\beta} + b(\delta_\gamma^\alpha \gamma_\delta^{\beta\delta} - \delta_\gamma^\beta \gamma_\delta^{\alpha\delta})]\xi^\gamma\}_{;\beta}\delta)_{,\alpha} \equiv 0. \tag{9.2}$$

If one takes $\xi^\alpha = \delta_4^\alpha$ and choses the $h_A^\alpha = (\sqrt{g_{AA}})^{-1}\delta_A^\alpha$ it is easily shown that $U_\gamma^{[\alpha\beta]} \equiv |h|[a\gamma_\gamma^{\alpha\beta} + b(\delta_\gamma^\alpha \gamma_\delta^{\beta\delta} - \delta_\gamma^\beta \gamma_\delta^{\alpha\delta})]$ reduces to the Freud superpotential for $\tfrac{1}{2}a = b = 1$. Thus, with this choice of the tetrad field and an arbitrary $\xi^\alpha(x)$ one can write

$$(U_\gamma^{\alpha\beta}\xi^\gamma)_{,\beta} = U_{\gamma,\beta}^{\alpha\beta}\xi^\gamma + U_\gamma^{\alpha\beta}\xi_{,\beta}^\gamma, \tag{9.3}$$

where $U_\gamma^{\alpha\beta} = (\sqrt{-g})^{-1} g_{\gamma\delta} [(-g)(g^{\alpha\lambda}g^{\beta\delta} - g^{\beta\lambda}g^{\alpha\delta})]_{,\lambda}$ is one form of the Freud superpotential.

The tetrad superpotential with $\frac{1}{2}a = b = 1$ is precisely the expression found by Møller[63] from a tetrad formulation of general relativity based on a variational principle of the form[64]

$$\delta \int [|h|(h_{;\beta}^{A\alpha}h_{A;\alpha}^\beta - h_{;\alpha}^{A\alpha}h_{A;\beta}^\beta)] \, d_4x = 0. \tag{9.4}$$

where $|h| = \det(h_{A\alpha}) = \pm\sqrt{-g}$. However, Møller does not use $U_\gamma^{[\alpha\beta]}$ in a tensor formulation; [i.e., in the form of (9.2)] but, rather, takes its ordinary divergence to form

$$U_{\gamma,\beta}^{[\alpha\beta]} = T_\gamma^\alpha(h_A^\mu, h_{A,\delta}^\mu, h_{A,\delta,\lambda}^\mu), \tag{9.5}$$

so that a tetrad gravitation pseudotensor is obtained. In addition, Møller[63] proposed to add supplementary conditions to fix the arbitrariness in the choice of tetrad field.

Here we propose that the arbitrariness in the choice of tetrad fields is quite in order[61,62] and that it should be fixed only when a specific local frame of reference has been chosen in which the conservation laws are to be applied. Of course, this is to be taken to include the possibility that the choice of a certain class of reference frames admitted by the given spacetime fixes the choice of tetrads. Also, it is proposed that this tetrad formulation should only be applied in accord with the symmetry method (see Sections VII and VIII) of obtaining conservation laws from tensorial generators, where the ξ^α must represent symmetry properties (e.g., r-parameter group of motions) admitted by the given spacetime.

When one considers tetrads and TTCLG locally determined by particular observers (defined, for example, along certain timelike trajectories) or matter "streamlines", it becomes clear that tetrad formulations become involved with the properties of curve congruences and transport processes. In this connection it is of interest to note that for a tetrad h_A^α suitably defined along a normal curve congruence u^α with ξ^α a Killing vector $(u^\alpha\xi_\alpha = \text{constant})$, the integral form of the TTCLG (9.2) reduces to Pirani's energy expression[22] provided $h_{A\alpha;\beta}\xi^\alpha = 0 (A = 0,1,2,3)$ where $h_{0\alpha} = u_\alpha$. The intrinsic formulation of the results of this section in terms of tetrad components gives results that do not give a conservation law in the expected form[61] in the case of tetrads that are anholonomic[65] (i.e., $\gamma_{[AB]}^C \neq 0$). A simple physical example of this type is provided by the case of an observer comoving with rotating "dust" sources[66]. More generally, anholonomic frames would play a fundamental role in investigating conservation laws of spacetimes with matter tensors that possess an intrinsically twisting timelike principal direction. Here, the natural local comoving frame is anholonomic and tetrad conservation law generators in the intrinsic form allow precisely the needed freedom in specifying such local observers.

X. CONSERVATION LAWS FOR CURVE CONGRUENCES WITH SPECIAL KINEMATICAL AND SYMMETRY PROPERTIES

Particular types of curve congruences are known to provide an important type of invariant characterization of the physical properties of Riemannian spacetimes[67]. We will refer to these invariant properties as kinematical properties. In general, these "kinematical" properties provide significant information (constituting an invariant characterization) in addition to that provided by symmetry group properties[68] which may be admitted. The simplest examples of these kinematical quantities specifying any curve congruence[69] are provided by the following familiar definitions given for the case of a timelike congruence defined by the vector field $u^\alpha(x)$ (where $u^\alpha u_\alpha = +1$ and the given metric is of signature -2):

$$\theta = u^\alpha_{;\alpha} \quad \text{(expansion)}, \tag{10.1a}$$

$$\sigma_{\alpha\beta} = u_{(\alpha;\beta)} - u_{(\alpha;|\gamma|u_\beta)}u^\gamma - (1/3)\theta h_{\alpha\beta} \quad \text{(shear)}, \tag{10.1b}$$

$$\omega_{\alpha\beta} = u_{[\alpha;\beta]} - u_{[\alpha;|\gamma|u_\beta]}u^\gamma \quad \text{(rotation)}, \tag{10.1c}$$

where $h_{\alpha\beta} = g_{\alpha\beta} - u_\alpha u_\beta$ is the projection operator for the given congruence.

Clearly, any conservation laws obeyed by these fundamental (or principal) types of curve congruences are bound to be of interest. We will see that conservation expressions of this type are especially interesting when considered in conjunction with particular symmetry properties (represented by symmetry groups) admitted by the given spacetime. Of course, it is to be recognized that, in general, such kinematical and symmetry demands are not unrelated. Indeed, by considering both kinematical properties and symmetry groups we are in a better position to interpret physically these conservation expressions in the broader context of the physical situation constituted by the given Riemannian spacetime. It follows that we should be able to learn how to make various types of useful application of such expressions. Of course, in the simplest cases, we could expect little beyond more or less obvious results. Here again, the subject that we propose to briefly discuss in this section is quite large and the best we can hope to do is to give some simple examples[70]. Also, we will try to make some comments that may be of value in finding and applying more significant examples in concrete cases of physical interest.

For our first example of a conservation expression associated with a curve congruence we consider the following very simple example of an expression involving a timelike geodesic congruence u^α and a Killing vector ξ^α ($u^\alpha = dx^\alpha/ds$):

$$(\sqrt{-g}\,\rho u^\alpha u^\beta \xi_\beta)_{;\alpha} = (\sqrt{-g}\,\rho u^\alpha u^\beta \xi_\beta)_{,\alpha}$$

$$= \sqrt{-g}\left(\frac{d\rho}{ds} + \theta\rho\right)u^\alpha \xi_\alpha + \sqrt{-g}\,\rho\,(u^\beta_{;\alpha}u^\alpha \xi_\beta + u^\alpha u^\beta \xi_{\alpha;\beta})$$

$$(10.2)$$

where ρ is a scalar function. Clearly, since $u^\alpha \xi_\alpha = \text{constant}$ ($\neq 0$ in general), the expression (10.2) leads to a weak conservation law generator for ρ satisfying $d\rho/ds + \theta\rho = 0$ and the ξ^α representing a group of motions (isometries). This conservation law generator can be immediately seen to be a special case of the generator (7.3) for a "dust" spacetime where $T^{\alpha\beta} = \rho u^\alpha u^\beta$. This example could be immediately extended to the case of a perfect fluid where one would find conditions representing the familiar hydrodynamical equations[71]. The value of this example is that it serves to point out that a number of the interesting expressions obtained will involve familiar tensors written in terms of the fundamental (or principal) curve congruences admitted in a form that may be less familiar[72].

For Riemannian spacetimes admitting symmetry properties leading to m-th order first integrals (FI) of geodesics (see Sections V and VI) $A_{\alpha_1 \ldots \alpha_m}u^{\alpha_1} \cdots u^{\alpha_m} = \text{constant}$ (where $A_{(\alpha_1 \ldots \alpha_m;\beta)} = 0$) we can write the following simple type of field conservation law generator for a timelike geodesic congruence ($u^\alpha = dx^\alpha/ds$):

$$(\sqrt{-g}\,\rho u^{\alpha_1} \cdots u^{\alpha_m}u^\beta A_{\alpha_1 \ldots \alpha_m})_{,\beta} = 0 \qquad (10.3)$$

for $d\rho/ds + \theta\rho = 0$. Obviously, the conservation expression (10.3) involving $m + 1$ u^α's is the field (geodesic curve congruence) analogue of the m-th order (FI) for the geodesic differential equations. Affine, projective, and special curvature collineations provide immediate simple examples of symmetry properties required for field conservation laws of this type. It will be noted that the generator (10.3) is equally valid for spacetimes admitting m-th order (FI) due to symmetry properties that cannot be characterized as point deformations (see Sections IV and VI).

The two simplest conservation expressions depending on the timelike geodesic congruence being shear free with symmetry defined by $\xi_{(\alpha;\beta);\gamma} = 0$ and rotation free with symmetry defined by $\xi_{[\alpha;\beta];\gamma} = 0$ are the following:

$$\{\sqrt{-g}\,[\tfrac{1}{2}(u^\alpha g^{\beta\gamma} + u^\beta g^{\alpha\gamma}) - \tfrac{1}{3}u^\gamma h^{\alpha\beta}]\xi_{\alpha;\beta}\}_{,\gamma} = 0 \qquad (10.4)$$

and

$$[\sqrt{-g}\,(u^\alpha g^{\beta\gamma} - u^\beta g^{\alpha\gamma})\xi_{\alpha;\beta}]_{,\gamma} = 0. \qquad (10.5)$$

For geodesic congruences a number of other simple expressions can be easily constructed using the rotation[73] $\omega_{\alpha\beta}$ or the dual of the rotation $*\omega^{\alpha\beta}$ where $(*\omega^{\alpha\beta}_{;\beta} = 0)$ and one or more vectors representing symmetry properties admitted by the given Riemannian spacetime. For example, in the case of a spacetime admitting a symmetry $\xi_{[\alpha;\beta];\gamma} = 0$ and a particular timelike congruence, where the absolute derivative of $\omega^{\alpha\beta}$ vanishes along the congruence with $d\rho/ds + \theta\rho = 0$, one easily finds

$$(\sqrt{-g}\,\rho\omega^{\alpha\beta}u^\gamma\xi_{[\alpha;\beta]})_{,\gamma} = 0. \qquad (10.6)$$

Also, for a geodesic congruence in the special case $\xi_{[\alpha;\beta]} = 0$ one can write

$$(\sqrt{-g}\,*\omega^{\alpha\beta}\xi_\alpha)_{,\beta} = 0. \qquad (10.7)$$

Aside from considerations of nongeodesic curve congruences, one can obtain expressions that appear to be potentially far more interesting than those given above (and their immediate generalizations) by working with expressions that involve one or more of the quantities θ, $\sigma_{\alpha\beta}$, $\omega_{\alpha\beta}$ and their absolute derivatives and one or more symmetry vectors that may be independent (corresponding to the same type of symmetry or a different symmetry). Of course, integral formulations of conservation expressions based on the above considerations could be of particular interest. Clearly, a general systematic classification of all the simpler types of potential conservation expressions for curve congruences is suggested by the above results and comments. In the near future we plan to report on some preliminary results of this type of investigation.

XI. OTHER INTEGRAL CONSERVATION EXPRESSIONS

It has been pointed out by Synge[74] that Stokes' Theorem and its extensions[75] are just as valuable as the familiar hypersurface integral expression $\oint \Lambda^\alpha d\Sigma_\alpha = 0$ (leading to $\int_1 \Lambda^\alpha d\Sigma_\alpha = \int_2 \Lambda^\alpha d\Sigma_\alpha$) in formulating integral conservation laws. For example, consider a closed 2-space V_2 spanned by an open 3-space V_3 with

$$\oint_{V_2} \Lambda_{\alpha\beta}\,d\tau^{[\alpha\beta]} = \int_{V_3} \Lambda_{\alpha\beta;\gamma}\,d\tau^{[\alpha\beta\gamma]} \qquad (11.1)$$

where it is noted that the integral on the right has the same value for all 3-spaces spanning the given closed V_2. In many respects this constitutes an integral conservation law in the same sense as the more familiar expression, except, unlike the Λ^α there are no conditions on the $\Lambda^{\alpha\beta}$.

Clearly, we will only be interested in antisymmetric $\Lambda_{\alpha\beta}$ (11.1). Synge suggested that (11.1) may be regarded as a "factory for making integral conservation laws" and that "our task is to select an $\Lambda_{\alpha\beta}$ to yield a law of physical interest"[74]. Here again, it would appear that symmetry considerations would be fundamental to this selection of physically interesting examples of $\Lambda_{\alpha\beta}$.

Of course, symmetry demands require special properties of the $\Lambda_{\alpha\beta}$. One important antisymmetric tensor of this type, which radically simplifies (11.1), is the harmonic tensor[76] defined by

$$\phi_{[\alpha\beta];\gamma} + \phi_{[\gamma\alpha];\beta} + \phi_{[\beta\gamma];\alpha} = 0 \tag{11.2a}$$

$$g^{\alpha\gamma}\phi_{[\alpha\beta];\gamma} = 0. \tag{11.2b}$$

Clearly, in this case, (11.1) leads to the following simple integral conservation law

$$\oint_{V_2} \phi_{[\alpha\beta]} \, d\tau^{[\alpha\beta]} = 0. \tag{11.3}$$

A concrete example of (11.3) is provided by the antisymmetric Killing tensor $\xi_{\alpha;\beta}$ (harmonic tensor second order) formed from a given Killing vector. The above example of harmonic tensors is very special and it should in no way prejudice the general merit of the expression (11.1) in its role as an integral conservation law generator.

Synge suggested that the following form of (11.1) may be of particular interest[74]

$$\oint_{V_2} F^{\alpha\beta}R_{\alpha\beta\gamma\delta} \, d\tau^{[\gamma\delta]} = \int_{V_3} F^{\alpha\beta}_{;\lambda}R_{\alpha\beta\gamma\delta} \, d\tau^{[\gamma\delta\lambda]} \tag{11.4}$$

where $\int F^{\alpha\beta}R_{\alpha\beta\gamma\delta;\lambda} \, d\tau^{[\gamma\delta\lambda]}$ vanishes on account of the Bianchi identities. Part of the motivation for interest in this particular expression comes from the observation that we need a nonvanishing conservation law for spacetimes where $R_{\alpha\beta}$ vanishes in certain regions (or everywhere)[77]. Again, the task is to pick out physically significant $F^{\alpha\beta}$. One choice suggested by Synge[74] is that the $F^{\alpha\beta}$ should be an eigentensor of the Riemann tensor (i.e., $F^{\alpha\beta}R_{\alpha\beta\gamma\delta} = \phi F_{\gamma\delta}$). Other choices made by Synge[74] were selected so as to yield a law having some resemblance to Newtonian conservation laws. We will not consider this matter further here; however, we wish to point out that there are numerous interesting possibilities for expressions like (11.4) suggested by the results of Section IX and other combinations involving the Riemann tensor and its duals (i.e., $R^{\alpha\beta}_{\mu\nu}R_{\alpha\beta\gamma\delta}$, $R^{\overset{*}{\alpha\beta}}_{\mu\nu}R_{\alpha\beta\gamma\delta}$, $R^{\overset{*}{\alpha\beta}}_{\overset{*}{\mu\nu}} R_{\alpha\beta\gamma\delta}$). Of course, if one has in mind essentially local applications one would expect that expressions constructed from (or involving) principal curve congruences and tetrads would be the most appropriate.

XII. CONCLUSIONS

(1) The conservation law generators (7.2) ($\{[\sqrt{-g}(\xi^{\alpha;\beta} - \xi^{\beta;\alpha})]_{,\alpha}\}_{;\beta} = 0$), (7.3) ($(\sqrt{-g}T_\beta{}^\alpha\xi^\beta)_{,\alpha} = 0$), and (9.2) ($[U_\gamma{}^{[\alpha\beta]}\xi^\gamma)_{;\beta}]_{,\alpha} = 0$) can provide local tensor conservation laws in differential form that are reasonably analogous to those of Lorentz-covariant field theories (in accord with the symmetry method) provided the given Riemannian spacetimes admit groups of motions (isometries). The particular conservation laws admitted ($\leqq 10$) depend on the physical identification of the particular group of motions admitted as a proper subgroup of the full Lorentz group.

(2) For asymptotically flat Riemannian spacetimes (physically closed systems) admitting asymptotic Killing vectors one can find (global) integral conservation laws (using the generators mentioned in (1) above) analogous to those of Lorentz-covariant field theories. Again, the particular conservation laws admitted ($\leqq 10$) depend on the physical identification of the particular group of motions admitted as a proper subgroup of the full Lorentz group.

(3) Symmetry properties (not necessarily isometries) are just as important in the Riemannian spacetime of Einstein's theory as they are in all other physical theories, the primary difference being that there are no built-in guaranteed symmetries in any given case. When they exist, they represent important physical properties of the spacetime (with important physical consequences) as evidenced in part by the first integrals they lead to for geodesic differential equations. Thus, in general relativity our interest in symmetry properties cannot be limited to those particular symmetries that preserve the field equations.

(4) The demand of general coordinate covariance in field theory leads to the formulation of strong conservation expressions independent of the question of particular symmetry properties being admitted. In the case of general relativity this means that the strong conservation expressions must include all the special conservation expressions admitted by particular Riemannian spacetimes (corresponding to particular symmetries that are a subgroup of $G_{\infty 4}$) in all their isometric representations. The viewpoint of strong generators embracing the totality of possible conservation expressions (corresponding to a given type of generator) for all spacetimes is of very limited value in practice. In any case, the actual physical conservation laws follow as a consequence of the symmetry properties that are admitted by the given Riemannian spacetime. The strong conservation expressions and identities following from general covariance are useful for mathematical reformulation even in the absence of symmetries.

(5) The general identities that follow from invariance demands under

infinite groups $G_{\infty n}$ (including general coordinate covariance in spacetime $G_{\infty 4}$) have another important area of application in field theories outside their use in mathematical reformulations and the construction of super-potentials. Namely, they can be used as a systematic method of identifying the consequences of symmetry demands and to help determine suitable modifications of field structures admitting particular symmetry properties.

(6) The conservation law generators (7.2) and (9.2) also lead to tensor conservation laws in the case of symmetry properties more general than motions. The number depends on the number of independent parameters involved in the vector ξ^α representing the symmetry property admitted. Here the main difficulty is in interpreting physically these more general symmetries and the conservation expressions they provide. Any integral conservation expressions formed from (7.2) and (9.2) will vanish for a closed universe.

(7) A tetrad formulation of the conservation laws in conjunction with the symmetry method appears to be essential to all strictly local applications involving particular observers and related transport processes. Thus, no general *a priori* conditions are to be imposed to remove the usual arbitrariness of the tetrad field. The arbitrariness in the orthonormal tetrad field (or quasi-orthonormal tetrad field for some radiation problems) is naturally determined by the observer, his transport process, and the physical application to be made.

(8) There appear to be many interesting possibilities for useful conservation expressions associated with the fundamental (or principal) curve congruences (with special kinematical properties) of Riemannian space-time together with various possible symmetry group properties. For essential local applications a tetrad formulation of such expressions could be required.

(9) Integral expressions based on Stokes' Theorem and its extensions appear to offer the possibility of very important applications, particularly, when used in conjunction with the above mentioned kinematical and symmetry considerations.

ACKNOWLEDGMENT

The author would like to take this occasion to express his deep appreciation to Professor Cornelius Lanczos for many helpful and stimulating discussions spanning a number of years. In regard to the present paper the author wishes to thank Dr. James York, Jr. for his helpful comments.

REFERENCES

1. By the differential form of a field conservation law we mean simply a local conservation law in the form of a vanishing ordinary four divergence (i.e., $T^{\alpha_1 \cdots \alpha \rho \beta}{}_{,\beta} = 0$). In this paper we do not refer to vanishing covariant divergences (i.e., $T^{\alpha_1 \cdots \alpha \rho \beta}{}_{;\beta} = 0$) as "conservation expressions" unless they happen to be equivalent to an ordinary divergence in some special case. Clearly, for any vector density of weight one (relative tensor of first order weight one) $T^{\beta}{}_{;\beta} \equiv T^{\beta}{}_{,\beta}$. Throughout this paper the operation of covariant differentiation will be denoted by a semicolon and partial differentiation by a comma. Brackets, [ ], and parentheses, ( ), are used to denote antisymmetrization and symmetrization of indices, respectively.

2. Earlier, two excellent survey articles, which also contain new results, were written covering this period: J. G. Fletcher, *Rev. Mod. Phys.* **32**, 65 (1960) and A. Trautman, article in *Gravitation: An Introduction to Current Research*, Ed. L. Witten (John Wiley and Sons, New York, 1962) p. 169. Also, R. Arnowitt, S. Deser, and C. Misner, *Phys. Rev.*, **122**, 997 (1961) considered most of the expressions for energy and momentum that are mentioned in Sections I and II in terms of invariance properties, physical boundary conditions, and the properties of the Hamiltonian in a canonical formulation.

3. For example, the original form of the energy-momentum conservation law proposed by Einstein [A. Einstein, *Preuss. Akad. Wiss. Sitz.*, 778 (1916b pt. 2) 1111 (1916c)] involving his gravitational pseudotensor leads to quite different values of the total energy for a physically closed system (e.g., Schwarzschild metric) just by transforming to a new coordinate system as first discussed by Schrödinger [E. Schrödinger, *Physik. Z.*, **19**, 4 (1918)] and Bauer [H. Bauer, *Physik. Z.* **19**, 163 (1918)]. In particular, Schrödinger found a coordinate frame in which all the energy components vanish identically for the field outside a spherically symmetric mass distribution. See W. Pauli, (translated by G. Field) *The Theory of Relativity* (Pergamon Press, New York, 1958) pp. 175–178 for further comments on this early history.

4. The superpotential $U_\alpha^{[\beta\gamma]}$ of a given "strong" conservation expression $T^\beta_{\alpha,\beta} \equiv 0$ is defined by the relation $T^\beta_\alpha \equiv U^{[\beta\gamma]}_{\alpha,\gamma}$. Here the term strong refers to the fact that the conservation expression vanishes as an identity. One can define a multiple infinity of other distinct identically vanishing divergence expressions [i.e., $(U_\alpha^{[\beta\gamma]}\phi^\alpha)_{,\gamma,\beta} \equiv 0$] corresponding to the possible distinct choices for the vector ϕ^α. (See Section III).

5. See, for example, L. P. Einsenhart, *Riemannian Geometry* (Princeton Univ. Press, Princeton, 1926) p. 233 and J. A. Schouten, *Ricci-Calculus* (Springer-Verlag, Berlin 1954) IV.

6. Here we refer to the fact that Lorentz-covariance (corresponding to the G_{10} group of motions admitted by flat spacetime) leads directly to the conservation laws of energy, momentum, and angular momentum. It should be noted that there are six angular momentum conservation laws. In addition to the

three familiar laws, one finds [see C. Lanczos, *Z. Physik*, **59**, 514 (1930)] three laws relating to the position of the center of inertia.

7. E. Noether, *Nachr. Akad. Wiss. Göttingen Math. Phys. Kl.*, 235 (1918).

8. Actually, one usually refers to the first and second Noether theorems. The first theorem relates to the weak conservation laws that follow when symmetry properties are admitted (*cf.* Section IV) and the equations of motion are satisfied. The second theorem refers to the identities that follow when the theory is invariant under a general group $G_{\infty n}$ (*cf.* Sec. III). For a more precise statement of the Noether theorems see, for example, A. Trautman[2] p. 178. More recently Trautman [A. Trautman, *Commun. Math. Phys.* **6**, 248 (1967)] has given a rigorous derivation of the Noether theorems in terms of modern mathematical methods. Further comments and references relating to the Noether theorem and its inversion can be found in a more recent paper by E. Candotti, C. Palmieri, and B. Vitale, *Nuovo Cimento*, **70A**, 233 (1970).

9. The Riemannian spacetimes referred to here as physically closed systems are asymptotically flat, thus, one can always introduce coordinate systems that would become Cartesian asymptotically.

10. Here it should be mentioned that Einstein [A. Einstein, *Preuss. Akad. Wiss Sitz.*, 448 (1918f)] and Klein [F. Klein, *Nachr. Akad. Wiss. Göttingen*, 394 (1918)] found further clarification of the significance of the expression $P_\alpha = \int (T_\alpha{}^4 + t_\alpha{}^4) \, dx^1 \, dx^2 \, dx^3$ called the total energy and momentum expression for a closed system where $T_\alpha{}^\beta$ is the usual matter tensor density and $t_\alpha{}^\beta$ is the Einstein pseudotensor. In particular, for reasonable assumptions defining the closed system they showed that the values of energy and momentum are independent of the choice of coordinates in the interior of the system provided they go into a quasi-Galilean system outside and that the components P_α behave like the components of a covariant vector under linear coordinate transformations. In this early period it was recognized [H. A. Lorentz *Amst. Akad. Versl.* **25**, 468, 1380 (1916)] that the expressions t_α^β were only unique to the extent that one would insist that they contain only up to first derivatives of the metric tensor. Also, Rosenfeld [L. Rosenfeld, *Acad. Roy. Belg.*, **18**, No. 6 (1940)] provided results and discussions contributing to the above considerations. In connection with the problem of the nontensorial character of Einstein's conservation law we should mention the works of N. Rosen (1940, 1963), A. Papapetrou (1948), and M. Kohler (1952, 1953) who have shown that a tensor formulation of the conservation laws can be obtained by introducing into the usual formalism of general relativity a second flat spacetime metric tensor $\bar{g}_{\alpha\beta}$. We do not consider these results here because of the arbitrary elements this formulation involves and because a more fundamental type of tensor formulation can be achieved by the tetrad formulation of general relativity which includes the second flat-metric formulation as a very special case. For a discussion of this matter and references to the above mentioned works see W. R. Davis, *Nuovo Cimento*, **43**, 200 (1966).

In connection with the above mentioned definition of total mass it should be stressed that the concept of mass depends crucially on the notion of a physically closed system and an asymptotic Newtonian framework (see, for example, C. Misner[56]). Of course, the mass so computed is also the "tidal" mass [i.e., the mass which measures the relative acceleration of particles (geodesic deviation)] corresponding to the fundamental tidal nature of the gravitational field of any source [in this connection see F. A. E. Pirani, *Acta Phys. Pol.*, **XV**, 389 (1956)].

11. Lanczos [C. Lanczos, *Z. Physik* **59**, 514 (1930)] viewed the condition $\phi_{\alpha;\beta} + \phi_{\beta;\alpha} = 0$ that he required in the asymptotic region as a necessary boundary condition to obtain covariant integral conservation laws. In the interior region he found that the ϕ_α would have to satisfy $g^{\beta\gamma}\phi_{\alpha;\beta;\gamma} + R^\gamma_\alpha \phi_\gamma = 0$. Also, it might be noted that Lanczos' notation for covariant differentiation in the above mentioned paper could easily be confused with ordinary partial differentiation.

12. L. Landau and E. Lifschitz (translated by M. Hamermesh), *The Classical Theory of Fields* (Addison-Wesley Pub. Co., Reading, Mass., 1951).

13. J. N. Goldberg, *Phys. Rev.*, **111**, 315 (1958).

14. P. G. Bergmann, *Phys. Rev.*, **112**, 287 (1958).

15. At this point we will adopt the term *complex* to describe the nontensorial geometric objects such as the various gravitational pseudotensors and their superpotentials that have been or will be introduced.

16. C. Møller, *Ann. Phys.*, **4**, 347 (1958).

17. A. Komar, *Phys. Rev.*, **113**, 934 (1959).

18. P. A. M. Dirac, *Phys. Rev. Letters*, **2**, 368 (1959). In this paper Dirac stresses that the primary requirement for conservation laws is that they provide *useful* integrals of the equations of motion. His approach [like that of Arnowitt, Deser, and Misner, *Phys. Rev.* **118**, 1100 (1960)] is to solve the energy problem in terms of a proper canonical formulation of Einstein's theory. In the linear approximation essentially the same results are found as when one uses the Einstein complex $t_\alpha{}^\beta$.

19. P. Freud, *Ann. Math.*, **40**, 417 (1939).

20. For example, using the Freud superpotential one can immediately write Einstein's total energy quantity (see footnote 10) in the form $P_4 = \oint U_4{}^{[4i]} dS_i$, $(i = 1, 2, 3)$. Also, because the superpotential makes it possible to write hypersurface integrals in terms of two-dimensional surface integrals, we can conclude that the total energy in any closed universe will vanish. Of course, we must have tensor expressions to make this statement rigorous.

21. As noted by Møller the integral expressions following from his complex are, in general, hypersurface dependent. This can be understood, in part, by noting that it has a different $1/r$ dependence than the Einstein complex for isolated systems. For a discussion of these and other related problems see C. Møller, *Ann. Phys.*, **12**, 118 (1961). Of course, even if the Møller expression had no defects one could still not deal with conserved quantities in any given physical frame of reference labelled with general coordinates. In particular, Møller's expression depends on one being able to define a frame at "rest".

22. F. E. A. Pirani, paper presented at the conference on Relativistic Theories of Gravitation, l'Abbaye de Royaumont, France, 1959. A. Trautman[2] p. 195 gives further comments on this work. For applications of this approach and further related developments see H. Dehnen, *Z. Physik*, **179**, 76, 96 (1964).

23. E. Schrödinger, *Proc. Roy Irish Acad.*, **52**, 1 (1948); **54**, 79 (1951).

24. P. G. Bergmann, *Phys. Rev.*, **75**, 680 (1949) and (with R. Schiller) *Phys. Rev.* **89**, 4 (1953); W. R. Davis, *Z. Physik*, **148**, 1 (1957). For a general discussion of the identities (including uses and examples) that follow for any kind of general invariance group see W. R. Davis and J. W. York, *Nuovo Cimento*, **65B**, 1 (1970).

25. Here it will be assumed that the field variables Ψ_A and Φ_Ω are defined over an n-dimensional manifold so that the lower case Greek letters run from 1 to n.

26. These conditions are found from the requirement that the commutator of two such transformations be a transformation of the same type.

27. J. Heller, *Phys. Rev.*, **81**, 946 (1951).

28. This simple identically vanishing divergence expression can also be obtained by straight-forward tensor operators starting with the general invariance demand on the variational principle [W. R. Davis and M. K. Moss, *J. Math. Phys.*, **7**, 975 (1966)].

29. For a more comprehensive discussion of symmetry properties and invariance principles and additional references see E. P. Wigner, *Proc. Nat. Acad. Sci. (U.S)*, **41**, 956 (1964), R. M. F. Houtappel, H. Van Dam, and E. P. Wigner, *Rev. Mod. Phys.*, **37**, 595 (1965) and W. R. Davis and James W. York, Jr., *Nuovo Cimento*, **65B**, 1 (1970); (with M. K. Moss) *Nuovo Cimento*, **65B**, 19 (1970). Houtappel, Van Dam and Wigner point out that invariance principles are used in two distinct ways in physics. First, they are used as superlaws which may guide research toward the unknown laws of nature at any given time. Second, they serve as tools that help provide information relating to the properties of the solutions of equations representing the laws of nature. They are particularly concerned with the problem of giving invariances a formulation directly in terms of observations or measurements and their results. They refer to invariances that can be so formulated as geometric invariances.

30. Here we use methods and notations that are most closely related to those of J. A. Schouten, *Ricci Calculus*, 2nd ed., (Springer-Verlag, Berlin, 1954) IV-VI; VII and K. Yano, *The Theory of Lie Derivatives and Its Applications* (North-Holland Pub. Co., Amsterdam, 1957). In particular, the operation of Lie differentiation with respect to the vector ξ^α is denoted by the symbol $\mathscr{L}$.

31. A symmetry property (that may be characterized as a continuous group) represented by an infinitesimal mapping (or transformation) can be viewed as a generalized deformation of an object of physical interest characterized by a δ-operation on the given object such that it is left invariant. [For example, in the case of Riemannian spaces admitting isometries $\delta\gamma_{\alpha\beta} = \epsilon\mathscr{L}g_{\alpha\beta} = \epsilon(\xi_{\alpha;\beta} + \xi_{\beta;\alpha}) = 0$.] This general δ operator can usually be thought of as a generalized Lie differential in some space where again the idea of

a symmetry property corresponding to the "non-distortion" of some physical object is helpful. However, there are many instances where a symmetry property can be defined by requiring that a given physical object obey a specific special infinitesimal transformation law (under the δ-"deformation") rather than requiring it to vanish. In most cases of this type one can find some other object with a vanishing δ-"deformation". As an example of these last remarks, consider Riemannian spaces admitting projective collineations defined by $(\mathscr{L}g_{\alpha\beta})_{;\gamma} = 2g_{\alpha\beta}\psi_{,\gamma} + g_{\gamma\alpha}\psi_{,\beta} + g_{\beta\gamma}\psi_{,\alpha}$ or by $\mathscr{L}\Pi^{\gamma}_{\alpha\beta} = 0$ where $\Pi^{\gamma}_{\alpha\beta}$ is the projective connection (see Eisenhart[33] pp. 98).

32. L. P. Eisenhart, *Riemannian Geometry*, (Princeton Univ. Press., Princeton, N. J., 1926) (Sections 40–41.

33. L. P. Eisenhart, *Non-Riemannian Geometry* (American Math. Soc., New York, 1927) Sections 46–49.

34. For a detailed discussion of symmetry properties of this type with some examples provided see Davis-Moss-York[29]. Often in the case of these general mappings, not representable as coordinate point deformations, it is assumed that two spaces have coordinate patches that can be put in one-to-one correspondence. One does not actually gain in generality when one adjoins coordinate point deformations to these general mappings because their group structures are distinct and they involve distinct sets of arbitrary parameters.

35. Figure 1 is essentially the same diagram that appeared in an earlier paper [G. H. Katzin, J. Levine, and W. R. Davis, *J. Math. Phys.*, **10**, 617 (1969)]. This diagram, which includes minor corrections and two new symmetry definitions, is from a paper by G. H. Katzin and J. Levine, *Colloquium Mathematicum* (Poland) XXVI, 21 (1972). This paper should be consulted for more details relating to these minor corrections and theorems relating to the symmetry proporties defined in the figure.

36. L. P. Eisenhart[32], p. 128.

37. W. R. Davis and M. K. Moss, *Nuovo Cimento*, **38**, 1558 (1965).

38. This survey paper (G. H. Katzin and J. Levine[35]) which includes a number of new theorems relating to null geodesics should be consulted for further examples and additional references to the relevant literature in this area.

39. In the case of null geodesics results of the same form as those given below hold with the affine parameter replaced by some other suitable invariant parameter.

40. Actually, to obtain the r independent (LFI) in practice one must identify the r independent arbitrary parameters in the ξ^i and proceed to eliminate them systematically from the generator $(d/ds)[g_{\alpha\beta}(dx^{\alpha}/ds)\xi^{\beta}] = 0$ thus obtaining the desired (LFI). Similar remarks hold for the (FI) generators given below corresponding to more general symmetries.

41. Several references relating to this problem can be found in Katzin-Levine[35].

42. Here we shall call any tensor of mth order that is completely symmetric in its indicies and satisfies $P(A_{\alpha_1\alpha_2\cdots\alpha_m;\beta}) = A_{(\alpha_1\alpha_2\cdots\alpha_m;\beta)} = 0$ a symmetric Killing tensor of m-th order. Thus, a second order symmetric

Killing tensor satisfies by definition $A_{\alpha\beta;\gamma}+A_{\gamma\alpha;\beta}+A_{\beta\gamma;\alpha}=0$. This terminology seems natural enough when one considers that the first order case of this definition is a Killing vector. However, the term Killing tensor is used, for example, [J. L. Synge, *Relativity: The General Theory* (North-Holland Pub. Co., Amsterdam, 1960) p. 236] for the second order tensor $\xi_{\alpha;\beta}$ which is formed from a Killing vector which is, of course, antisymmetric and satisfies $P(\xi_{\alpha;\beta;\gamma}) = 0$. In this paper we will refer to this type of Killing tensor as an antisymmetric second order Killing tensor.

43. This relation is most easily obtained by writing out $A_{\alpha\gamma;\beta}+A_{\beta\gamma;\alpha}-A_{\alpha\beta;\gamma}$ in full and introducing the identity $A_{\alpha\beta;\gamma} = 0$ where (;) denotes covariant differentiation with respect to $A_{\alpha\beta}$.

44. Conformal mappings are a very special case of this type of mapping. This can be immediately seen by substituting $A_{\alpha\beta} = \sigma g_{\alpha\beta}[A^{\alpha\beta} = (1/\sigma)g^{\alpha\beta}]$ which gives $\left\{\begin{array}{c}\gamma\\\alpha\beta\end{array}\right\} = \left\{\begin{array}{c}\gamma\\\alpha\beta\end{array}\right\} + \frac{1}{2}\delta_\alpha^\gamma(\ln\sigma)_{,\beta} + \frac{1}{2}\delta_\beta^\gamma(\ln\sigma)_{,\alpha} - g^{\gamma\delta}g_{\alpha\beta}(\ln\sigma)_{,\delta}$.

45. This is most easily accomplished by writing out the $A_{\alpha\beta;\gamma}$ using the identity $A_{\alpha\beta;\gamma} = 0$ (see footnote 43).

46. Here it should be mentioned that Levi-Civita [T. Levi-Civita, *Annali di Matematica*, **24**, 255 (1896)] gives the $\bar{g}_{ij}$ and g_{ij} for spaces in geodesic correspondence (based on a classification of the roots of $|\bar{g}_{ij}-\lambda g_{ij}| = 0$) in the canonical systems of coordinates in which he was able to integrate the relevant equations. For a modern treatment of this problem in the case of Riemannian spaces with indefinite metric see C. F. Martin, Dissertation (1962), North Carolina State University, Raleigh, N. C.

47. As a physical example of interest we mention the (QFI) of the Friedmann-Lemaître cosmological spacetime first found by Tolman [R. C. Tolman, *Relativity, Cosmology and Thermodynamics* (Oxford Univ. Press, Oxford, 1934) Chapter X]. Martin[46] showed that this (QFI) actually follows because the Friedmann-Lamaître spacetime admits geodesic correspondence. Also, in this connection see Davis-Moss-York[29] (pp. 25–26).

48. This has been stressed in varying degrees in more recent times by several workers in this field see (including original references cited in these papers), for example, J. G. Fletcher[2], A. Trautman[2], and W. R. Davis and M. K. Moss, *Nuovo Cimento*, **27**, 1492 (1963).

49. In the absence of particular symmetries, these expressions, used in conjunction with simple group theoretical considerations involving the formal assumption of the corresponding pseudosymmetry property, can lead to natural definitions of physical quantities even when they are not actually conserved.

50. The symmetry method as described above was first introduced by Davis-Moss[48]. Working from tensor generators with the symmetries admitted (in accord with the symmetry method) means simply that we are using the "absolutely" defined geometric structure that exists in the Riemannian spacetime. Clearly, the coordinates used to label this structure can have nothing to do with the essential physical properties of this spacetime.

51. W. R. Davis and M. K. Moss, *Nuovo Cimento*, **38**, 1531 (1965).

52. Here it is of interest to note

$$[\sqrt{-g}(\xi^{\alpha;\beta}-\xi^{\beta;\alpha})]_{;\beta} = [\sqrt{-g}(\xi^{\alpha;\beta}-\xi^{\beta;\alpha})]_{,\beta}$$

$$= \chi^{\alpha\beta}_{\gamma,\beta}\xi^{\gamma} + (\chi^{\alpha\beta}_{\gamma}+\psi^{\alpha\beta\delta}_{\gamma,\delta})\xi^{\gamma}_{,\beta} + \psi^{\alpha\beta\delta}_{\gamma}\xi^{\gamma}_{,\beta,\delta},$$

where $\chi_{\gamma}^{\alpha\beta} = \sqrt{-g}\,(g_{\gamma\lambda,\delta}-g_{\gamma\delta,\lambda})g^{\lambda\alpha}g^{\delta\beta}$ is the Møller[16] superpotential (units $8\pi k = 1 = c$) $\psi_{\lambda}^{\alpha\beta\delta} = \sqrt{-g}\,g_{\alpha\lambda}(g^{\alpha\delta}g^{\beta\lambda} - g^{\alpha\lambda}g^{\beta\delta})$. Thus, for $\xi^{\alpha} = \delta^{\alpha} \times$ constant one obtains the Møller superpotential.

53. See Trautman[2] and the earlier references to his work that are cited.

54. It should be noted that (7.2) is initially a strong conservation law generator while (7.1) is a weak generator. Also, we should point out that (7.1) and (7.2) will always agree when the ξ^{α} are Killing vectors because in that case $\mathcal{L}(\sqrt{-g}\,R) = -(\delta_{\alpha}^{\beta}\sqrt{-g}\,R\xi^{\alpha})_{\beta} = 0$ and $[\sqrt{-g}\,(\xi^{\alpha;\beta}-\xi^{\beta;\alpha})]_{;\beta} = \sqrt{-g}\,R_{\beta}^{\alpha}\xi^{\beta}$.

55. Clearly, (7.2) holds identically for any ξ^{α} (hence, for an ξ^{α} representing any symmetry property) while (7.3) only holds of ξ^{α} that represent isometries. For the special spacetimes with $T_{\alpha}^{\alpha} = 0$ (7.3) also holds for spacetime admitting conformal motions (see Fig. 1).

56. For physically closed systems one can write $\int \sqrt{-g}\,\xi^{|\alpha;\beta|;\beta}d\Sigma_{\alpha} = \int \sqrt{-g}\,\xi^{4;i}d_{3}x = \int \sqrt{-g}\,\xi^{4;i}_{;i}dS_{i}$. This approach to the calculation of the total energy as more completely described below is in accord with the symmetry method[50] and it has been applied to several examples by Davis-Moss[51]. They also make calculations of total linear and angular momentum. For another approach to the identification of the proper total energy-momentum quantities in asymptotically flat spacetimes see C. W. Misner, article in *Conférence internationale sur les théories relativistes de la gravitation* Ed. L. Infield (Gauthier-Villars, Paris, 1964). Earlier, C. W. Misner, *Phys. Rev.*, **130**, 1590 (1963) observed that in general the following three properties for an energy formula are incompatible: (a) a surface integral or Gaussian flux formulation, (b) general covariance sufficient for applicability to closed spaces, (c) positive definiteness. In this paper Misner showed that Komar's proposed method [A. Komar, *Phys. Rev.* **127**, 1411 (1962); 1873 (1963)] of defining energy using ξ^{α} orthogonal to families of minimal surfaces in conjunction with (7.1) in its integral forms is defective in principle and for specific examples. For an application of the integral conservation law in the form $\oint \xi^{[\alpha;\beta]}d\tau_{[\alpha\beta]}$ specialized for radiation applications see L. T. Tamburino and J. H. Winicour, *Phys. Rev.*, **150**, 1039 (1966). See M. K. Moss, *Nuovo Cimento*, **57B**, 257 (1968) for further comments on the transformation properties of the total energy and momentum of physically closed systems. In regard to total energy calculations for physically closed systems we should also mention the early result of R. C. Tolman, *Phys. Rev.*, **35**, 875 (1930) $\mathscr{E} = \int (T_{4}^{4} - T_{1}^{1} - T_{2}^{2} - T_{3}^{3})d_{3}x$ holding for static metrics which would be used in conjunction with interior solutions. Such calculations could be of interest in attempting to determine the various contributions to the total mass.

57. J. M. Cohen, *J. Math. Phys.*, **9**, 905 (1968) and M. K. Moss and W. R. Davis, *Nuovo Cimento*, **11B**, 84 (1972). Cohen based his calculation on (7.3) using hypersurface theory to write his integral expression in terms of a surface integral. Moss and Davis made a straightforward integral application of (7.2) and give comparisons of the two methods.

58. A. V. Aminova, *Soviet Physics-Doklady*, **16**, 294 (1971).

59. G. H. Katzin, J. Levine, and W. R. Davis, *J. Math. Phys.*, **11**, 1578 (1970); P. C. Aichelberg, *J. Math. Phys.*, **11**, 2458 (1970). For other examples of curvature collineations see Katzin-Levine-Davis[35] and C. D. Collison, *J. Math. Phys.*, **11**, 818 (1970). In regard to the energy problem for gravitational radiation an important result was obtained earlier by D. R. Brill, *Ann. Phys.*, **7**, 466 (1959) who showed that a class of axially symmetric time symmetric wave solutions exist corresponding to wave packets of positive definite mass which can only vanish if the space is flat.

60. Orthonormal tetrads in Riemannian spacetime satisfy $g_{\alpha\beta} = h_{A\alpha}h_{\beta}^{A}$. For a discussion of orthonormal tetrads, see, for example, J. L. Synge, *Relativity: The General Theory* (North-Holland Pub. Co., Amsterdam, 1960).

61. W. R. Davis and J. W. York, Jr., *Nuovo Cimento*, **61B**, 271 (1969).

62. W. R. Davis, *Nuovo Cimento*, **43B**, 200 (1966).

63. C. Møller, *Mat. Fys. Dan. Vid. Selsk.*, **1**, No. 10 (1961); also see "Conservation Laws in the Tetrad Theory of Gravitation", in Report of the 1962 Warszaw Conference. For recent applications of these results see C. Møller, *Mat. Fys. Dan. Vid. Selsk.*, **34**, No. 3 (1964).

64. For a general tetrad formulation of the variational principle of Einstein's gravitational theory see W. R. Davis, *Nuovo Cimento*, **5B**, 153 (1971).

65. In particular, the frame is nonholonomic when the vector field $h_A{}^\alpha$, representing the four-velocity of the local observer, has intrinsic rotation. For a discussion of anholonomic frames see, for example, J. A. Schouten[30] Chapters II (Sec. 9) and III (Sec. 9).

66. G. F. Ellis, *J. Math. Phys.*, **8**, 1171 (1967).

67. Perhaps, the most straight-forward example of the type of fundamental curve congruence that is referred to are the principal congruences defined by the eigenvalue-eigenvector problem of the matter tensor. For example, one would refer to a principal time-like congruence defined in this way as a matter "streamline". This type of approach has been very important in studies of gravitational radiation where one is interested in the principal null congruences of the Riemannian tensor. For more details and other types of example curve congruences providing an invariant classification of the properties of Riemannian spacetimes see, for example, J. Ehlers, article in *Recent Developments in General Relativity* (Macmillan, New York, 1962) and the references given in this work including the papers by P. Jordan, J. Ehlers, W. Kundt, R. K. Sachs, and M. Trümper, *Akademie der Wissenschaften und der Literatur* (Franz Steiner Verlag, Wiesbaden, 1961).

68. Here we use the term "kinematical" recognizing, for example, that the quantities θ and $\sigma_{\alpha\beta}$ have definite dynamical implications for the time

development of the gravitational field (i.e., the proper initial value problem of Einstein's theory). Petrov (Dissertation, Moscow State Univ., 1957) first gave an invariant classification of Riemannian spacetimes based on the groups of motions they admit. For a brief discussion of this work see A. Z. Petrov, article in *Recent Developments in General Relativity* (Macmillan Co., New York, 1962) pp. 371–378. Also, see A. Z. Petrov, *Einstein Räume*, (Akademie-Verlag, Berlin, 1964). More recently A. Z. Aminora[58] has undertaken to extend this invariant classification of Riemannian space-times to include groups of projective collineations.

69. That is, properties in addition to those specifying the general type of congruence (e.g., geodesic or a particular type of trajectory) timelike, null, or spacelike. In addition to a specification of (10.1a)–(10.1c) one could also consider their absolute derivatives.

70. Several simple examples of conservation expressions have already appeared in the papers of Katzin-Levine-Davis[53,59]. Also, see G. H. Katzin, J. Levine and W. R. Davis, *Tensor N. S.*, **21**, 52 (1970).

71. See, for example, J. Ehlers[67] (1962).

72. In Katzin-Levine-Davis[53] an example is given involving the Bel-Robinson tensor $T^{\alpha\beta\gamma\delta}$ written in terms of a null congruence (Petrov type N Field). Also a conservation law generator of the form $(\sqrt{-g}T^{\alpha\beta\gamma\delta}\xi_\alpha\xi_\beta\xi_\gamma)_{,\delta} = 0$ is considered for the case of spacetimes admitting curvature collineations.

73. For the case of nongeodesic normal timelike congruences one is immediately led to Pirani's[22] strong conservation expression $\{[\sqrt{-g}(u^{\alpha;\beta} - u^{\beta;\alpha})]_{;\beta}\}_{,\alpha} = 0$. In this connection see the remarks in Sections II and IX.

74. J. L. Synge[10] Chapter VI.

75. J. L. Synge and A. Schild, *Tensor Analysis* [Toronto Univ. Press, Toronto (1949)] Chapter VII.

76. K. Yano, *The Theory of Lie Derivatives and Its Applications* (Interscience Pub. Co., New York, 1957) pp. 215–218.

77. Of course, this type of expression would be of particular interest for application involving gravitational radiation. In connection with gravitational radiation see footnotes 56, 59, 70, and 72 and the survey article by F. A. E. Pirani, (L. Witten ed.) *Gravitation: An Introduction to Current Research* (John Wiley and Sons, Inc., New York, 1962) pp. 199–226.

An elementary procedure for the evaluation of electric networks

L. JÁNOSSY

Central Research Institute for Physics, Budapest, Hungary

INTRODUCTION

When trying to obtain the numerical or analytical solution of complicated problems I have often met the following situation. In the first attempt to deal with the problem one makes simplifications. The simplifications, however, do not really help to solve the original problem but on the contrary they destroy the inherent symmetries and thus make the solution of the problem very difficult. Treating the same problem without the simplifications one may get—with the help of the inherent symmetries—the exact solution and one may simplify at the last stage only, namely when evaluating the formulae which are generally valid.

An example of this is the analysis of electric networks. There exist many procedures for particular networks. We show presently how the general formalism can be developed without regard to particular cases. There exists an extensive literature dealing with network analysis† and thus our procedure is not new. However, we extend our consideration physically to somewhat more general systems than is usually done— furthermore we give the arguments without the technicalities which are needed for the application of computer techniques, so we hope to make clearer the basic ideas involved.

---

† Compare e.g.: "*Computer Oriented Circuit Design*", Edited by Kuo and Magnuson, Prentice-Hall, Inc., Englewood Cliffs, N.J., 1969, p. 73, and in particular W. J. McCalla and D. O. Pederson, *IEEE Transactions* CT-18/1, 14, 1971.

I

Consider a circuit consisting of $N+1$ points $P_1, P_2, \cdots, P_{N+1}$. Pairs of points are connected by resistors with resistances $R_{kl} = R_{lk}$, $k \neq l = 1, 2, \cdots, N+1$. The conductance are denoted by

$$S_{kl} = 1/R_{kl}. \qquad (1)$$

If two points e.g. P_k and P_l are not connected directly then we take

$$S_{kl} = 0.$$

Thus we do not particularly distinguish between the many networks which differ from each other by their topological structures.

We denote the potentials of the points P_k by V_k thus we introduce an $N+1$ component vector

$$V = V_1, V_2, \cdots, V_{N+1}. \qquad (2)$$

Usually one considers networks which are fed through a pair of points. The current flows into the network, say through P_1 and comes out through P_{N+1}. We consider here the more general case in which each of the points P_k is fed from the outside by a current $J_k^{(0)}$, $k = 1, 2, \cdots, N+1$. Thus we introduce an $N+1$ component vector

$$J^{(0)} = J_1^{(0)}, J_2^{(0)}, \cdots, J_{N+1}^{(0)}.$$

Remembering Kirchoff's first law we see that the components are not independent but we have

$$\sum_{k=1}^{N+1} J_k^{(0)} = 0. \qquad (3)$$

The current flowing from P_k to P_l may be denoted by J_{kl}. Since $J_{kl} = -J_{lk}$ we can introduce an antisymmetric matrix J with components J_{kl} describing the currents flowing in the system.

Kirchhoff's laws may thus be written

$$\sum_{l}' J_{kl} + J_k^{(0)} = 0, \qquad (4)$$

where the Σ' signifies summation over all l except $l = k$. Further

$$J_{kl} = S_{kl}(V_l - V_k). \qquad (5)$$

Introducing (4) into (5) we find that

$$\overline{S}V = J^{(0)}, \qquad (6)$$

where $\bar{S}$ is a matrix with the following elements

$$\bar{S}_{kl} = \begin{cases} -S_{kl} & \text{if} \quad k \neq l \\ \Sigma'_m S_{km} & \text{if} \quad k = l. \end{cases} \tag{7}$$

If we were to take $\bar{S}$ as the reciprocal conductance matrix, then (6) would give the current-voltage distribution in the circuit in the form of Ohm's law – only the symbols represent vectors and matrices and not numbers as in the usual form of Ohm's law.

However, it follows from the definition (7) that

$$\det \bar{S} = 0$$

and thus $\bar{S}$ possesses no reciprocal.

The real significance of (5) becomes apparent, however, if we exclude in a quite arbitrary way one of the suffices. We can, for example, omit the suffix $N+1$ and introduce a matrix S' of the N-th order so that

$$S'_{kl} = \bar{S}_{kl} \qquad \text{if} \qquad k, l = 1, 2, \cdots, N$$

similarly we can also write

$$\left. \begin{array}{l} V'_k = V_k \\ J_k^{(0)'} = J_k^{(0)} \end{array} \right\} \qquad \text{if} \qquad k = 1, 2, \cdots, N, \tag{8}$$

where V'_k and $J_k^{(0)'}$ are the elements of the vectors $\mathbf{V}'$ and $\mathbf{J}^{(0)'}$ each having N components only. Doing so we have to be careful. Because of (3) it follows that the N components of J' can be given arbitrarily but this implies that $J_{N+1}^{(0)} = -\sum_{k=1}^{N} J_k^{(0)}$. Furthermore we suppose $V_{N+1} = 0$. Thus the current entering the point $N+1$ must vanish when taken together with those entering the other points. Further the potential of the $N+1$-th point is taken to be zero.

If we reduce the current vector to N components and the conductance matrix to one of N-th order, then we find that in general

$$\det S' \neq 0,$$

and thus we can introduce a resistance matrix

$$\mathbf{R}' = \mathbf{S}'^{-1}. \tag{9}$$

In place of (6) we find that

$$\mathbf{V}' = \mathbf{J}^{(0)'}\mathbf{R}'. \tag{10}$$

The above relation has exactly the form of Ohm's law. With the help of

(10) we can thus calculate the potentials of the points of the circuit if we are given the currents J_k, $k = 1,2, \cdots , N$ fed into the N points P_k, $k = 1,2, \cdots , N$.

The relation (10) gives N linear equations between the V_k, $k = 1,2, \cdots ,N$ and $J_l^{(0)}$, $l = 1,2, \cdots ,N$. It is not necessary to consider the V_k as the unknown quantities and to determine them in terms of the $J_k^{(0)}$. We can select, arbitrarily, N quantities out of the $2N$ quantities V_k and $J_l^{(0)}$ and consider the remaining ones as the unknowns. (10) provides in any case a system of linear equations suitable for the determination of the remaining unknowns.

It remains to investigate the conditions for the non-vanishing of det S'. For this purpose we investigate the network composed of three points. In this case

$$S' = \begin{pmatrix} S_{12} + S_{13} & -S_{12} \\ -S_{12} & S_{12} + S_{23} \end{pmatrix} \tag{11}$$

and thus

$$\det S' = S_{12}S_{23} + S_{23}S_{13} + S_{13}S_{12}. \tag{12}$$

Since $S_{kl} \geqslant 0$ we have also det S' $\geqslant 0$.

From (11) we see that det S' $= 0$ if at least two of the conductivities $S_{kl} = 0$. In a triangle this means that det S' $= 0$ if one of the three points remains unconnected with the other two. Thus in the case of the triangle det S' $= 0$ is a sign that the distribution of the potentials is not a function of arbitrarily imposed currents. Thus det S' gives a warning that the problem has no solution for physical reasons.

The inverse of S' is

$$S'^{-1} = \frac{1}{\det S'} \begin{pmatrix} S_{12} + S_{23} & +S_{12} \\ +S_{12} & S_{12} + S_{13} \end{pmatrix}.$$

Thus

$$V = S'^{-1}J^{(0)}$$

and

$$\left. \begin{aligned} V_1 &= \frac{(S_{12} + S_{23})J_1^{(0)} + S_{12}J_2^{(0)}}{\det S'} \\ V_2 &= \frac{S_{12}J_1^{(0)} + (S_{12} + S_{13})J_2^{(0)}}{\det S'} \end{aligned} \right\} \tag{13}$$

If we put

$$J_2^{(0)} = 0, \quad \text{and} \quad S_{13} = 0,$$

then

$$\frac{1}{\det S'} = R_{12}R_{23},$$

and

$$V_1 = (R_{12} + R_{23})J_1^{(0)} \qquad V_2 = R_{23}J_1^{(0)},$$

since

$$V_3 = 0, \qquad J_3^{(0)} = -J_1^{(0)}.$$

We see that (13) expresses Ohm's law for two resistors connected in series. It follows also that the effective resistance to the current between the points 1 and 3 is equal to

$$R_{\text{eff}} = R_{12} + R_{23}. \tag{14}$$

If we take $J_2^{(0)} = 0$ but $S_{13} \neq 0$ then we find similarly

$$R_{\text{eff}} = \frac{1}{R_{13}} + \frac{1}{R_{12} + R_{23}}. \tag{15}$$

(14) and (15) give the laws of how resistors combine if they are put in series and in parallel.

In case of $N \geq 3$ the determinant of S' may also be found in the form of a sum of non-negative terms.

Since all the terms in det S' are non-negative det S' vanishes only if every term contains at least one factor equal to zero. This is the case if for a fixed index l all the

$$S_{kl} = 0 \qquad k = 1, 2, \cdots, N+1. \tag{16}$$

The conditions (16) imply that the point P_l is not connected with any of the other points.

The determinant is also zero if we have

$$S_{kl} = 0 \quad \text{if} \quad \begin{matrix} k = K_1, K_2, \cdots, K_n \\ l = K_{n+1}, K_{n+2}, \cdots, K_{N+1} \end{matrix} \Big\} \tag{17}$$

where $K_1, K_2, \cdots, K_{N+1}$ is a permutation of the numbers $1, 2, \cdots, N+1$.

The latter configuration is one where the points $P_1, \cdots P_{N+1}$ can be divided into two groups: $P_{K_1} \cdots P_{K_n}$ and $P_{K_{n+1}} \cdots P_{K_{N+1}}$ so that there may be connections between points within a group but no connection between one group and another.

Thus the configurations obeying (17) are ones where the network consists of at least two independent parts.

II

The treatment can be extended to circuits containing capacitances and inductances as well as resistances. Consider first a network containing capacitances and resistances.

The capacity of the system can be introduced by a symmetric matrix C with elements C_{kl} so that

$$Q_k = \sum_{l=1}^{N} C_{kl} V_l \qquad k = 1, 2, \cdots, N, \tag{18}$$

where V_k is the potential of the point P_k if the electric charges upon the various points are equal to

$$Q = Q_1, Q_2, \cdots, Q_N.$$

In place of (18) we can also write

$$Q = CV \qquad \text{or} \qquad V = C^{-1}Q. \tag{19}$$

As can be shown from general considerations C is a symmetric matrix with positive elements and its eigenvalues are all positive real numbers.

The currents flowing in the above network are such that

$$\frac{1}{c}\frac{dQ_k}{dt} = \sum_l{}' J_{kl} + J_k^{(0)}.$$

Thus with the help of (5) we find in place of (6)

$$\frac{1}{c}\frac{dQ}{dt} + \bar{S}V = J^{(0)}.$$

Multiplying from the left with the matrix C^{-1} we find with the help of (19) that

$$\frac{1}{c}\frac{dV}{dt} = C^{-1}J^{(0)} - C^{-1}\bar{S}V. \tag{20}$$

Writing $S = R^{-1}$ we have $C^{-1}S = (RC)^{-1}$ the latter matrix is the analogue of $\tau = 1/RC$, the time constant of a simple circuit of capacitance C and resistance R.

Equation (20) gives the differential equation which determines the time rate of change of the potentials. (20) can be solved by taking the Laplace transform. The eigenvalues λ_n of the matrix $C^{-1}S = (RC)^{-1}$ gives the N-time constants of the system, namely

$$\tau_n = c/\lambda_n,$$

and one finds in the case of $J^{(0)} = 0$ that

$$V_k = \sum V_k(0)\, e^{-t/\tau}.$$

III

We may also consider a network where there are inductances placed between the points of the network. Let us denote by $L_{kk'll'}$ the coefficient of mutual inductance between the sections $P_kP_{k'}$ and $P_lP_{l'}$ (if $k = l$ $k' = l'$ then $L_{kk'll'}$ is the coefficient of self-induction). The voltage induced between P_l and $P_{l'}$ by the current $J_{kk'}$ flowing from P_k to $P_{k'}$ can thus be written

$$V_{kk'll'}^{(i)} = -\frac{1}{c} L_{kk'll'} \frac{dJ_{kk'}}{dt}$$

and thus the total induced voltage may be written

$$V_{ll'}^{(i)} = -\frac{1}{c} \sum_{k,k'} L_{kk'll'} \frac{dJ_{kk'}}{dt} - \frac{1}{c} \sum_{k} L_{kll'}^{(0)} \frac{dJ_k^{(0)}}{dt},$$

where the second term gives the induced voltage produced by the lines feeding the circuit.

Remembering that

$$J_{lk} = S_{kl}(V_l - V_k + V_{kl}^{(i)}),$$

and that the charge in the point P_k obeys

$$\frac{1}{c}\frac{dQ_k}{dt} = J_k^{(0)} + \sum_l J'_{lk}$$

we can eliminate the potentials and as a result obtain a system of differential equations for the currents J_{kl} which can be written

$$\frac{1}{c^2} \mathcal{M} \frac{d^2\mathcal{J}}{dt^2} + \frac{1}{c}\frac{d\mathcal{J}}{dt} + \mathcal{D}\mathcal{J} = \mathbf{D}^{(0)}\mathbf{J}^{(0)} + \frac{1}{c}\mathcal{M}^{(0)}\frac{d^2\mathbf{J}^{(0)}}{dt^2},$$

where the $\mathcal{M}$ and $\mathcal{D}$ are super matrices, the scalar product to be extended over a pair of suffices; for example;

$$\left(\mathcal{M}\frac{d^2\mathcal{J}}{dt^2} \right)_{kk'} = \sum_{ll'} M_{kk'll'} \frac{d^2J_{ee'}}{dt^2}$$

The elements of the various matrices are found as follows

$$M_{kk'll'} = S_{k'k}L_{ll'k'k}$$

$$D_{knln} = S_{lk}(C_{ln}^+ - C_{kn}^+)$$

$$C_{kl}^+ = (\mathbf{C}^{-1})_{kl}$$

further

$$D_{kk'l}^{(0)} = S_{k'k}D_{k'lkl}$$

and

$$M^{(0)}_{kk'l} = S_{k'k}L^{(0)}_{lk'k}.$$

We see thus that the differential equation of the network containing resistors, capacitors and inductances has the same form as the differential equation giving the circuit containing a resistor, a capacity and an inductance. In the case of the network the coefficients of the differential equation have to be replaced by matrices of the order of N^2.

Canonical Polynomials in
the Lanczos Tau Method

EDUARDO L. ORTIZ

Imperial College
University of London

I. INTRODUCTION

In a memoire which covered a whole issue of the Journal of Mathematics and Physics for the year 1938, C. Lanczos[11] dealt with the application of the Chebyshev polynomials in approximation problems.

The central theme of this memoire, which has become a classic in numerical mathematics, is an analysis of the interplay of *extrapolating* and *interpolating* expansions and their role in improving the accuracy of polynomial approximations of functions in a finite interval *J*.

Two of the methods developed in this paper are the so-called *economization* (or *telescoping*) *method* and the *tau method*. The first is designed to reduce the number of terms of a truncated power series expansion of a function $y(x)$, without sensibly damaging the original accuracy in the interval *J*. This method has been applied extensively in the construction of computer approximation of functions[7]. In the tau method we assume that $y(x)$ is defined implicitly in *J* by a linear differential operator with polynomial (or rational) coefficients and the economization is made without a direct reference to a power series expansion, which may not even exist in *J*. Instead, the operator which defines $y(x)$ is *modified* in such a way that it has an exact polynomial solution $y_n^*(x)$ with an error of an oscillatory character and a maximum amplitude much smaller, in general, than that of the error of a Taylor polynomial of the same degree.

In this contribution we discuss a systematization and some extensions of the tau method.

II. MATHEMATICAL TABLES AND MACHINE TABLES

By the end of the Second World War automatic electronic computers became available which were capable of performing arithmetical operations at very high speed with high precision. With the advent of such machines it became easier to calculate the values of various mathematical functions as required rather than consult bulky tables, when such were available.

In this new environment, it was more expedient to rely on polynomial approximations of a small degree requiring a minimum amount of stored information and capable of giving function values with a prescribed accuracy in a definite range. This approach by-passes the problems of storage of a large mass of information and subsequent retrieval and interpolation. Consequently, there was a shift in the mathematical interest in this field from the computation of tables to methods for the construction of various types of polynomial approximations.

This last problem had occupied the minds of some of the most remarkable mathematicians of the last century, when the basic theorems were proved. The possibility of finding a polynomial approximation of a continuous function defined in J with any prescribed degree of accuracy was first demonstrated by K. Weierstrass, in 1885[33]. Notwithstanding the several constructive proofs of Weierstrass' theorem which give a definite formula for the construction of the polynomial approximation, they do not give a practical answer† to the problem mentioned above as computers have a very definite limitation with regard to the *degree* of the approximation. For instance, to get an approximation to e^x in the interval $[0, 1]$ to an accuracy of $0 \cdot 2 \, 10^{-4}$ using Bernstein's polynomials[1], a polynomial of degree around 10^4 would be required.

What we aim to get—say for a degree not greater than n— is the exceptional polynomial $P_n(x)$ which minimizes

$$\max_{x \in J} |y(x) - P_n(x)|, \tag{1}$$

where the search is conducted among all polynomials of degree less than or equal to n. These polynomials of *best approximation* were studied by P. L. Chebyshev in 1853 and he discovered[4] some important general properties of these polynomials. For instance, the fact that the error

---

† Some of them, however, are rich in ideas of an enormous interest from the numerical point of view. See, for instance, in connection with the problem discussed in this contribution, Bernstein's proof[2] of the converse of Lebesgue's[15] theorem on the reduction of the approximation by trigonometric polynomials to an approximation by algebraic polynomials and notice the use Bernstein makes of approximations in terms of Chebyshev polynomials.

given by $P_n(x)$ has an oscillatory character and that its maximum $E_n(y)$ is attained at least $n + 2$ times with alternating signs[4]. However, except for a few but important cases, we know neither the amplitude $E_n(y)$ of the error nor the exact form of the polynomial $P_n(x)$ of best approximation in the sense indicated in (1). One of the cases in which both things are known is that of the approximation of x^n by a (non-trivial) polynomial of degree $n - 1$ or, what amounts to the same thing, the best approximation of degree n of the function zero in the interval J. This problem was solved by Chebyshev by means of the polynomials (for simplicity we assume that $J \equiv [-1, 1]$)

$$T_n(x) = \cos(n \arccos x), \quad x \in [-1, 1],$$

which bear his name.

There is still another classical result which I would like to mention as, in a way, it gives a clue to the problem of bridging the gulf between arbitrary polynomial approximations and those which may be of interest in numerical mathematics. In 1914, following a conjecture of Bernstein, Muntz[21] proved an important theorem in which he discusses the conditions under which a system of powers of x: $\{x^{\lambda_i}\}$, $i = 1, 2, \cdots$ is closed in the space of continuous functions defined in a finite interval J. His result indicates that it is possible to rarify the system of powers $\{1, x, x^2, \cdots\}$ without the closure properties being affected. For instance, if instead of the set of exponents $0, 1, 2, \cdots$ we take only the set of prime numbers plus zero, the resulting system will still be adequate for the representation of continuous functions defined in J.

As Lanczos said[14], this possibility of transfering the approximating roles of powers of x from one to the other shows the high degree of inefficiency of this system of representation. (In $L_2[J]$ we would argue in terms of skew angularity).

These results from classical analysis are closely related to Lanczos' central idea of economization of polynomial approximations, a procedure designed before the computer era and which made possible the construction of some of the most efficient *machine tables* or compact polynomial approximations of functions for computer use. The relative inefficiency of power series expansions for the approximation of functions in a finite interval J could be corrected and the expansion *contracted*, provided it is re-projected on an orthogonal system and especially if that system emphasizes the contribution of the elements of the basis with lower index, as is the case for the system of Chebyshev polynomials, where

$$x^n = \frac{1}{2^{n-1}} \left\{ T_n(x) + \binom{n}{1} T_{n-2}(x) + \binom{n}{2} T_{n-4}(x) + \cdots \right\} (x \in [-1, 1]).$$

This contraction is done at the expanse of a small error evenly distributed over the interval J†.

However, in this method we depend on the knowledge of a basic power series expansion. A step further is a method in which we get rid of this representation and go straight to the compact expansion: this is the tau method.

III. THE TAU METHOD

The idea of the tau method is essentially as follows: suppose that we want to solve by means of a power series expansion, say, the simple differential equation $y'(x) + y(x) = 0$, $y(0) = 1$, which defines the function $y(x) = e^{-x}$.

We take for $y(x)$ a formal power series expansion $a_0 + a_1 x + a_2 x^2 + \cdots + a_n x^n + \cdots$ and insert it in the differential equation. Hence, we get a system of linear algebraic equations

$$ja_j + a_{j-1} = 0, \qquad j = 1, 2, \cdots \tag{2}$$

and solve it in terms of a_0. Later on we adjust the value of a_0 so that the initial condition is satisfied and test the formal expansion for convergence.

The solution of this problem will be an infinite power series and no polynomial solution would be obtained unless the exact solution is a polynomial one. To get a polynomial approximation, say of order n, we could truncate the series defined by (2) but this is equivalent to solving only the first n equations in (2) with a perturbation term of the form τx^n in the right hand side, so that in the $(n+1)$th equation

$$(n+1)a_{n+1} + a_n = \tau$$

$a_{n+1} = 0$ and then the cancellation of coefficients $a_j, j > n$, propagates *downwards* instead of upwards, and then the solution is not destroyed.

The solution is a partial sum of the Taylor series for $y(x)$, i.e., a truncated power series expansion around $x = 0$ and therefore the accuracy it gives goes down as we go away from the point of expansion. At this stage Lanczos[11] goes back to his considerations on *point* and *interval* expansions and instead of τx^n he proposes a perturbation term which *distributes* the error more evenly over the interval J in which the solution is required.

If this interval is, say $[-1, 1]$, then it is natural to replace the original right hand side of the equation (the function zero) by its best algebraic approximation by a polynomial of degree n. That is, by the Chebyshev polynomial $T_n(x)$.

† On the optimal property of Chebyshev expansions see Lanczos[12]. Rivlin and Wayne Wilson has also considered the problem in a very interesting paper[32].

In this way there is an exact polynomial solution in the n-dimensional subspace in which the problem is now projected and the projection is done in such a way that the right hand side is minimized in the uniform sense in J. It will oscillate between $\pm\tau$ with $n+1$ consecutive extrema instead of growing with $|x|$.

IV. THE CANONICAL POLYNOMIALS

Lanczos' memoire of 1938 was followed by concrete applications of the tau method to the generation of polynomial approximations which were used in the construction of a number of mathematical tables[23-25].

This work was followed in 1952 by an extensive table of Chebyshev polynomials[12] computed at the National Bureau of Standards for $x = 0(0\cdot001)2$ and $n = 1(1)12$†, for which Lanczos wrote a detailed introduction in which he put forward the idea of expressing the tau method approximation as a weighted arithmetic mean of succesive partial sums of the power series solution. This idea, developed further in his book of 1956, "Applied Analysis"[13], is related to the so-called **canonical polynomials** $Q_n(x)$, $n \in N$, associated with a linear differential operator $D \equiv \sum_{k=0}^{\nu} P_k(x)\,\mathrm{d}^k/\mathrm{d}x^k$, with polynomial coefficients $P_k(x)$. The canonical polynomials are defined by the functional relationship

$$DQ_n(x) = x^n, \qquad n \in N.$$

Let us assume that $Dy(x) = 0$ is a proposed problem with initial conditions $y^{(j)}(\alpha) = y_\alpha^{(j)}$, $j = 0, \cdots, \nu - 1$; J the interval in which the solution is sought and α a point of J. For simplicity we will assume that D is a first order operator.

Therefore, if the canonical polynomials were known for all non-negative integers $n \in N$, then, on account of the linearity of D, the solution of the perturbed problem

$$Dy_n^*(x) = \tau T_n(x) = \tau \sum_{k=0}^{n} c_k^{(n)} x^k$$

would simply be

$$y_n^*(x) = \tau \sum_{k=0}^{n} c_k^{(n)} Q_k(x)$$

† Customarily Chebyshev polynomials $T_n(x) = \cos{(n \arc\cos x)}$ and $U_n(x) = \{\sin{[(n+1) \arc\cos x]}\}/\sin x$ are defined in the interval $[-1, 1]$ with amplitude ± 1. The polynomials $C_n(x)$ and $S_n(x)$ tabulated in [12] are defined in the interval $[-2, 2]$ and oscillate with an amplitude ± 2. The two types of polynomial are simply related by $T_n(x) = C_n(2x)/2$ and $U_n(x) = S_n(2x)$. The ± 2 normalization was advocated by J. C. P. Miller[19], who wrote the Foreword of the NBS tables and previously computed an early table of Chebyshev polynomials with Jones, Conn and Pankhurst[20].

where the parameter τ is adjusted to match the initial condition $y(\alpha) = y_\alpha$. Therefore,

$$y_n^*(x) = y_\alpha \frac{\sum\limits_{k=0}^{n} c_k^{(n)} Q_k(x)}{\sum\limits_{k=0}^{n} c_k^{(n)} Q_k(\alpha)} \tag{3}$$

The canonical polynomial $Q_i(x)$ associated with D is computed by solving a system of linear equations like (2) for $0 \leqslant j \leqslant i$ with a 1 in the right hand side of the i-th equation. However, the actual solution of a series of such systems may not be a trivial problem as the system may be overdetermined and the canonical polynomials may present multiplicities or be undefined for some subset of index i. This is all reflected in decisions to be taken when the tau method is automatized for use in a computer. In the next paragraphs we will discuss an alternative to this approach which renders the procedure recursive and throws some light on the problems of multiplicity[26–28].

The expression of the approximate solution in terms of canonical polynomials offers several advantages: they neither depend on the initial, or boundary, conditions of the problem we want to solve nor on the interval in which the solution is sought.

When an approximation of a higher degree $n + s$, $s \geqslant 1$, is required, it is only necessary to compute one or more canonical polynomials and weight these and the ones already computed with a different set of coefficients $c_k^{(n+s)}$ ($k = 0, \cdots, n + s$) to get the desired approximation.

Even more, they can also be used to generate expansions in arbitrary systems of polynomials[31] and to solve eigenvalue problems where the parameter may enter either linearly or non-linearly[3].

V. THE RECURSIVE FORM OF THE TAU METHOD

Let us consider again the simple equation

$$y'(x) + y(x) = 0, \quad y(0) = 1, \quad x \in [-1, 1]. \tag{4}$$

If we set $Dy(x) \equiv y'(x) + y(x)$, our first task is to try to find the elements $Q_n(x)$ of the sequence Q of canonical polynomials associated with D and such that D maps $Q_n(x)$ in x^n for all $n \in N$.

As $Q_n(x)$ is by definition a polynomial and D is a linear operator which maps polynomials into polynomials, it seems reasonable to start considering the effect of D on the *monomial x^n*. This is the *polynomial*

$$Dx^n = \sum_{r=0}^{m} a_r^{(n)} x^r \tag{5}$$

of degree $m \geqslant n$. Then,

$$\frac{1}{a_m^{(n)}} Dx^n = x^m + \frac{1}{a_m^{(n)}} \sum_{r=0}^{m-1} a_r^{(n)} x^r$$

If we assume that at the stage m we know all $Q_r(x)$ with $r < m$, on account of the linearity of D we can write

$$\frac{1}{a_m^{(n)}} D \left[x^n - \sum_{r=0}^{m-1} a_r^{(n)} Q_r(x) \right] = x^m \qquad (6)$$

Therefore, recalling the definition of $Q_m(x)$:

$$Q_m(x) = \frac{1}{a_m^{(n)}} \left[x^n - \sum_{r=0}^{m-1} a_r^{(n)} Q_r(x) \right] \qquad (7)$$

we find in our particular case, as $m = n$, $a_m^{(n)} = 1$, $a_{m-1}^{(n)} = n$ and $a_r^{(n)} = 0$ for $0 \leqslant r < m - 1$ and $m \geqslant 2$, that

$$Q_n(x) = x^n - nQ_{n-1}(x)$$

which successively gives (without a starting value) $Q_0(x) = 1$, $Q_1(x) = x - 1$, $Q_2(x) = x^2 - 2x + 2$, $Q_3(x) = x^3 - 3x^2 + 6x - 6$, etc., and solves the problem of finding the sequence Q, with complete independence from the Taylor expansion.

There are, however, some difficulties in applying this approach, as described, to the general case. On the one hand m need not be equal to n, in general it will be *greater* (for instance if $Dy(x) \equiv y'(x) + xy(x)$), leaving a *gap* between the exponent of x^n and the leading one of Dx^n, on the other hand, $a_m^{(n)}$ (which is an algebraic function of n) may be equal to zero for some values of n ($a_1^{(1)} = 0$ if $Dy(x) = xy'(x) - y(x)$).

In both cases there will also be gaps in the sequence Q, i.e., there will be some indices $v \in S$ for which the corresponding canonical polynomials will remain *undefined*. The existence of undefined canonical polynomials affects the possibility of generating all the canonical polynomials by means of (7), as not all the $Q_r(x)$, $0 \leqslant r \leqslant m - 1$ may be defined. In turn it affects the possibility of getting a solution at all of the perturbed problem $Dy_n^*(x) = \tau T_n(x)$, as there may be no canonical polynomials available to generate the powers x^v, $v \in S$, in the expression of $T_n(x)$.

This fact poses a new problem: the characterization of the set S of indices of undefined canonical polynomials. Finally, it is easy to produce examples (such as $Dy(x) = x^2 y''(x) + 2(x-1)y'(x) - 2y(x)$) where there are *multiple canonical polynomials* associated with a certain index ($Q_0(x)$, in the last example, is either $-\frac{1}{2}$ or $-x/2$).

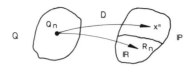

FIGURE 1. Representation of the space of polynomials $\mathbb{P}$ by means of D applied to the canonical polynomials $Q_n(x)$.

In order to go around these difficulties we introduce a more workable definition of $Q_n(x)$:

$$DQ_n(x) = x^n + R_n(x) \qquad (8)$$

where $R_n(x)$ is a polynomial generated by $\{x^v\}$, where $v \in S$, i.e., the *residual* polynomial $R_n(x)$ belongs to the subspace $\mathbb{R}$ generated by the powers of x which are *unattainable* by means of the operator D acting upon polynomials.

Then, although we cannot generate x^v, $v \in S$, with the $Q_n(x)$'s, their residual polynomials $R_n(x)$, which belong to $\mathbb{R}$, will take care of that segment of the perturbation polynomial.

Let us consider now the meaning of multiple canonical polynomials associated with a given operator D. If there were no residuals and $Q_n(x)$, $\bar{Q}_n(x)$ were two canonical polynomials of order n associated with D, then $Q_n(x) - \bar{Q}_n(x)$ would be a polynomial solution of $Dy(x) = 0$. If residuals are allowed in the definition of the canonical polynomials, then $D[Q_n(x) - \bar{Q}_n(x)]$ would either be zero or in $\mathbb{R}$. The last possibility is absurd as $Q_n(x) - \bar{Q}_n(x)$ is again a polynomial. Then multiple canonical polynomials can only differ by a multiple of a polynomial solution of $Dy(x) = 0$.

Therefore the multiplicity is a consequence of the algebraic kernel U_D of the operator D being non-empty. A computational consequence of this fact is the possibility of using multiple canonical polynomials to characterize the set U_D.†

If we classify the canonical polynomials associated with D, multiple or not, in classes of equivalence $\mathscr{L}_n(x)$ $n \in N - S$ with modulo $U_D(x)$, then a one-to-one correspondence can be established between every operator D and a sequence $\mathbb{L} = \{\mathscr{L}_n(x)\}$ and, with it, the possibility of inverting (5) to get (7), i.e. the recursive expression for the canonical polynomials is justified (details of the proof are given in[28]).

To generate the sequence of canonical polynomials associated with an operator D we start with a provisional *finite* set (see[28]) $Z \supset S$, such that for all $n \in N - Z$, $a_m^{(n)} \neq 0$, and we use (7) to generate $Q_m(x)$ for

---

† Other approaches are discussed in[27] and[17].

all $n \in N - Z$. (The residuals of the polynomials in this list belong to the subspace generated by $\{x^j\}, j \in Z$. Z in due course will be reduced to S). For the remaining set of indices we use (5): if we get new canonical polynomials, we add them to the list (and reduce the residuals, i.e., the dimension of Z). If we get a canonical polynomial of some order t which was already in the list, by linear combinations with the ones already in the list we either generate a canonical polynomial which was not already in the list (and then reduce the dimension of Z) or a polynomial solution of $Dy(x) = 0$ (and then start generating U_D). When all indices have been used, the process stops and we have reduced Z to S, generated U_D and produced a list of canonical polynomials and residuals for $n \in N - S$. For details of this procedure see Algorithm 6.1 in our paper[28] and for the programme for the University of London CDC 6000 computer, see the paper by Purser and Rodriguez L.-Canizares[30].

VI. ALGEBRAIC FORMULATION OF THE TAU METHOD

The ideas we have just discussed only depend on the fact that D maps polynomials into polynomials and can be extended in a purely algebraic form in which D is an endomorphism of a vector space Y, such that its restriction D' to the vector space $\mathbb{P}[x] \subset Y$ of polynomials in x is an injective endomorphism of $\mathbb{P}[x]$, the range, I, of which has a *finite* codimension.

Given the equation

$$Dy(x) = 0, \quad x \in J,$$

where $y(x) \in Y$ and J is a compact interval in the reals, we try to obtain, in J, an approximate solution of order n with elements of $\mathbb{P}[x]$.

In the tau method we do so by replacing the right hand side of our problem by an approximation $\tau \rho_n(x)$ of order n and in $\mathbb{P}[x]$:

$$D'y_n(x) = \tau \rho_n(x), \quad x \in J,$$

and solving this problem exactly in $\mathbb{P}[x]$.

As D' is injective we can consider its inverse $(D')^{-1}$. We then try to find a "useful" basis $\mathbb{B}$ in I in which $\rho_n(x)$ has a simple expression. The image of the elements of $\mathbb{B}$ by $(D')^{-1}$ determine a base Q in $\mathbb{P}[x]$ in which the expression of $y_n(x)$ is known.

The elements of the base Q, which solve our approximation problem, are called **Lanczos' canonical polynomials.**

This approach is discussed by Llorente and Ortiz in (16) where it is shown that every endomorphism D of the type indicated above is associated with a uniquely determined sequence of linearly independent canonical polynomials. On the other hand, a sequence Q defines a family of endomorphisms D'_a of $\mathbb{P}[x]$ such that every endomorphism D of Y with its restriction to $\mathbb{P}[x]$ in the family D'_a, has Q as its sequence of canonical polynomials. Two elements of such a family differ by an endomorphism of $\mathbb{P}[x]$ of a finite dimensional range. Finally, we have shown that the use of residuals is numerically equivalent to the formal use of undefined canonical polynomials and the subsequent cancelation of their coefficients.

VII. INTEGRATED FORMS OF THE TAU METHOD

From the point of view of this method it is clear that integrated forms will give in general, more accurate results for approximations $y_n^*(x)$ of a given degree n, as an integrated form **raises** the order of $y_n^*(x)$ more than the corresponding differential form does. Then $Iy_n^*(x)$ will satisfy a perturbation term of a higher order than $Dy_n^*(x)$.

There is no essential difficulty in the extension of the recursive form of the tau method to integrated forms. We define the canonical polynomials $q_n(x)$ associated with the integrated operator I as before:

$$Iq_n(x) = x^n + r_n(x), \qquad r_n(x) \in \mathbb{R},$$

where $\mathbb{R}$ has the same meaning as in the differential case. In the construction of this method we can either introduce a notion analogous to that of the algebraic kernel U, used in Section V, or dispense of this idea at the price of having $\nu - 1$ (ν is the order of the operator) more undefined canonical polynomials. Details of the first procedure are given in our paper[29].

We have also two options as regards the definition of I: we can introduce the initial conditions in the definition of I or leave them free, in which case we have more flexibility when the same problem has to be solved for different sets of initial conditions.

For simplicity, let us consider again (more elaborate examples are given in[29]) equation (4):

$$Dy(x) = y'(x) + y(x) = 0, \qquad y(0) = 1, \qquad x \in [-1, 1].$$

Method 1

We will consider first an *indefinite integrated* form I_i:

$$I_i y(x) = y(x) + \int y(t)\,dt = C \tag{9}$$

where C is an arbitrary constant.

We then form

$$I_i x^n = \frac{x^{n+1}}{n+1} + x^n, \qquad n \in N.$$

Therefore

$$q_{n+1}(x) = (n+1)[x^n - q_n(x)], \qquad n \in N,$$

where $q_0(x)$ remains undefined ($\nu = 1$), then $r_0(x) = 1$ and†

$$q_1(x) = 1; \qquad\qquad\qquad r_1(x) = 1$$
$$q_2(x) = 2(x-1); \qquad\qquad r_2(x) = -2$$
$$q_3(x) = 3(x^2 - 2(x-1)); \qquad\qquad r_3(x) = 6$$
$$q_4(x) = 4(x^3 - 3(x^2 - 2(x-1))); \qquad r_4(x) = -24$$
$$q_5(x) = 5(x^4 - 4(x^3 - 3(x^2 - 2(x-1)))); r_5(x) = 120$$

and so on.

The solution of the perturbed problem associated with (9), namely

$$I_i y_n^*(x) = C + \tau T_n(x)$$

is, formally,

$$y_n^*(x) = C q_0(x) + \tau \sum_{k=0}^{n} c_k^{(n)} q_k(x)$$

As $q_0(x)$ remains undefined and the residuals of $q_k(x)$, $k \geq 0$, are constants, we will match C with the weighted sum of residuals:

$$C r_0(x) = C = -\tau \sum_{k=0}^{n} c_k^{(n)} r_k(x)$$

(which makes the coefficient of the undefined canonical polynomial $q_0(x)$ equal to zero) and adjust the constant τ so that the initial condition $y(0) = 1$ is satisfied:

$$y_n^*(0) = \tau \sum_{k=0}^{n} c_k^{(n)} q_k(0) = 1$$

It is clear that we do not need to compute the actual value of C. The approximate solution of degree 4 is then:

$$y_5^*(x) = \frac{1}{1805}[80x^4 - 320x^3 + 900x^2 - 1800x + 1805]$$

and it satisfies (4) with a perturbation term equal to $\tau T_5(x)$.

---

† Obviously the residuals satisfy a recurrence relation easily deducible from that for $q_n(x)$. In this case $r_{n+1}(x) = -(n+1) r_n(x)$.

Method 2

We consider now a *definite integrated* form I_d of the operator D:

$$I_d y(x) = y(x) + \int_{-1}^{x} y(t)\,dt = e \qquad (10)$$

which incorporates the initial condition of the problem. Then

$$I_d x^n = \frac{x^{n+1}}{n+1} + x^n - \frac{(-1)^{n+1}}{n+1}$$

and therefore

$$q_{n+1}(x) = (n+1)\left[x^n - q_n(x) + \frac{(-1)^{n+1}}{n+1} q_0(x) \right]$$

as $q_0(x)$ remains undefined we transfer the last term to the residual of $q_{n+1}(x)$. Therefore,

$$
\begin{aligned}
q_1(x) &= 1; & r_1(x) &= 2 \\
q_2(x) &= 2(x-1); & r_2(x) &= -5 \\
q_3(x) &= 3(x^2 - 2(x-1)); & r_3(x) &= 16 \\
q_4(x) &= 4(x^3 - 3(x^2 - 2(x-1))); & r_4(x) &= -65 \\
q_5(x) &= 5(x^4 - 4(x^3 - 3(x^2 - 2(x-1)))); & r_5(x) &= 326
\end{aligned}
$$

and so on. The perturbed problem associated with (10) is

$$I_d y_n^*(x) = e + \tau T_n(x)$$

and its solution is

$$y_n^*(x) = e q_0(x) + \tau \sum_{k=0}^{n} c_k^{(n)} q_k(x).$$

In this case we have only one parameter, τ, and it is used to ensure that the weighted sum of residuals ($q_0(x)$ is again undefined and the residuals of all $q_n(x)$, $n > 1$, are again constants) matches with the number e (formally, the coefficient of $q_0(x)$, i.e., we again cancel the coefficient of the undefined canonical polynomial):

$$-\tau \sum_{k=0}^{n} c_k^{(n)} r_k(x) = e.$$

Then

$$y_n^*(x) = e \frac{\displaystyle\sum_{k=0}^{n} c_k^{(n)} q_k(x)}{\displaystyle -\sum_{k=0}^{n} c_k^{(n)} r_k(x)}$$

Again, an approximation of degree n satisfies a perturbation term of degree $n + 1$. The approximate solution of degree 4 is:

$$y_5^*(x) = \frac{e}{4906}[80x^4 - 320x^3 + 900x^2 - 1800x + 1805]$$

Notice that now $y_n^*(-1) \neq e = y(-1)$, but to $e + \tau T_n(-1)$, which is the initial condition of the perturbed equation. It is clear that once we know the form of the canonical polynomials in one of these forms we can immediately get the corresponding form (and residuals) in the others.

In Figure 2 we show the error curves of approximate solutions of degree 4 obtained with the differential, the definite integrated and the indefinite integrated forms. In agreement with our initial remark in this paragraph, the errors given by the integrated forms have a smaller amplitude.

The definite integrated approximation shows an even more striking feature: it has $n + 2$ extrema with alternating sign and amplitudes which differ only in the fifth decimal place (see Figure 3). If we now take into account de la Vallée Poussin's theorem (Natanson[22]), we can deduce very close bounds for $E_n(e^{-x})$. For $n = 4$:

$$0.516 \; 10^{-3} < E_4(e^{-x}) < 0.579 \; 10^{-3}, \qquad -1 \leqslant x \leqslant 1,$$

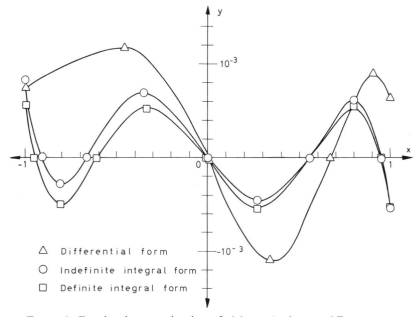

△ Differential form

○ Indefinite integral form

□ Definite integral form

FIGURE 2. Fourth order approximations of $y(x) = e^{-x}, -1 \leqslant x \leqslant 1$ Error curves.

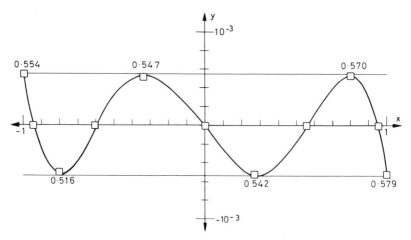

FIGURE 3. Fourth order approximation of $y(x) = e^{-x}$, $-1 \leqslant x \leqslant 1$. Error curve corresponding to the definite integrated form.

which shows how close to the best are the approximations obtained with this method.† Approximations of order higher than 20 are generated with the computer program described in[30] in tenths or hundreds of a second.

VIII. EXPANSION OF THE APPROXIMATE SOLUTION IN MORE GENERAL BASES

The numerical solution of ordinary differential equations in Chebyshev series was first considered by Lanczos in 1938 (see[11] pp. 177–78). In the same paper he pointed out the advantages of arranging the solution in Chebyshev polynomials from the point of view of judging the convergence.

Since then, several authors have considered this problem, either from the point of view of the tau method or trying direct substitution in the differential equation or in its integrated form. An account of these schemes is given by Luke[18] in his most interesting book on the approximation of the special functions.

The problem can also be considered from the point of view of the recursive form of the tau method, with the advantage that the solution obtained is recursive (Ortiz[29] and [31]).

---

†These bounds are in complete agreement with those given by Meinardus in his book *Approximation von Funktionen und ihre numerische Behandlung*, Springer-Verlag (1964), pp. 93–94. I am grateful to J. Freilich (Imperial College) for pointing it out to me.

In the first case we apply the operator D to $T_n(x)$ instead of x^n, as in (5):

$$DT_n(x) = \sum_{r=0}^{m} a_r^{(m)} T_r(x)$$

From it we deduce a recursive formula for canonical polynomials $Q_n(x)$ such that $DQ_n(x) = T_n(x) + R_n(x)$, in complete analogy with the procedure outlined in Section V. The obvious advantage is that the solution of the perturbed problem

$$Dy_n^*(x) = \tau T_n(x)$$

in Chebyshev polynomials is simply given by

$$y_n^*(x) = \tau Q_n(x)$$

Let us now solve equation (4) in terms of Chebyshev polynomials:

$$DT_n(x) = T_n(x) + 2n \sum_{k=0}^{[(n-1)/2]} T_{n-1-2k}(x)$$

where $T_0(x)$ should be taken in the summation with a weight equal to $\frac{1}{2}$. Therefore

$$Q_n(x) = T_n(x) - 2n \sum_{k=0}^{n-1/2} Q_{n-1-2k}(x)$$

and we get:

$$Q_0(x) = T_0(x)$$
$$Q_1(x) = T_1(x) - T_0(x)$$
$$Q_2(x) = T_2(x) - 4T_1(x) + 4T_0(x)$$
$$Q_3(x) = T_3(x) - 6T_2(x) + 24T_1(x) - 27T_0(x)$$
$$Q_4(x) = T_4(x) - 8T_3(x) + 48T_2(x) - 200T_1(x) + 224T_0(x)$$

and so on. Hence the solution of the perturbed problem

$$Dy_4^*(x) = \tau T_4(x)$$

is simply given by

$$y_4^*(x) = \frac{1}{177} Q_4(x)$$

after adjusting τ to the initial condition $y(0) = 1$.

We can also apply this method to the solution of integrated forms. As an example let us solve again equation (9), now in terms of Chebyshev polynomials. The indefinite integrated form gives:

$$I_i y(x) = y(x) + \int y(t)\,dt = C, \qquad y(0) = 1, \qquad x \in [-1, 1]$$

Then

$$I_i T_n(x) = \frac{T_{n+1}(x)}{2(n+1)} + T_n(x) - \frac{T_{n-1}(x)}{2(n-1)}, \qquad n > 1$$

$$I_i T_1(x) = \tfrac{1}{4} T_2(x) + T_1(x) + \tfrac{1}{4} T_0(x),$$

$$I_i T_0(x) = T_0(x) + T_1(x).$$

Therefore

$$q_1(x) = T_0(x)$$

$$q_2(x) = 4[T_1(x) - T_0(x)]$$

$$q_{n+1}(x) = 2(n+1)\left[T_n(x) - q_n(x) + \frac{1}{2(n-1)}\, q_{n-1}(x) \right]$$

for $n > 1$. Similarly, $r_0(x) = -1$; $r_1(x) = 1$; $r_2(x) = -3$; ..... The solution of the perturbed problem

$$y_n^*(x) + \int y_n^*(t)\,dt = C + \tau T_n(x)$$

is obtained in a way entirely similar to the one described in Section VII: we adjust τ and C (which need not be computed explicitly) so that the residual polynomial $q_n(x)$ matches C (and then the coefficient of the undefined canonical polynomial $q_0(x)$ is equal to zero) and the initial condition is satisfied in

$$y_n^*(x) = \tau q_n(x),$$

with $\tau = 1/q_n(0)$.

An approximate solution in terms of Chebyshev polynomials of orders not greater than four is given by

$$y_5^*(x) = \frac{1}{1805}\left[10T_4(x) - 80T_3(x) + 490T_2(x) - 2040T_1(x) + 2285T_0(x) \right]$$

If we make use of the definite integrated form, we get instead

$$y_5^*(x) = \frac{e}{4906}\left[10T_4(x) - 80T_3(x) + 490T_2(x) - 2040T_1(x) + 2285T_0(x) \right]$$

The following table shows the difference (Δ) between the coefficients of these approximations obtained with the differential (D), the indefinite (I_i) integrated and the definite (I_d) integrated forms and those of high precision truncated Chebyshev series expansions given in[5].

Table I

a_i	ΔD	ΔI_i	ΔI_d
a_0	0·000529	0·000138	0·000009
a_1	0·000375	0·000124	0·000009
a_2	0·000309	0·000027	0·000006
a_3	0·000861	0·000015	0·000012
a_4	0·000175	0·000066	0·000066

$$y_n^* (x) = \sum_{i=0}^{n} a_i T_i(x)$$

The last application shows that the recursive form of the tau method contains, as a particular case in which the *test function* in (5) is $T_n(x)$ instead of x^n, the method of Clenshaw[5] and Fox[9]. Besides, the solution obtained with this approach is purely recursive.

Using other systems of polynomials instead of x^n or $T_n(x)$ we can find approximate expansions in arbitrary systems of polynomials $\lambda = \{\lambda_n(x)\}$ $n \in N$ of a function $y(x)$ provided that it can be defined implicitly by means of an operator Φ which maps polynomials into polynomials. In this case we use as a perturbation an expansion in terms of the polynomials of the system λ of the best approximation of order n of zero in J. We then generate a sequence of canonical polynomials $\sigma_n(x)$: $\Phi\sigma_n(x) = \lambda_n(x) + R_n(x)$ where R_n has the same meaning as before. An approximation of $y(x)$ in terms of the polynomials of the system λ is given by

$$y_n^*(x) = \tau \sum_{k=0}^{n} b_k^{(n)} \sigma_k(x)$$

where $b_k^{(n)}$ are the coefficients of the expansion of the polynomial of order n of best approximation of zero in the system λ.

In this case we have a margin of choice for the operator Φ, and we try to select one with an empty algebraic kernel and such that the number of conditions to be satisfied by the solution is as small as possible.

If, for instance, we want to find an approximate expansion of $y(x) = e^{-x}$ in the interval $-1 \leq x \leq 1$ in terms of, say, Legendre polynomials, we take as an operator Φ:

$$\Phi y(x) = y'(x) + y(x) = 0$$

with the supplementary condition $y(0) = 1$.
Then form

$$\Phi P_n(x) = P_n(x) + \sum_{k=1}^{[(n+1)/2]} [2(n-2k)+3] P_{n+1-2k}(x),$$

from which we deduce that

$$\sigma_n(x) = P_n(x) - \sum_{k=1}^{[(n+1)/2]} [2(n-2k)+3]\sigma_{n+1-2k}(x)$$

for all $n \in N$.

With this choice of $\Phi, S = \phi$ and only one free parameter must be adjusted in order to get the solution of the perturbed problem:

$$\Phi y_n^*(x) = \tau E_n(x).$$

If we take as E_n the expansion in Legendre polynomials of the best approximation of zero in $[-1,1]$ in the Chebyshev sense, and call $b_k^L, k = 0(1)n$, its coefficients, then

$$\tau = 1/\sum_{k=0}^{n} b_k^L \sigma_k(0).$$

Details and examples of this method can be found in our paper[31].

IX. EIGENVALUE PROBLEMS

Fox[8-9] and Fox and Parker[10] have discussed the application of the original formulation of the tau method to the numerical solution of eigenvalue problems for linear differential equations. The same problem was considered by Chaves and I using the recursive approach outlined in Section V (see[3]).

In this case the differential operator depends on a parameter λ and so do the canonical polynomials. The extension is not trivial because of the fact that the algebraic kernel of D_λ depends on the spectrum and is *movable* in the sense that it may be empty for some eigenvalues and not for others. However, there are several advantages in using this approach: exact polynomial solutions satisfying the boundary conditions are immediately detected; the basis in which the eigensolutions are represented is generated recursively, the order of the λ-determinant is *independent* of the degree of the desired approximation; as the order of approximation increases, the lower eigenvalues are obtained with rapidly increasing accuracy and the eigenvalues of higher order give a wide image of the spectrum of D_λ and, finally, most of the computational effort to get an approximation of degree n is useful if an approximation of higher order or at different boundary points is required.

We will illustrate these considerations with an example.

Let D_λ be

$$D_\lambda y(x) = x(3x^2 - 1)y''(x) - 2y'(x) - \lambda xy(x) = 0, \qquad (11)$$

with the boundary conditions $y(\pm 1) = 0$.

If we apply D_λ to x^n we get†

$$D_\lambda x^n = [3n(n-1)-\lambda]x^{n+1} - n(n+1)x^{n-1}$$

and consequently,

$$Q_{n+1}(x,\lambda) = \frac{1}{3n(n-1)-\lambda}\,[x^n + n(n+1)Q_{n-1}(x,\lambda)]$$

for $n \geq 1$. Therefore the set of indices of undefined canonical polynomials $Q_n(x,\lambda)$ is $S_\lambda = \{0\}$.

In order to satisfy the three conditions of our problem: two boundary conditions and one undefined canonical polynomial, we use a three term perturbation, for instance, a linear combination of Chebyshev polynomials:

$$H_n(x) = \tau_0 T_n(x) + \tau_1 T_{n-1}(x) + \tau_2 T_{n-2}(x).$$

Therefore,

$$y_n^*(x,\lambda) = \sum_{i=0}^{2} \tau_i \sum_{k=0}^{n-1} c_k^{(n-i)}Q_k(x,\lambda) = \tau_0 A(x,\lambda) + \tau_1 B(x,\lambda) + \tau_2 C(x,\lambda).$$

The approximate solution must then satisfy the three homogeneous conditions:

$$\begin{cases} \tau_0 A(-1,\lambda) + \tau_1 B(-1,\lambda) + \tau_2 C(-1.\lambda) = 0 \\ \tau_0 A(+1,\lambda) + \tau_1 B(+1.\lambda) + \tau_2 C(+1,\lambda) = 0 \\ \tau_0 \alpha \qquad\quad + \tau_1 \beta \qquad\quad + \tau_2 \gamma \qquad\quad = 0, \end{cases}$$

where α, β, γ are the sum of residuals in the first, second and third terms of y_n^* respectively.

In order to get a non-trivial solution, the λ-determinant must vanish:

$$\begin{vmatrix} A(-1,\lambda) & B(-1,\lambda) & C(-1,\lambda) \\ A(+1,\lambda) & B(+1,\lambda) & C(+1,\lambda) \\ \alpha & \beta & \gamma \end{vmatrix} = 0 \qquad (12)$$

The roots of the algebraic equation (12) give the eigenvalues of problem (11).

In particular, the first eigenvalue is $\lambda = 6$ and, as $a_m^{(n)} = 3n(n-1)-\lambda$, the set $S_6 = \{0,3\}$ and therefore it contains S_λ. Also $U_{D_6} \supset U_{D_\lambda}$. The eigensolution corresponding to this eigenvalue is the exact polynomial solution

$$y_n^*(x,6) = k(x^2-1),$$

which satisfies both boundary conditions. More details and examples are given in our paper[3].

† The extension of this procedure to other bases is obvious.

FINAL REMARKS

Canonical polynomials can also be defined in such a way that it is possible to *focus* them on particular intervals of the real axis. The advantage of using this type of polynomial is that the approximate solution needs no modification when the interval in which it was constructed is changed. The canonical polynomials can also be defined, subject to certain constraints, to satisfy prescribed conditions in particular points.

REFERENCES

1. Bernstein, S., Démonstration du théorème de Weierstrass, fondée su le calcul des probabilités, *Soobshch. Protok. Khar'kov. mat. Obshch.* (2), **13** (1912–13), 1–2.

2. ———, Sur la meilleure approximation des fonctions continues, *Mém. Acad. r. Belg. Cl. Sci.* 4 (1912).

3. Chaves, T. and Ortiz, E. L., On the Numerical Solution of two point boundary value problems for linear differential equations, *Z. angew. Math. Mech.*, **48** (1968) 415–18.

4. Chebyshev, P. L., *Oeuvres Vol. I* St. Petersburg, 1899.

5. Clenshaw, C. W., Chebyshev series for mathematical functions, *Math. Tab. Nat. Lab*, 5, London H.M. Stationary Office (1962).

6. Cody, W. J., A Survey of practical rational and polynomial approximations of functions, *SIAM Rev.*, **12**, 3, (1970), 400–23.

7. Fike, C. T., *Computer evaluation of mathematical functions*, Prentice-Hall, Englewood Cliffs, New Jersey, (1968).

8. Fox, L., *The numerical solution of two point boundary value problems, in ordinary differential equations*, Oxford, Pergamon Press, (1957).

9. ———Chebyshev methods for ordinary differential equations, *Comput. J.*, 4, (1962), 318–31.

10. ——— and Parker, I. B., *Chebyshev polynomials in numerical analysis*, Oxford, University Press, (1968).

11. Lanczos, C., Trigonometric interpolation of empirical and analytical functions, *J. Math. Phys.* **17**, (1938), 123–199.

12. ——— *Introduction, Tables of Chebyshev polynomials*, Appl. Math. Ser. U.S. Bur. Stand., 9, Washington, Government Printing Office, (1952).

13. ——— *Applied Analysis*, Prentice-Hall, Englewood Cliffs, New Jersey, (1956).

14. ——— *Uses of Chebyshev polynomials*, lecture given on Nov. 20, 1962 (unpublished) at Prof. J. A. P. Hall's Visiting Lectureship Programmes. Hatfield Polytechnic.

15. Lebesgue, H., Sur l'approximation des fonctions, *Bull. Sci. math.* **22** (1898). Reproduced in *Oeuvres Scientifiques, Vol. III*, pp. 11–20, Genève (1972).

16. Llorente, P. and Ortiz, E. L., Sur quelques aspects algébriques d'une méthode d'approximation de M. Lanczos, *Math. Notae,* **21,** (1968), 17–23.

17. ———— ———— On the existence and construction of polynomial solutions of differential equations, *Revta Un. mat. argent.,* **23,** (1968), 183–89.

18. Luke, Y. L., *The special functions and their approximation,* 2 vols., New York, Academic Press, (1969).

19. Miller, J. C. P., Two numerical applications of Chebyshev polynomials, *Proc. Roy. S. Edinb.* **62,** (1946) 204–210.

20. ————, Jones, C. W., Conn, J. F. C. and Pankhurst, R. C., Tables of Chebyshev polynomials, *Proc. Roy. S. Edinb,* **62** (1946), 182–203.

21. Müntz, Ch., *Über den Approximationssat von Weierstrass,* in *Schwarz-Festschrift,* Springer, Berlin, (1914), 303–312.

22. Natanson, I. P., *Konstruktive Funktionentheorie,* Berlin, Akademie-Verlag, (1955).

23. *National Bureau of Standards, Tables of sine, cosine and exponential integral, Appl. Math. Ser.,* Government Printing Office, Washington, (1941).

24. ———— *Tables of probability functions, Appl. Math. Ser.,* Washington Government Printing Office, (1941).

25. ———— *Tables of spherical Bessel functions,* New York, Columbia University Press, (1947).

26. Ortiz, E. L., On the generation of the canonical polynomials associated with certain linear differential operators, *Res. Rep., London, Imperial College* (1964), 1–22.

27. ———— Polynomlösungen von Differentialgleichungen, *Z. angew. Math. Mech.* **46,** (1966), 394–95.

28. ———— The tau method, *SIAM J. Numer. Anal.* **6,** (1969), 480–92.

29. ———— A unified recursive approach to the Lanczos and Clenshaw-Fox approximation methods (to appear elsewhere) (Presented at the Symposium on Approximation Theory organized by A. Talbot at the University of Lancaster, Lancaster, (1969)).

30. ————, W. F. C. Purser and F. J. Rodriguez L.-Canizares, Automation of the tau method, *Res. Rep., London, Imperial College* (1972), 1–58 (Presented at the Conference on Numerical Analysis organized by the Royal Irish Academy, Dublin, (1972)).

31. ———— A recursive method for the approximate expansion of functions in a series of polynomials, *Comp. Phys. Comm.* **4** (1972) 151–156.

32. Rivlin, T. and M. Wayne Wilson, An optimal property of Chebyshev expansions, *J. Approx. Theory,* **2** (1969), 312–317.

33. K. Weierstrass, Über die analytische Darstellbarkeit sogennannter willkürlicher Funktionen einer reellen Argumente, *Sber. preuss. Akad. Wiss.,* (1885), 633–39; 789–805. Reproduced in *Mathematische Werke,* **3,** 1–37, Berlin (1903).

Rational Approximations
From Chebyshev Series

C. W. CLENSHAW and K. LORD

University of Lancaster, Lancaster, England

I. INTRODUCTION

We introduce our subject, and motivate our discussion, by considering two distinct but related problems in numerical approximation. The first is the approximation of a function f of a complex variable $z = x + iy$ in the neighbourhood of some given point, while the second is the approximation of f on a given closed interval of a line. Without loss of generality, we take the given point to be $z = 0$, and the given interval to be $-1 \leq x \leq 1$. In each case we concentrate on the non-singular situation: that is, f is assumed to be regular in a circle of radius $r > 0$, and also in an ellipse of nonzero area, with foci ± 1. Equivalently, the Taylor series expansion of $f(z)$ about $z = 0$ converges uniformly in a neighbourhood of the origin, while the Chebyshev series expansion of $f(x)$ converges uniformly on $[-1, 1]$. We denote the Taylor series expansion by

$$f(z) = \sum_{r=0}^{\infty} c_r z^r, \tag{1}$$

and the Chebyshev series expansion on $[-1, 1]$ by

$$f(x) = \sum_{r=0}^{\infty} {}' a_r T_r(x). \tag{2}$$

Here $T_r(x) = \cos(r \cos^{-1} x)$, and $\sum_{r=0}^{\infty} {}' u_r$ denotes the sum $\frac{1}{2}u_0 + u_1 + u_2 + \cdots$. The properties of Chebyshev polynomials, of such great power and convenience in problems of numerical approximation, have been systematically exploited during the past 35 years by many numerical analysts, following the trail blazed by Lanczos in a number of very

95

significant publications, notably [8], [9] and [10]. In particular, the introduction to [9] contains a very useful and concise collection of those properties which are of greatest relevance to numerical analysis. A similar collection may be found in [2], and a general survey of the applications of Chebyshev polynomials to numerical problems has been given by Fox and Parker[4]. We shall assume a familiarity with the basic properties of the polynomials.

If we seek a polynomial approximation of degree n to $f(z)$ valid near the origin, it is natural to consider using a truncation of the Taylor series. That is, $f(z)$ is approximated by

$$f_1(z) = \sum_{r=0}^{n} c_r z^r. \tag{3}$$

If we next seek a polynomial approximation of degree n to $f(x)$ on $[-1, 1]$, then of course $f_1(x)$, given by (3), is at once a candidate. However, a more accurate approximation is usually given by

$$f_2(x) = \sum_{r=0}^{n}{}' a_r T_r(x). \tag{4}$$

Justification of this statement will be found in the references cited, but intuitive support is supplied immediately by consideration of the regions of convergence of the two series.

In the complex case, where every direction in the complex plane is of equal interest, we use the Taylor series, with its *circle* of convergence. In the real case, however, we are prepared to sacrifice accuracy (or rapidity of convergence) in the imaginary direction in order to favour the real axis; in the Chebyshev case we accordingly have an *ellipse* of convergence, with the major axis lying on the real line.

Next we examine the approximation of f by rational functions. An effective means of producing rational approximations in a complex variable is to develop elements of the Padé table from the Taylor series. This table is a two-dimensional array whose (m, n) element is defined as that rational function of degree m in the numerator and n in the denominator whose own Taylor series expansion, $\sum_{r=0}^{\infty} C_r z^r$ say, agrees with that of $f(z)$ up to and including the term in z^{m+n}. Discussion of the Padé table from a numerical viewpoint will be found in [5]. (This reference also contains descriptions of devices such as the Q-D algorithm of Rutishauser and the ϵ-algorithm of Wynn, which can be used to facilitate the construction of a Padé table.) For simplicity, we assume that the rational function is finite at $z = 0$, and that its numerator and denominator have no common factors.

We write

$$R_{m,n}(z) = \frac{U_m(z)}{V_n(z)} = \sum_{r=0}^{\infty} C_r z^r, \qquad (5)$$

and we require to find the coefficients in $U_m(z) = \sum_{r=0}^{m} u_r z^r$ and in $V_n(z) = \sum_{r=0}^{n} v_r z^r$. We give v_0 the value of unity; this is permissible because our simplifying assumptions ensure that $V_n(0) \neq 0$. It then follows that

$$\sum_{r=0}^{m} u_r z^r = \left\{ \sum_{r=0}^{n} v_r z^r \right\}\left\{ \sum_{r=0}^{\infty} C_r z^r \right\} \qquad (6)$$

whence

$$u_r = \sum_{s=0}^{n} v_s C_{r-s}, \qquad r = 0, 1, 2, \cdots, \qquad (7)$$

where $u_r = 0$ for $r > m$, $c_k = C_k = 0$ for $k < 0$ and where $v_0 = 1$. Thus, identification of the C_r with the c_r in (1) for $r = 0, 1, 2, \cdots, m+n$ yields

$$u_r = \sum_{s=0}^{n} v_s c_{r-s}, \qquad r = 0, 1, 2, \cdots, m+n. \qquad (8)$$

This system of $m+n+1$ linear simultaneous equations may usually be solved for the unknowns u_r $(r = 0, 1, \cdots, m)$ and v_r $(r = 1, 2, \cdots, n)$. The system divides naturally into two subsystems; the n equations

$$0 = \sum_{s=0}^{n} v_s c_{r-s}, \qquad r = m+1, m+2, \cdots, m+n, \qquad (9)$$

may be solved first, to give the unknown v_r, and then the $m+1$ equations

$$u_r = \sum_{s=0}^{n} v_s c_{r-s}, \qquad r = 0, 1, \cdots, m, \qquad (10)$$

yield the u_r directly.

Our principal aim in the present paper is to extend the Padé idea to the case of Chebyshev series, in order to obtain readily an accurate rational approximation on $[-1,1]$ to a function with a convergent Chebyshev series expansion. In our next section we describe one such extension, due to Maehly[11]. In Section III we suggest an alternative, which seems to preserve the Padé analogy more closely, and which also offers computational advantages. This is followed in Sections IV and V by an account of the application of this alternative, with an example. Section VI is devoted to a discussion of a modification designed to deal with slowly-convergent Chebyshev series, while Section VII contains a treatment of some of the computational details of the various approaches. Numerical results for several functions are presented in Section VIII, and we conclude in Section IX with a mention of the scope of the proposed method.

II. THE GENERALIZED PADÉ APPROXIMATION

If we seek rational approximations to f on $[-1,1]$, it now seems natural to apply to the Chebyshev series a technique analogous to that applied to the Taylor series about $z = 0$ in producing the Padé table. One such approach was suggested by Maehly[11] and has been subsequently described in numerical terms, with examples, by Ralston[12] and by Fike[3]. An extended account of the theory was given by Cheney[1], who dealt with expansions in *any* functions (ϕ_k, say) which satisfy relations of the form

$$\phi_i\phi_j = \sum_k A_{ijk}\phi_k.$$

The method is basically as follows. On the assumption that a formal series expansion for f in terms of the ϕ_k is available, the rational function U_m/V_n is obtained by equating to zero the leading terms in the series expansion of

$$V_n f - U_m,$$

this expansion again being expressed in terms of the ϕ_k. Cheney calls a rational function derived in this way a "generalized Padé approximation", which we shall henceforth abbreviate to "GP". We see that the GP procedure presents its unknown parameters in linear form, and so leads to a generally straightforward computing problem. We also note, however, that the annihilation of the leading terms in

$$V_n f - U_m \tag{11}$$

does not lead to the same result as annihilating the leading terms in

$$f - (U_m/V_n) \tag{12}$$

in general, though it does so when $\phi_k = z^k$. The form (12) is, of course, usually less attractive computationally than (11), since its parameters occur nonlinearly. In the special case where $\phi_k = T_k$ however, the form (12) proves to be tractable, and indeed appears to offer significant advantages both in computational convenience and accuracy of approximation.

III. THE CHEBYSHEV-PADÉ TABLE

We first define a new table, which we call the "Chebyshev-Padé" (or "CP") table. By analogy with the Padé case, its (m,n) element is that rational function whose formal Chebyshev series expansion agrees with that of f up to and including the term in T_{m+n}. As far as we are aware, the construction of such a table has not been discussed hitherto in the literature, though Hornecker[7] described a rational approximation which is in fact an element of this table. This important work seems to have been overlooked to a large extent, and Maehly's approach, leading to the GP approximation, has attracted far more attention. Hornecker's method for the calculation of the error of the approximation is particularly elegant, and we follow it closely. We approach the calculation of the coefficients somewhat differently however, and we describe in Section VII a technique for building up the CP table in a systematic and simple manner.

We now investigate the nature of the table. The (m,n) element is written as

$$S_{m,n}(x) = \frac{P_m(x)}{Q_n(x)} = \sum_{r=0}^{\infty}{}' A_r T_r(x), \tag{13}$$

and we require to find the coefficients in $P_m(x) = \sum_{r=0}^{m}{}' p_r T_r(x)$ and in $Q_n(x) = \sum_{r=0}^{n}{}' q_r T_r(x)$, assuming as before that they have no common factors and that $Q_n(x)$ has no zeros on $[-1,1]$. (The consequences of this latter assumption being false for some elements of the CP table will be considered later.) We give to q_0 the value of 2, in order to maintain the analogy with the Padé table. This normalization might appear to be restrictive, in the sense that it requires the coefficient of T_0 in the expansion of Q_n to be nonzero. However, we appeal to the following theorem, which is a special case of one proved by Cheney [1, p. 110].

Theorem A: The function continuous on $[-1, 1]$, represented by the formal Chebyshev series $\sum_{r=N}^{\infty} a_r T_r(x)$, has at least N changes of sign on $[-1, 1]$ (if it does not vanish identically). Application of this theorem shows that since Q_n has no zeros on $[-1, 1]$, its Chebyshev series must have $N = 0$; that is, its coefficient of T_0 is nonzero.

In an attempt to find the coefficients p_r and q_r, an obvious step is to write

$$\sum_{r=0}^{m}{}' p_r T_r = \left(\sum_{r=0}^{n}{}' q_r T_r \right)\left(\sum_{r=0}^{\infty}{}' A_r T_r \right). \tag{14}$$

The right hand side of this equation may then be expanded in Chebyshev series by a term-by-term multiplication, followed by use of the fundamental relation

$$T_r T_s = \tfrac{1}{2} (T_{r+s} + T_{|r-s|}). \tag{15}$$

Coefficients of each Chebyshev polynomial may then be equated. But this, of course, is precisely the GP process. The weakness of its analogy with the classical Padé table is evident if we proceed further. Our equation corresponding to (7) will be

$$p_r = \tfrac{1}{2} \sum_{s=0}^{n}{}' q_s (A_{r+s} + A_{|r-s|}), \qquad r = 0, 1, 2, \cdots, \tag{16}$$

where $p_r = 0$ when $r > m$ and where $q_0 = 2$. We should now identify the leading coefficients A_r with the corresponding a_r from (2), so as to produce the system

$$p_r = \tfrac{1}{2} \sum_{s=0}^{n}{}' q_s (a_{r+s} + a_{|r-s|}), \qquad r = 0, 1, \cdots, m+n, \tag{17}$$

which is to be solved for the unknown coefficients, just as in the Padé case. The weakness to which we have referred relates to the presence (in the system (17)) of coefficients a_r for values of r up to $m + 2n$, whereas we seek agreement between (2) and (13) up to $r = m+n$ only. Thus, our solution should be independent of the a_r for $r > m+n$.

In order to achieve such a solution, we first regard (16) as a difference equation which might be used to calculate the coefficients A_r in the Chebyshev series expansion of a known rational function, regular near $[-1, 1]$. In particular, we examine (16) for $r > m$, so that we have

$$0 = \sum_{s=0}^{n}{}' q_s (A_{r+s} + A_{|r-s|}), \qquad r = m+1, m+2, \cdots, \tag{18}$$

where A_{-k} may be replaced by A_k $(k > 0)$ wherever a negative suffix occurs. This is a symmetric linear difference equation in A_r of order $2n$ with constant coefficients. Its general solution may therefore be expressed as a linear combination of the r-th powers of the roots of the polynomial equation

$$0 = \sum_{s=0}^{n}{}' q_s (t^s + t^{-s}). \tag{19}$$

Now to find these roots, we first halve the degree of the polynomial equation by writing

$$x = \tfrac{1}{2}(t + t^{-1}).$$

This is a particularly appropriate transformation here since $x = \cos\theta$ implies $t = e^{i\theta}$, so that

$$T_s(x) = \tfrac{1}{2}(t^s + t^{-s}).$$

Thus t is a root of (19) if and only if x is a root of

$$0 = \sum_{s=0}^{n}{}' q_s T_s(x),$$

that is, if and only if x is a zero of Q_n. Clearly, if t_i satisfies (19) then so does t_i^{-1}; also $|t_i| \neq 1$ since Q_n has no zeros on $[-1, 1]$. Thus we may factorize our polynomial in t in the following manner

$$\sum_{s=0}^{n}{}' q_s(t^{n+s}+t^{n-s}) = \mu \left(\sum_{s=0}^{n} \gamma_s t^{n-s} \right) \left(\sum_{s=0}^{n} \gamma_s t^s \right), \qquad (20)$$

where each zero of $\sum_{s=0}^{n} \gamma_s t^{n-s}$ is less than unity in modulus, and where we choose γ_0 to be unity. (μ is merely a scaling factor.) Since $A_r \to 0$ as $r \to \infty$, it must be expressible in terms of those roots of modulus less than unity; we conclude that the A_r satisfy the difference equation of order n given by

$$\sum_{s=0}^{n} \gamma_s A_{|r-s|} = 0, \qquad r = m+1, m+2, \cdots. \qquad (21)$$

By equating the leading coefficients A_r with the corresponding a_r, we are now able to solve for the γ_s. These γ_s are in turn related to the coefficients q_s (from (20)) by the equation

$$q_s = \mu \sum_{i=0}^{n-s} \gamma_i \gamma_{s+i}. \qquad (22)$$

To accord with our convention that $q_0 = 2$, we find (setting $s = 0$)

$$\mu^{-1} = \tfrac{1}{2} \sum_{i=0}^{n} \gamma_i^2. \qquad (23)$$

IV. THE CALCULATION OF $S_{m,n}(x)$

We now outline a computational procedure for finding $S_{m,n}(x)$, given the coefficients a_r in (2) for $r = 0, 1, \cdots, m+n$. This is not the procedure which we recommend for practical use (that follows in Section 7), but it serves to clarify the numerical situation.

As in the Padé case, we equate A_r to a_r for $r = 0, 1, \cdots, m+n$, and we then calculate the coefficients γ_s for $s = 1, 2, \cdots, n$ by solving the system of n simultaneous equations given by

$$\sum_{s=0}^{n} \gamma_s a_{|r-s|} = 0, \qquad r = m+1, m+2, \cdots, m+n, \qquad (24)$$

remembering that $\gamma_0 = 1$. The γ_s enable us to find the coefficients q_s of Q_n directly from equation (22), and we can then calculate the coefficients of P_m from

$$p_r = \tfrac{1}{2} \sum_{s=0}^{n}{}' q_s(a_{r+s}+a_{|r-s|}), \qquad r = 0, 1, \cdots, m. \qquad (25)$$

The error E, given by

$$E(x) = f(x) - S_{m,n}(x) \qquad (26)$$

is readily expanded in a Chebyshev series. For when the γ_s have been found, we can generate A_r for $r = m+n+1, m+n+2, \cdots$ from equation (21). Then we use

$$E(x) = \sum_{r=m+n+1}^{\infty} (a_r - A_r) T_r(x) \qquad (27)$$

to give $E(x)$; we see that $|E(x)|$ is bounded in $[-1, 1]$ by

$$\sum_{r=m+n+1}^{\infty} |a_r - A_r|,$$

and this bound is readily calculated, and often realistic in that it is attained (or very nearly attained) at some point of $[-1, 1]$. In application, examination of the "tail" of the series for $S_{m,n}(x)$ provides a reliable indication of the existence or otherwise of the (m, n) element. (In this context, "non-existence" implies that the formal Chebyshev series expansion of the element is divergent, which in turn implies the presence of a pole of $S_{m,n}(x)$ on $[-1, 1]$. Where it occurs, divergence of the expansion is suggested by the numerical growth of the coefficients A_r for $r > m+n$.)

V. AN EXAMPLE

To illustrate the results that can be obtained by application of the various methods which are available, we now discuss the specific example of rational approximations for e^x on $[-1, 1]$, quadratic in both numerator and denominator. We select this case because it is the one treated by Ralston[12] in some detail; he finds the Padé approximant $R_{2,2}(x)$, an "economized" Padé approximant $R_{2,2}^{(E)}(x)$ (obtained by a refinement process analogous to the Lanczos economization of a polynomial[8]), a rational approximation obtained by the Maehly method $R_{2,2}^{(M)}$ and the minimax rational approximation $R_{2,2}^*(x)$. (It should be noted that our notation for these rational approximations does not coincide in every case with that of Ralston.) $R_{2,2}^*(x)$ is, of course, that approximation of the desired form whose error has the minimum uniform norm; in that sense, this approximation is the best possible, and is effectively the standard against which the others are measured. Essentially, we seek an approximation whose precision bears comparison with that of $R_{2,2}^*(x)$, but which is much easier to compute in the general case. We give in Table 1 the values of the error norms for the various approximations, including that for $S_{2,2}(x)$, the CP approximant. It is a characteristic and scarcely surprising feature of this function that the maximum error is attained in each case at the point $x = 1$.

Table I

Approximation	max $\mid e^x -$ Approximation $\mid$ $[-1, 1]$
$R_{2,2}(x)$	4.00×10^{-3}
$R_{2,2}^{(E)}(x)$	2.39×10^{-4}
$R_{2,2}^{(M)}(x)$	1.89×10^{-4}
$R_{2,2}^{*}(x)$	0.87×10^{-4}
$S_{2,2}(x)$	1.45×10^{-4}

It will be observed that the CP approximant is rather more accurate than that obtained by the Maehly method. (That both are significantly better than those derived from the Padé table is not unexpected, since they aim at the comparatively limited target of approximation on $[-1, 1]$, whereas the Padé table offers approximation in a circle in the complex plane.) This pattern appears to be quite typical of the general situation: where the CP element (m, n) exists (that is, where the Chebyshev series expansion of $S_{m,n}(x)$ is uniformly convergent on $[-1, 1]$) we expect it to be the best of the rational approximations which are obtainable by simple direct methods.

To further illustrate the resemblance between $S_{2,2}(x)$ and $R_{2,2}^{*}(x)$, we give in Figure 1 a plot of the error curve for the former.

It is a general and valuable result that where $S_{m,n}(x)$ exists, the error $E(x)$, as given by equation (27), has at least $m + n + 1$ sign changes, by

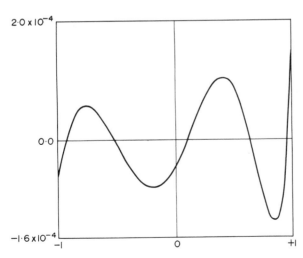

FIGURE 1   Plot of the error $[e^x - S_{2,2}(x)]$ as a function of x.

Theorem A. Thus $E(x)$ has the appropriate number of oscillations for a minimax error curve, and is therefore an effective starting point for an iterative procedure for determining a minimax rational approximation.

VI. CASE OF SLOW CONVERGENCE

A practical difficulty of some importance arises when the convergence of the Chebyshev series (2) is slow. The foregoing procedure is by no means invalidated here; the problem is that the coefficients a_r, required for $r = 0, 1, \cdots, m + n$, may not be easily calculated. If we use the procedure advocated in [6], which would seem to be the most appropriate available in the general case, we shall find coefficients $a_r^{(N)}$ for $N = 2, 4, 8, 16, \cdots$, which are related to the required coefficients a_r by the equation

$$a_r^{(N)} = \sum_{k=0}^{\infty} {}' \left(a_{2kN+r} + a_{|2kN-r|} \right). \tag{28}$$

Under the assumptions which we have made, $a_r^{(N)} \to a_r$ as $N \to \infty$, and in many examples of interest $a_r^{(N)}$ will be an adequate approximation to a_r (for all r) for a quite moderate value of N. In these cases the construction of the CP elements can proceed in the manner indicated above.

It will be appreciated, however, that the very motivation for seeking a rational (rather than polynomial) approximation may be the experimental discovery, during the computation of the $a_r^{(N)}$, that convergence is slow, and that $a_r^{(N)}$ can be an adequate approximation to a_r only if a very large value of N is adopted. The question arises—can we obtain a rational approximation from the $a_r^{(N)}$, even if these do not closely approximate the a_r? The answer to this is affirmative, though we cannot at present suggest how an element of the CP table can be computed in this case. For we have, using equation (28),

$$\sum_{s=0}^{n} {}' q_s (a_{r+s}^{(N)} + a_{|r-s|}^{(N)})$$

$$= \sum_{s=0}^{n} {}' q_s \left\{ \sum_{k=0}^{\infty} {}' \left(a_{2kN+r+s} + a_{|2kN-r-s|} \right) + \sum_{k=0}^{\infty} {}' \left(a_{|2kN+r-s|} + a_{|2kN-r+s|} \right) \right\}$$

$$= \sum_{s=0}^{n} {}' q_s \left\{ \sum_{k=0}^{\infty} {}' \left(a_{2kN+r+s} + a_{|2kN+r-s|} \right) + \sum_{k=0}^{\infty} {}' \left(a_{|2kN-r+s|} + a_{|2kN-r-s|} \right) \right\}.$$

Hence if

$$\sum_{s=0}^{n} {}' q_s (a_{r+s} + a_{|r-s|}) = 0, \qquad r = m+1, m+2, \cdots,$$

then

$$\sum_{s=0}^{n}{}' q_s(a_{r+s}^{(N)}+a_{|r-s|}^{(N)}) = 0, \qquad r = m+1, m+2, \cdots, m+n, \qquad (29)$$

for $N > m+\frac{1}{2}n$.

Thus we can obtain a GP approximation, using the coefficients $a_r^{(N)}$ in place of the a_r. This requires solution of the equations (29) for the q_s, and then finding p_r directly from

$$p_r = \tfrac{1}{2} \sum_{s=0}^{n}{}' q_s(a_{r+s}^{(N)}+a_{|r-s|}^{(N)}), \qquad r = 0, 1, \cdots, m. \qquad (30)$$

Thus the procedure requires knowledge of the coefficients $a_r^{(N)}$ for $r = 0, 1, 2, \cdots, m+2n$. Accordingly, we shall choose N to be the smallest number of the form 2^k (k being an integer) which exceeds $m+2n$. (The reasons for choosing an N of this form are given in [6].)

VII. COMPUTATIONAL ASPECTS

There is a practical drawback to the method of Section 6 which may often be serious; moreover, the disadvantage may extend to the GP procedure in general. We now discuss this point, and compare the GP case with the CP table, where the difficulty is readily avoided.

We consider the application of Maehly's method, which is the GP method in the Chebyshev polynomial case. The first step is the calculation of the $q_s(s = 1, 2, \cdots, n)$ from the system

$$\sum_{s=0}^{n}{}' q_s(a_{r+s}+a_{|r-s|}) = 0, \qquad r = m+1, m+2, \cdots, m+n \qquad (31)$$

given $q_0 = 2$.

Now let us suppose for simplicity that the function $f(x)$, which we are attempting to approximate by the rational function P_m/Q_n, is exactly a rational function of the required degrees m and n. The linear system (31) then tends to be ill-conditioned (since the coefficient of q_s is dominated by $a_{r-s} = A_{r-s}$ which satisfies (21), a difference equation of order n with constant coefficients). The ill-conditioning becomes more severe when some zeros of $\sum_{s=0}^{n} \gamma_s t^{n-s}$ are appreciably smaller in modulus than others.

The unfortunate conclusion may be inferred, that when the greatest gains are likely to accrue from rational approximation, that is when $f(x)$ has a Chebyshev series expansion which is very close to that of a rational function, the conditioning of the GP method is at its worst. Experience tends to confirm this expectation, and there remains a severe computational obstacle to obtaining GP approximations of high accuracy.

In contrast, the CP table can be generated without the embarrassment of ill-conditioning. We recall that the classical Padé table may be built up using the Q-D algorithm[13] or the ϵ-algorithm[14], avoiding the necessity for solving a new linear system for each element. It is possible in the CP case again to construct the elements in a systematic, step-by-step manner.

Let us first suppose that we have computed the (m, n) and $(m+1, n)$ elements of the CP table. That is to say, we know the coefficients $\gamma_s^{(m,n)}$ in the equation

$$\sum_{s=0}^{n} \gamma_s^{(m,n)} a_{|r-s|} = 0, \qquad r = m+1, \cdots, m+n, \qquad (32)$$

and similarly the coefficients $\gamma_s^{(m+1,n)}$. (The extended notation for the coefficients γ_s needs no explanation.) Suppose we now wish to compute the coefficients $\gamma_s^{(m+1,n+1)}$.

First let

$$\gamma_s^{(m+1,n+1)} = \gamma_s^{(m+1,n)} + \sigma^{(m,n)} \gamma_{s-1}^{(m,n)}. \qquad (33)$$

Then

$$\sum_{s=0}^{n+1} \gamma_s^{(m+1,n+1)} a_{|r-s|} = \sum_{s=0}^{n+1} \gamma_s^{(m+1,n)} a_{|r-s|} + \sigma^{(m,n)} \sum_{s=0}^{n+1} \gamma_{s-1}^{(m,n)} a_{|r-s|}$$

$$= \sum_{s=0}^{n} \gamma_s^{(m+1,n)} a_{|r-s|} + \sigma^{(m,n)} \sum_{s=0}^{n} \gamma_s^{(m,n)} a_{|r-s-1|},$$

by virtue of the fact that $\gamma_{n+1}^{(m+1,n)} = 0$ and $\gamma_{-1}^{(m,n)} = 0$. Now each sum on the right vanishes for $r = m+2, \cdots, m+n+1$. Therefore we choose $\sigma^{(m,n)}$ so that

$$\sum_{s=0}^{n} \gamma_s^{(m+1,n)} a_{|m+n+2-s|} + \sigma^{(m,n)} \sum_{s=0}^{n} \gamma_s^{(m,n)} a_{|m+n+1-s|} = 0, \qquad (34)$$

and then equation (33) gives the required coefficients. In order to complete the CP table by a "staircase" scheme of computation, we need other relations of similar type. For instance, one is furnished by the following equations, corresponding to (33) and (34),

$$\gamma_s^{(m+1,n+1)} = \gamma_s^{(m,n+1)} + \rho^{(m,n)} \gamma_{s-1}^{(m,n)}, \qquad (35)$$

where $\rho^{(m,n)}$ is given by

$$\sum_{s=0}^{n+1} \gamma_s^{(m,n+1)} a_{|m+n+2-s|} + \rho^{(m,n)} \sum_{s=0}^{n} \gamma_s^{(m,n)} a_{|m+n+1-s|} = 0. \qquad (36)$$

Where the γ_s are known for any particular element (m,n), then the required parameters p_s and q_s can be obtained directly as in Section IV.

It is of course possible that a similar procedure could be devised for the more effective production of the GP approximations. We have not pursued this, because for practical application, the CP would appear to be preferable, both aesthetically and on grounds of accuracy. As far as the theory is concerned, the GP covers a much wider field, and the possible illconditioning of its linear systems is largely irrelevant.

VIII. NUMERICAL RESULTS

CP tables have been derived for a number of common functions. The techniques described in the last section were applied to build up the elements from the polynomial case ($n = 0$). Perhaps the most interesting information to be gleaned from these tables is the accuracy with which the elements represent the given function. Accordingly we now present the leading sector of a table of the error bound $E_{m,n}$, given by

$$E_{m,n} = \sum_{r=m+n+1}^{\infty} |a_r - A_r|,$$

for each of several functions. Each element is written as $A(-B)$, to be interpreted as $A.10^{-B}$, with two significant figures in A. These tables are believed to be generally correct to within a unit of the second figure, though there may be slightly larger inaccuracies towards the lower right corner in those cases where the full working precision is demanded. The coefficients a_r for these functions were taken from (2), where they are given to a precision of 20 decimal places, and the results were produced from a double precision FORTRAN program.

It is of some interest that the most accurate approximation with a given number of parameters, that is with $m + n$ fixed, is usually on or adjacent to the principal diagonal ($m = n$); this feature is exemplified by the table for e^x in $[-1, 1]$. For easy identification, this "best" element is under-lined for each $m + n$. In the analysis of each example, we have of course used Chebyshev expansions in a variable appropriate to that example: that is to say, x for the normal situation in $[-1, 1]$, $2x^2 - 1$ for an even function in $[-1, 1]$, and $2x - 1$ for the interval $[0, 1]$. (The corresponding Chebyshev polynomials are T_r, T_{2r} and T_r^* respectively.) Tables such as those given for e^x, 2^{-x}, $\ln(1+x)$ and $(1/x)\tan^{-1} x$ have been found for other functions; they usually present a similar overall appearance. The pattern is, however, violated to a significant extent by two functions, namely $\Gamma(1+x)$ in $[0, 1]$ and $\cos \pi x$ in $[-1, 1]$. The "best" element for the former seems to wander randomly as $m + n$ varies, though the first column of the table is favoured. In the case of $\cos \pi x$, the best element

e^x in $[-1,1]$. (T_r)

m \ n	0	1	2	3	4	5	6
0	*1·5(0)*	*3·2(−1)*	5·0(−2)	6·1(−3)	5·9(−4)	4·8(−5)	3·4(−6)
1	5·9(−1)	*3·5(−2)*	2·6(−3)	1·8(−4)	1·1(−5)	6·3(−7)	3·4(−8)
2	9·4(−2)	3·3(−3)	*1·4(−4)*	6·7(−6)	3·0(−7)	1·3(−8)	5·2(−10)
3	1·2(−2)	2·6(−4)	7·8(−6)	*2·6(−7)*	8·7(−9)	2·9(−10)	9·2(−11)
4	1·3(−3)	1·8(−5)	3·9(−7)	9·8(−9)	*2·6(−10)*	6·8(−12)	1·8(−13)
5	1·1(−4)	1·1(−6)	1·8(−8)	3·5(−10)	7·4(−12)	*1·6(−13)*	3·5(−15)
6	7·7(−6)	6·2(−8)	7·9(−10)	1·2(−11)	2·1(−13)	3·8(−15)	*7·0(−17)*

2^{-x} in $[0,1]$. (T_r^*)

m \ n	0	1	2	3	4	5	6	7
0	*2·7(−1)*	*2·3(−2)*	1·3(−3)	5·5(−5)	1·9(−6)	5·5(−8)	1·4(−9)	2·9(−11)
1	2·8(−2)	*7·3(−4)*	2·0(−5)	5·1(−7)	1·2(−8)	2·4(−10)	4·4(−12)	7·3(−14)
2	1·6(−3)	2·2(−5)	*3·7(−7)*	6·2(−9)	9·9(−11)	1·5(−12)	2·2(−14)	2·9(−16)
3	7·2(−5)	5·9(−7)	6·5(−9)	*7·8(−11)*	9·5(−13)	1·1(−14)	1·3(−16)	1·4(−18)
4	2·5(−6)	1·4(−8)	1·1(−10)	9·9(−13)	*9·3(−15)*	8·8(−17)	8·3(−19)	
5	7·3(−8)	2·9(−10)	1·7(−12)	1·2(−14)	9·1(−17)	*7·0(−19)*		
6	1·8(−9)	5·4(−12)	2·6(−14)	1·4(−16)	8·8(−19)			
7	4·0(−11)	2·9(−13)	7·8(−16)	2·6(−16)				

m \ n	0	1	2	3	4	5	6	7
0	*3·8 (−1)*	*3·3 (−2)*	3·9 (−3)	5·0 (−4)	6·9 (−5)	1·0 (−5)	1·5 (−6)	2·2 (−7)
1	3·1 (−1)	*1·1 (−3)*	6·8 (−5)	5·5 (−6)	5·1 (−7)	5·4 (−8)	6·0 (−9)	7·1 (−10)
2	2·3 (−1)	*9·3 (−5)*	*2·6 (−6)*	1·5 (−7)	1·0 (−8)	8·0 (−10)	7·1 (−11)	6·8 (−12)
3	1·9 (−1)	9·9 (−6)	1·7 (−7)	*5·9 (−9)*	*3·2 (−10)*	2·0 (−11)	1·4 (−12)	1·2 (−13)
4	1·6 (−1)	1·2 (−6)	1·4 (−8)	3·5 (−10)	*1·3 (−11)*	*6·9 (−13)*	4·2 (−14)	2·8 (−15)
5	1·4 (−1)	1·5 (−7)	1·3 (−9)	2·5 (−11)	7·5 (−13)	*3·0 (−14)*	*1·5 (−15)*	8·8 (−17)
6	1·2 (−1)	2·1 (−8)	1·3 (−10)	2·0 (−12)	4·9 (−14)	1·6 (−15)	*6·6 (−17)*	*3·4 (−18)*
7	1·0 (−1)	2·9 (−9)	1·4 (−11)	1·8 (−13)	3·6 (−15)	1·0 (−16)	3·5 (−18)	*1·8 (−19)*

$\ln(1+x)$ in $[0,1]$. (T_r^*)

m \ n	0	1	2	3	4	5	6	7
0	*1·2 (−1)*	1·3 (−2)	1·6 (−3)	2·2 (−4)	3·1 (−5)	4·5 (−6)	6·7 (−7)	1·0 (−8)
1	*6·0 (−3)*	*2·9 (−4)*	2·1 (−5)	1·8 (−6)	1·9 (−7)	2·0 (−8)	2·3 (−9)	2·8 (−10)
2	5·5 (−4)	*1·4 (−5)*	*6·6 (−7)*	4·1 (−8)	3·1 (−9)	2·6 (−10)	2·4 (−11)	2·4 (−12)
3	6·1 (−5)	9·5 (−7)	*3·8 (−8)*	*1·5 (−9)*	8·6 (−11)	5·9 (−12)	4·5 (−13)	3·7 (−14)
4	7·6 (−6)	8·0 (−8)	1·9 (−9)	*6·9 (−11)*	*3·3 (−12)*	1·8 (−13)	1·2 (−14)	
5	1·0 (−6)	7·7 (−9)	1·4 (−10)	4·0 (−12)	*1·5 (−13)*	*7·3 (−15)*		
6	1·4 (−7)	8·1 (−10)	1·2 (−11)	2·7 (−13)	8·6 (−15)			
7	2·0 (−8)	9·0 (−11)	1·0 (−12)	2·3 (−14)				

$1/x \tan^{-1} x$ in $[-1,1]$. (T_{2r})

n \ m	0	1	2	3	4	5	6	7
0	6·7(−2)	6·3(−2)	5·8(−3)	1·6(−3)	2·3(−4)	4·3(−5)	7·3(−6)	1·3(−6)
1	6·3(−2)	*	1·2(−3)	3·8(−4)	1·2(−5)	1·2(−6)	5·2(−8)	4·1(−9)
2	5·1(−3)	8·8(−4)	1·5(−5)	3·9(−6)	5·0(−8)	2·2(−8)	1·2(−9)	1·7(−10)
3	4·7(−4)	1·1(−4)	3·3(−6)	7·7(−7)	2·2(−8)	5·7(−8)	1·0(−10)	1·0(−11)
4	7·8(−5)	1·6(−5)	2·3(−7)	2·1(−8)	5·5(−10)	3·8(−11)	8·8(−13)	7·4(−14)
5	3·0(−6)	3·9(−7)	1·8(−9)	6·2(−10)	6·8(−12)	1·5(−12)	4·2(−14)	5·4(−15)
6	2·9(−7)	5·6(−8)	6·0(−10)	2·2(−10)	1·4(−12)	4·2(−13)	2·8(−15)	2·0(−16)
7	3·2(−8)	3·5(−9)	3·4(−11)	2·1(−12)	3·8(−14)	1·9(−15)	3·5(−17)	2·0(−18)

$\Gamma(1+x)$ in $[0,1]$. (T_r^*)

n \ m	0	1	2	3	4	5	6	7
0	1·3(0)	3·3(−1)	3·1(−2)	1·4(−3)	4·1(−5)	7·9(−7)	1·1(−8)	1·1(−10)
1	*	1·1(−1)	1·7(−3)	2·9(−5)	4·1(−7)	4·4(−9)	3·8(−11)	2·7(−13)
2	*	1·5(−2)	8·6(−5)	7·0(−7)	5·3(−9)	3·5(−11)	2·0(−13)	9·6(−16)
3	*	2·3(−3)	4·3(−6)	1·7(−8)	7·5(−11)	3·2(−13)	1·2(−15)	4·1(−18)
4	*	3·3(−4)	2·0(−7)	4·1(−10)	1·1(−12)	3·0(−15)	7·9(−18)	
5	*	4·9(−5)	9·2(−9)	9·6(−12)	1·5(−14)	2·8(−17)		
6	*	7·1(−6)	4·1(−10)	2·2(−13)	2·1(−16)			
7	*	1·0(−6)	1·8(−11)	4·9(−15)				

$\cos \pi x$ in $[−1,1]$. (T_{2r})

n \ m	0	1	2	3	4	5	6	7
0	7·5 (−2)	7·0 (−2)	5·6 (−3)	5·0 (−4)	8·4 (−5)	3·3 (−6)	3·0 (−7)	3·5 (−8)
1	7·0 (−2)	*	9·3 (−4)	1·1 (−4)	1·7 (−5)	4·2 (−7)	5·9 (−8)	3·7 (−9)
2	6·4 (−3)	1·4 (−3)	1·8 (−5)	3·5 (−6)	2·4 (−7)	2·1 (−9)	6·3 (−10)	3·6 (−11)
3	1·8 (−3)	4·0 (−4)	4·3 (−6)	8·2 (−7)	2·2 (−8)	6·6 (−10)	2·3 (−10)	2·3 (−12)
4	2·6 (−4)	1·3 (−5)	5·3 (−8)	2·3 (−8)	5·9 (−10)	7·2 (−12)	1·5 (−12)	
5	4·9 (−5)	1·2 (−6)	2·4 (−8)	6·1 (−8)	4·0 (−11)	1·6 (−12)		
6	8·3 (−6)	5·6 (−8)	1·2 (−9)	1·1 (−10)	9·4 (−13)			
7	1·4 (−6)	4·4 (−9)	1·9 (−10)	1·1 (−11)				

$\{\Gamma(1+x)\}^{-1}$ in [0,1]. (T_r^*)

seems to demand a denominator of low degree. It is tempting to suggest that this behaviour for the cosine function is a consequence of its lack of singularity for any finite argument, which implies good approximation by polynomials. However, similar remarks apply to e^x and 2^{-x}, whose CP tables exhibit a normal character. We conclude that the striking nature of the CP table for $\cos \pi x$ derives from the fact that its behaviour on $[-1, 1]$ (on which the table is, of course, based) does not even *suggest* a singularity for any finite argument.

It is also noteworthy that $\Gamma(1 + x)$ and $\cos \pi x$ are the only functions in our sample whose CP tables have nonexistent elements (indicated by an asterisk).

For interest, we include a CP error table for the reciprocal of one of the functions already treated, namely $\Gamma(1 + x)$. One might expect the table for the reciprocal to resemble the transpose of the table for the function, and indeed a striking similarity is observed.

CONCLUSION

We have described a well-conditioned method for constructing the CP table. This table furnishes rational functions which may be used as effective approximations in their own right, or as starting points in an iterative procedure designed to produce optimal rational approximations.

It is natural to investigate the possibility of extending the method to deal with a more general class of expansions, but this possibility seems remote. The method depends essentially upon the fact that the coefficients in the expansion of the rational function (A_r in the foregoing discussion) satisfy a linear difference equation of order n, where n is the degree of the denominator.

This will indeed be true if the functions of the expansion themselves satisfy a first-order difference equation or (if the expansion is convergent) a symmetric second-order equation. In the first case, we have a classical Padé table, and in the second a CP table; in each, the argument will be determined by the coefficient of the difference equation.

The method has been presented in its simplest form, with little or no attention paid to such matters as reducibility and singularity. These and other questions will be explored in a future paper.

ACKNOWLEDGEMENT

We wish to thank Dr. D. Kershaw for his helpful comments on the presentation of this material.

REFERENCES

1. Cheney, E. W. (1966). *Introduction to Approximation Theory.* McGraw-Hill, New York.
2. Clenshaw, C. W. (1962). *Chebyshev Series for Mathematical Functions.* NPL Math. Tables Vol. 5, HMSO, London.
3. Fike, C. T. (1968). *Computer Evaluation of Mathematical Functions.* Prentice-Hall, New Jersey.
4. Fox L. and Parker, I. B. (1968). *Chebyshev Polynomials in Numerical Analysis.* O.U.P., London.
5. Handscomb, D. C. (ed.) (1966). *Methods of Numerical Approximation.* Pergamon, Oxford.
6. Hayes, J. G. (ed.) (1970). *Numerical Approximation to Functions and Data.* Athlone, London.
7. Hornecker, G. (1960). Méthodes practiques pour la détermination approchée de la meilleure approximation polynômiale ou rationelle. *Chiffres*, **3**, 193–228.
8. Lanczos, C. (1938). Trigonometric interpolation of empirical and analytical functions. *J. Math. Phys.*, **17**, 123–199.
9. Lanczos, C. (1952). *Tables of Chebyshev polynomials.* NBS Appl. Math. Series 9. Government Printing Office, Washington.
10. Lanczos, C. (1957). *Applied Analysis.* Pitman, London.
11. Maehly, H. J. (1960). Rational approximations for transcendental functions. Proceedings of the International Conference on Information Processing, UNESCO. Butterworths, London.
12. Ralston, A. (1965). *A First Course in Numerical Analysis.* McGraw-Hill, New York.
13. Rutishauser, H. (1957). *Der Q-D Algorithmus.* Birkhäuser, Basel.
14. Wynn, P. (1956). On a Procrustean technique for the numerical transformation of slowly-convergent sequences and series. *Proc. Camb. Phil. Soc.*, **52**, 665–671.

Accuracy of Computed Eigensystems
and Invariant Subspaces

G. Peters and J. H. Wilkinson

National Physical Laboratory, Teddington, Middlesex, England

I. INTRODUCTION

One of the most fundamental problems in computational linear algebra is the determination of the eigenvalues and eigenvectors of a matrix. Over the last thirty years this problem has been the subject of intensive research and algorithms now exist which are remarkably 'accurate' and reliable. A pleasing feature of research in this area is the way in which the error analysis of the underlying causes of instability in computational algebra has concentrated research in the most profitable directions.

In spite of the success of the error analysis and the effectiveness of modern algorithms, a proper appreciation of the current position is not as widespread as might have been expected. It is not uncommon to see criticisms of the accuracy of results obtained even when these may well be described as "almost best possible" for the precision of computation being used. In this paper we attempt a brief survey of the current position.

II. SUMMARY OF PERTURBATION THEORY FOR NORMAL MATRICES

Only in rare cases can eigenvalue computations be carried out exactly. Usually rounding errors are involved and as a result we can at best hope to find the eigensystem of some matrix $A + E$ rather than of A, where $\|E\|/\|A\|$ is related to the machine precision. The relationship of the eigensystem of $A + E$ to that of A is therefore of fundamental importance.

Results for the perturbation of the eigenvalues are particularly satisfactory when $A + E$ and A are both normal. The Wielandt-Hoffman

115

theorem then states that if the eigenvalues of $A + E$ are λ_i' and those of A are λ_i, they can be ordered so that

$$\left\{ \sum |\lambda_i' - \lambda_i|^2 \right\}^{1/2} \leq \|E\|_F = \epsilon \|A\|_F \quad \text{say} \tag{2.1}$$

where $\|\cdot\|_F$ denotes the Frobenius norm. This result holds for all ϵ but in practice we are interested in the case when ϵ is small. Notice that the result is true independent of the multiplicity of the eigenvalues of $A + E$ and A; *multiple roots of normal matrices are no more sensitive than well separated roots*. Since

$$\|A\|_F = \left\{ \sum |\lambda_i|^2 \right\}^{1/2} \tag{2.2}$$

equation (2.1) may be written

$$\left\{ \frac{\sum |\lambda_i' - \lambda_i|^2}{\sum |\lambda_i|^2} \right\}^{1/2} \leq \frac{\|E\|_F}{\|A\|_F} = \epsilon \tag{2.3}$$

a particularly cogent form of the result.

When ϵ is small (2.3) shows that the root mean square of perturbations is *relatively small*, but of course the relative perturbation of a 'small' eigenvalue may be high. In particular A may have a zero eigenvalue while $A + E$ has not and vice versa.

So far we have interested ourselves in the case when $\|E\|_F$ is small relative to $\|A\|_F$. In some circumstances we shall be interested in more specialised perturbations for which

$$|e_{ij}| \leq \epsilon |a_{ij}| \tag{2.4}$$

i.e. to cases when we have small relative perturbations in each individual element. For matrices of a special type perturbations of this kind *may* lead to small relative perturbations in each individual eigenvalue and in practice this is sometimes important. We return to this point again in the final section.

Most normal matrices in which we are interested are, in fact, Hermitian, and the most important algorithms effectively produce eigenvalues which correspond to Hermitian perturbations of the given matrix. Hence the Wielandt-Hoffman theorem is relevant. Moreover we have the simplification that the λ_i' and λ_i are real and the orderings which give (2.3) are the monotonic orderings of both the λ_i' and λ_i. We may summarise this by saying that small Hermitian perturbations in the norm of A make small perturbations in all eigenvalues relative to $\|A\|_F$; the problem of computing the eigenvalues must always be regarded as *well conditioned* unless we are interested in high relative accuracy of each individual eigenvalue and some of the eigenvalues are small relative to $\|A\|_F$.

There is a second theorem which is of value in connexion with Hermitian perturbations of Hermitian matrices. This states that if A and E are Hermitian, λ'_i are the eigenvalues of $A + E$ are λ_i are those of A both sets being in decreasing order then

$$|\lambda'_i - \lambda_i| \leq \|E\|_2 \tag{2.5}$$

However $\|E\|_2 = \max |\mu_i|$ where the μ_i are the eigenvalues of E and the determination of $\|E\|_2$ is therefore a major computation. In practice one usually obtains some bound for $\|E\|_2$. We can, of course, use the result $\|E\|_2 \leq \|E\|_F$ but if we do this we have a weaker result than the Wielandt–Hoffman theorem. In practice error analysis of an algorithm often gives a sharp bound for $\|E\|_F$ rather than $\|E\|_2$ and the Wielandt–Hoffman theorem is then the more useful result.

Results for the eigenvectors are much less satisfactory. Eigenvectors corresponding to eigenvalues with separations which are small relative to $\|A\|$ are extremely sensitive. Some of the better eigenvalue algorithms for Hermitian matrices produce vectors which are almost exactly orthogonal (even when they correspond to close eigenvalues) and which are 'close' to the exact eigenvectors of $A + E$ where E is small and Hermitian. We are therefore interested in the eigenvectors of $(A + E)$.

Suppose λ and y are an eigen-pair of $A + E$, that the eigenvalues of A are λ_i and that x_i are a complete set of orthogonal eigenvectors. We assume that y and the x_i are normalised so that

$$\|y\|_2 = \|x_i\|_2 = 1 \tag{2.6}$$

We have

$$(A + E)y = \lambda y \tag{2.7}$$

and writing

$$y = \sum \alpha_i x_i \tag{2.8}$$

$$Ey = \sum_1^n (\lambda - \lambda_i)\alpha_i x_i \tag{2.9}$$

From the Wielandt-Hoffman theorem we know that there must be at least one λ_j such that $|\lambda - \lambda_j| \leq \|E\|$; let us choose λ_j to be the nearest eigenvalue. From equation (2.9)

$$\|E\|_2 = \|E\|_2\|y\|_2 \geq \|Ey\|_2 = \left[\sum (\lambda - \lambda_i)^2 \alpha_i^2\right]^{1/2} \geq \left[\sum_{i \neq j} (\lambda - \lambda_i)^2 \alpha_i^2\right]^{1/2} \tag{2.10}$$

Suppose λ_j is a well-isolated eigenvalue and

$$|\lambda - \lambda_i| \geq b \qquad (i \neq j) \tag{2.11}$$

then (2.10) gives

$$\left[\sum_{i \neq j} \alpha_i^2\right]^{1/2} \leq \|E\|_2/b \tag{2.12}$$

In this case y will be a good approximation to x_j and the component of y which is orthogonal to x_j will be of length less than $\|E\|_2/b$. As b becomes smaller this result becomes progressively weaker; when b is of the order of magnitude of $\|E\|_2$, (2.12) finally fails to give any useful bound for the error in y as an approximation to x_j. This deterioration is inevitable because eigenvectors corresponding to close eigenvalues are indeed extremely sensitive to perturbations in the matrix.

However we may still get a very valuable result if we are prepared to lower our sights a little. Let us consider the case when A has an eigenvalue of multiplicity k. The corresponding eigenvectors span a subspace of dimension k and *any orthogonal basis of this subspace is equally acceptable.* This immediately focuses attention on orthogonal bases of invariant subspaces rather than on individual eigenvectors. Suppose A has a group of k close eigenvalues which are well separated from the remaining $(n-k)$. Will $(A+E)$ have an invariant subspace close to that corresponding to the k close eigenvalues of A? It is easy to see that it will. Let us assume for simplicity that the eigenvalues are numbered so that the cluster of eigenvalues are $\lambda_1, \lambda_2, \ldots \lambda_k$ and that

$$|\lambda_i - \lambda_j| \geq b \qquad i \leq k, j > k \tag{2.13}$$

Let λ be an eigenvalue of $A + E$ "associated with" a member of the cluster and let y be a corresponding eigenvector. Then from (2.10)

$$\|E\|_2 \geq \left[\sum_{i>k} (\lambda - \lambda_i)^2 \alpha_i^2 \right]^{1/2} \tag{2.14}$$

Now since λ is related to an eigenvalue of the cluster

$$|\lambda - \lambda_i| \geq b - \|E\| \qquad i > k \tag{2.15}$$

and hence

$$\left[\sum_{i>k} |\alpha_i^2| \right]^{1/2} \leq \|E\| / (b - \|E\|) \tag{2.16}$$

showing that y contains only a small component orthogonal to the subspace spanned by the k eigenvectors of A corresponding to the cluster.

The result is best illustrated by an example. Suppose A is of order five, $\|E\| = 10^{-10}$ and

$$\lambda_1 = 1, \quad \lambda_2 = 1 - 10^{-6}, \quad \lambda_3 = 1 - 10^{-4}, \quad \lambda_4 = 0.5, \quad \lambda_5 = 0.3$$

The corresponding eigenvalues λ_i' of $A + E$ are such that $|\lambda_i' - \lambda_i| \leq 10^{-10}$. The eigenvalues λ_4 and λ_5 are well isolated and if x_4' and x_5' are the eigenvectors of $A + E$ then these will be good approximations to x_4 and x_5 respectively. For example x_4' will contain a component orthogonal to x_4 bounded by $10^{-10}/(0.2 - 10^{-10})$. The situation as regards x_1' and x_2' and x_3' is less satisfactory; x'_1 may have a component orthogonal to x_1 of magnitude up to $10^{-10}/(10^{-6} - 10^{-10})$. Roughly speaking we may say that x'_1 gives x_1 correct to about 4 decimals, x'_2 gives x_2 to about 4 decimals, x'_3 gives x_3 correct to about 6 decimals while x'_4 and x'_5 give x_4 and x_5 correct to about 10 decimals. *However x'_1, x'_2, x'_3 give an orthogonal basis for the invariant subspace spanned by x_1, x_2 and x_3 which is correct almost to 10 decimals* as we see by applying (2.16) with $k = 3$, while x'_1 and x'_2 give an orthogonal basis of the invariant subspace spanned by x_1 and x_2 which is correct to about 6 decimals. By taking larger groups of close eigenvalues we can make greater claims for the accuracy of the orthogonal basis obtained from $(A + E)$ as invariant subspaces of A. However the larger the group, the less information we are giving. Indeed if we place all the eigenvalues in one group an exact orthogonal basis is always provided by the columns of the identity matrix! Given a tolerance η which is not too small, we may determine the smallest groups which give orthogonal bases for the corresponding subspaces all having errors less than η. For example if $\eta = 10^{-5}$ we must group λ_1 and λ_2, while λ_3, λ_4 and λ_5 may each be taken separately. If $\eta = 10^{-7}$ however λ_1, λ_2 and λ_3 must be grouped together. The situation is comparatively simple for Hermitian matrices mainly because $A + E$ has orthogonal eigenvectors in this case.

We have not yet said much about the case when A has true multiple eigenvalues. If we have *a priori* knowledge that certain eigenvalues of A are truly multiple the situation is surprisingly different. Suppose, for example, A has the eigenvalues 1, 1, 0·7, 0·5, 0·3 then we do not need to group the first two eigenvalues in spite of their zero separation. Any two orthogonal vectors in the invariant subspace corresponding to λ_1 and λ_2 are acceptable and x'_1 and x'_2 will give approximations to an x_1 and x_2 correct to ten decimals.

III. PERTURBATION THEORY FOR NON-NORMAL MATRICES

Perturbation theory for eigenvalues of non-normal matrices is essentially less satisfactory and bounds almost inevitably involve some quantities for which no estimates are available until at least an approximate eigensystem has been determined. As far as eigenvectors are concerned we have the further complication that A may be defective in which case there will not be a complete set of eigenvectors.

The distinction between defective and non-defective matrices is not nearly so clear-cut from a computational point of view, since even if A is defective, rounding errors generally lead to a non-defective matrix and since the corresponding multiple eigenvalues of A are extremely sensitive

$A + E$ may not have any particularly 'close' eigenvalues even when E is 'small'. For this reason it is less of a restriction to limit oneself to the case of non-defective matrices than it might appear to the theorist.

Typical of the results available is that of the Bauer-Fike theorem. Let A be a non-defective matrix and X be a non-singular matrix such that

$$A = X \operatorname{diag}(\lambda_i) X^{-1} \tag{3.1}$$

An immediate deduction is that

$$A + E - \lambda I = X [\operatorname{diag}(\lambda_i - \lambda) + X^{-1} E X] X^{-1} \tag{3.2}$$

If λ is an eigenvalue of $A + E$ then $A + E - \lambda I$ is singular and hence $\operatorname{diag}(\lambda_i - \lambda) + X^{-1} E X$ is singular, implying that λ lies in at least one of the discs with centre λ_i and radius $\|X^{-1} E X\|$, where $\|\cdot\|$ is any norm for which $\|\operatorname{diag}(\alpha_i)\| = \max |\alpha_i|$. Hence we have the following theorem (Bauer-Fike[2]).

All eigenvalues of $A + E$ lie in the discs with centres λ_i and radii equal to $\|X\| \|X^{-1}\| \|E\|$. By a simple continuity argument one may deduce further that if any r of the discs are isolated from the remainder their union contains precisely r eigenvalues. In particular a single isolated disc contains precisely one eigenvalue. The sensitivity of the whole set of eigenvalues is therefore related to $\|X\| \|X^{-1}\|$ which is the condition number of the matrix X of eigenvectors with respect to inversion, usually denoted by $\kappa(X)$. The matrix X is not, of course, unique and clearly it is $\min \|X\| \|X^{-1}\|$ which is relevant.

When A is normal X may be chosen to be unitary and $\|X\|_2 = 1$; hence each eigenvalue of A lies in a disc of radius $\|E\|_2$ centred on a λ_i. This shows once again that all eigenvalues of normal matrices are well-conditioned and notice that the result now applies to perturbations for which $A + E$ is no longer normal.

For non-normal matrices the theorem gives information on the factors affecting the sensitivity of eigenvalues but since we shall seldom have *a priori* information about $\kappa(X)$ it does not provide information which can be used. When some calculation has been performed and we have information about X we can usually say much more about the sensitivity than can be deduced from the Bauer-Fike theorem. An outstanding weakness of the Bauer-Fike theorem is that the discs are all of the same size even though some eigenvalues may be quite insensitive.

In The Algebraic Eigenvalue Problem[3] Wilkinson has analysed the perturbation of the eigensystem of a non-normal matrix by applying Gerschgorin's theorem to the perturbed diagonal matrix $\operatorname{diag}(\lambda_i) + X^{-1} E X$. This is a valuable technique since it leads to a particularly useful method of deriving *a posteriori* error bounds for a computed eigensystem.

However, it is more convenient for our present purposes to give perturbation results based on the calculus. Such an analysis is 'unfashionable' at present and is often criticised because it gives only asymptotic results. In our experience the insight provided by this analysis is severely underestimated.

The impact of the results is somewhat sharper if one assumes that A is scaled so that $\|A\|_2 = 1$ and we consider a perturbed matrix $A + \epsilon E$ where we also have $\|E\|_2 = 1$. (It should be emphasized that no real restriction is involved). We assume that A has a complete set of left-hand eigenvectors y_i and right hand eigenvectors x_i, normalised so that

$$\|x_i\|_2 = \|y_i\|_2 = 1 \qquad i = 1, \ldots, n \qquad (3.3)$$

If λ_i is a simple eigenvalue of A, then $(A + E)$ has an eigenvalue λ_i' such that

$$\lambda_i' - \lambda_i = \epsilon y_i^H E x_i / y_i^H x_i + O(\epsilon^2) \quad \text{as} \quad \epsilon \to 0. \qquad (3.4)$$

Since $|y_i^H E x_i| \leq \|y_i\| \|E\| \|x_i\| = 1$ the sensitivity of λ_i is primarily dependent on $(y_i^H x_i)^{-1}$ and this can be arbitrarily large even when λ_i is an isolated root. When x_i corresponds to a non-linear elementary divisor $y_i^H x_i = 0$ and the result is no longer true. The perturbation of λ_i then depends on a fractional power of ϵ and its rate of change when $\epsilon = 0$ is infinite.

Although one does not know how small ϵ must be before the higher order terms in (3.4) are negligible there is a great deal of useful information in this relation. In particular it shows the main factors affecting the sensitivity of individual eigenvalues. The quantities $(y_i^H x_i)$ are referred to as s_i in[3] and this notation is now quite widely used. There is no loss of generality in assuming that $y_i^H x_i$ is positive. Notice that if Q is unitary

$$QAQ^H Q x_i = \lambda_i Q x_i \quad \text{and} \quad (y_i^H Q^H) QAQ^H = \lambda_i y_i^H Q^H \qquad (3.5)$$

and hence $Q y_i$ and $Q x_i$ are normalized left-hand and right-hand eigenvectors of QAQ^H. Since $y_i^H Q^H Q x_i = y_i^H x_i$ the s_i are invariant with respect to unitary similarity transformations.

An *a priori* error analysis of the QR algorithm shows that it produces a unitary reduction of a general matrix A to upper triangular form T such that

$$T = Q(A + E)Q^H \qquad (3.6)$$

where E is directly related to the computer precision. It is easy to find an 'accurate' value of an s_i corresponding to a triangular matrix, and hence effectively of $A + E$. Hence when the QR algorithm is completed one effectively knows the s_i corresponding to some $A + E$. If any of these s_i is small, the corresponding eigenvalue of A is certainly sensitive.

Notice that there are always perturbations for which (3.4) has its full significance. Indeed if we take $E = y_i x_i^H$ (so that $\|E\|_2 = 1$) we have

$$\lambda_i' - \lambda_i = \epsilon/s_i + O(\epsilon^2) \tag{3.7}$$

On the other hand there are perturbations for which the term in ϵ vanishes e.g. $E = x_j y_j^H (j \neq i)$. Indeed from the Schur triangular canonical form it is obvious that there are perturbations in A which leave all eigenvalues unaltered. At this stage it is salutary to be reminded that there are diagonal similarity transforms of A, of the form $D^{-1}AD$, having arbitrarily small values of s_i. The matrix

$$A = \begin{bmatrix} 3 & 1 \\ 1 & 3 \end{bmatrix} \tag{3.8}$$

for example, has eigenvalues $\lambda_1 = 4$ and $\lambda_2 = 2$ and we have

$$2^{1/2}x_1 = \begin{bmatrix} 1 \\ 1 \end{bmatrix}, \quad 2^{1/2}y_1 = \begin{bmatrix} 1 \\ 1 \end{bmatrix}, \quad 2^{1/2}x_2 = \begin{bmatrix} 1 \\ -1 \end{bmatrix}, \quad 2^{1/2}y_2 = \begin{bmatrix} 1 \\ -1 \end{bmatrix} \tag{3.9}$$

so that $s_1 = s_2 = 1$ as we would expect from a symmetric matrix. However the matrix

$$\bar{A} = \begin{bmatrix} 3 & 10^{10} \\ 10^{-10} & 3 \end{bmatrix} \tag{3.10}$$

obtained by a diagonal similarity of A has eigenvectors given by

$$x_1 = \begin{bmatrix} 1 \\ 10^{-10} \end{bmatrix} \bigg/ a, \quad y_1 = \begin{bmatrix} 10^{-10} \\ 1 \end{bmatrix} \bigg/ a, \quad x_2 = \begin{bmatrix} 1 \\ -10^{-10} \end{bmatrix} \bigg/ a, \quad y_2 = \begin{bmatrix} -10^{-10} \\ 1 \end{bmatrix} \bigg/ a \tag{3.11}$$

where $a = (1 + 10^{-20})^{1/2}$ so that

$$s_1 = s_2 = 2(10^{-10})/a^2 \doteq 2(10^{-10}) \tag{3.12}$$

It is true that small perturbations in $\bar{A}$ relative to its norm can make much larger perturbations in the eigenvalues. For example with

$$E = \begin{bmatrix} 0 & 0 \\ -10^{-10} & 0 \end{bmatrix} \text{ for which } \|E\|_2 \doteq 10^{-20}\|A\| \tag{3.13}$$

the eigenvalues of $A + E$ are $\lambda_1 = \lambda_2 = 3$. Clearly the small s_i and the sensitivity of the eigenvalues of $\bar{A}$ are an 'artificial' phenomenon. It is good practice when computing the eigenvalues of a matrix to start by performing a diagonal similarity, chosen in such a way as to make $D^{-1}AD$ as 'balanced' as possible. This was first discussed by Osborne[4] and has been extended recently by Parlett, Reinsch and Wilkinson (see e.g.[5]). In all further discussion we shall assume that the given A has been balanced in this way and that any small s_i are therefore 'genuine'.

Since a zero s_i can certainly not arise if all the eigenvalues of A are simple it is natural to conjecture that if any s_i is small A must be 'near' to a matrix having a multiple eigenvalue. Recently Wilkinson[7] has shown that this is indeed true and that there is a matrix $A + E$ with

$$\|E\| \leq s_i\|A\|/(1 - s_i^2)^{1/2} \tag{3.14}$$

having a multiple eigenvalue. It is interesting that the authors' conjecture was motivated by the relation (3.4) using the following argument. Suppose A has just two ill-conditioned eigenvalues λ_1 and λ_2 and the others are well conditioned, so that s_1 is small. (In fact in this case s_2 will be almost equally small but we do not need to assume this). Let us consider a perturbation ϵE with $E = y_1^H x_1 \arg (\lambda_2 - \lambda_1)$. From (3.4) we have

$$\frac{d\lambda_1}{d\epsilon} = \arg (\lambda_2 - \lambda_1)/s_1 \tag{3.15}$$

Hence small changes proportional to $y_1^H x_1 \arg (\lambda_2 - \lambda_1)$ make rapid changes in λ_1 in the direction of λ_2. The trace is changing only slowly and eigenvalues $\lambda_3, \ldots, \lambda_n$ are by our assumption also changing only slowly. Hence λ_2 must be changing fast, the change in it being almost equal and opposite to that in λ_1. For such perturbations E, λ_1 and λ_2 therefore move rapidly together and as they get closer s_1 will become even smaller. One would therefore expect that for a perturbation smaller than $|\frac{1}{2}(\lambda_1 - \lambda_2)s_1|$ we will obtain a double eigenvalue roughly at $\frac{1}{2}(\lambda_1 + \lambda_2)$. Since one would expect that $\lambda_1 - \lambda_2$ would be small compared with $\|A\|$ in this case, A will usually be much closer to a matrix having a multiple eigenvalue than the rigorous bound (3.14) leads one to expect. In fact the estimate $|\frac{1}{4}(\lambda_1 - \lambda_2)s_1|$ is usually remarkably good, which will not surprise anyone who has not reacted in an irrational way to the use of the elementary calculus for such purposes!

IV. EIGENVECTOR PERTURBATIONS

We now turn to the eigenvector perturbations analogous to (3.4). If A has simple eigenvalues λ_i, and y_i and x_i are a complete set of left-hand and right-hand eigenvectors then $A + \epsilon E$ has an eigenvector x_i' such that

$$x_i' = x_i + \epsilon \sum_{j \neq i} \frac{y_j^H E x_i}{(\lambda_i - \lambda_j)s_j} x_i + O(\epsilon^2) \quad \text{as} \quad \epsilon \to 0 \tag{4.1}$$

This gives the perturbation in x_i of order ϵ resolved in the directions of the remaining x. It shows that the perturbations may be large in the direction of any x_j for which $\lambda_i - \lambda_j$ is small. This had already been observed even in the case of normal matrices. However the quantities $1/s_j$ also enter into the expression for the perturbation but this can be misleading. Suppose A is close to a matrix having a quadratic elementary divisor and λ_k and λ_{k+1} are the relevant pair of eigenvalues. In this case x_k and x_{k+1} will be nearly parallel. The vectors x_j ($j \neq i$) will be a very poor basis in which to express the perturbation; x_i' may well have comparatively large perturbations in the directions of x_k and x_{k+1} which almost cancel one another out and together give quite a normal contribution to the perturbation in x_i.

For normal matrices we showed that it is more natural to think in general of determining orthogonal bases of invariant subspaces rather than individual eigenvectors, and it was easy to see how to group computed eigenvectors together in the optimum way to give invariant subspaces of prescribed accuracy. Provided we have an algorithm which gives accurate eigenvectors of some $A + E$ where E is small, the separation of the *computed* eigenvalues gives sufficient information to determine the grouping.

For non-normal matrices the situation is incomparably more difficult. Even if we had an algorithm which gave eigenvectors of A itself correct to working accuracy, this would not necessarily enable us to give an accurate orthogonal basis for a prescribed invariant subspace. Suppose for example that A is close to a matrix $(A + F)$ having a cubic elementary divisor, the corresponding eigenvalues of A being λ_1, λ_2 and λ_3. First it should be observed that λ_1, λ_2 and λ_3 may not be unduly close because the multiple eigenvalues of $(A + F)$ will be very sensitive. The corresponding eigenvectors x_1, x_2, x_3 will all be approximations to the one eigenvector which $(A + F)$ possesses corresponding to the triple eigenvalue. If we had the *exact* x_1, x_2 and x_3 and performed *exactly* the Schmidt orthogonalization process on them we would obtain an orthogonal basis for the relevant invariant subspace of A. However because x_1, x_2 and x_3 will be nearly parallel (on a computer using a t digit word, x_1, x_2 and x_3 will probably agree in their first $t/3$ digits!) mere end figure errors in x_1, x_2 and x_3 will lead to a poor determination of the two directions orthogonal to x_1 in the invariant subspace. Hence even if we know how we wish to group the λ_i it will be inadequate to compute the full set of eigenvectors and then to attempt to produce orthogonal bases from them.

V. COMPUTATION OF ORTHOGONAL BASES
FOR AN INVARIANT SUBSPACE

Suppose we have an algorithm which gives an exact unitary reduction of some matrix $A + E$ to upper-triangular form T so that

$$Q^H(A+E)Q = T \tag{5.1}$$

for some unitary Q; the eigenvalues of $A + E$ are of course the diagonal elements μ_i (say) of T. There are two main problems to be solved.

(i) Given a group of eigenvalues $\mu_{i_1}, \mu_{i_2}, \ldots, \mu_{i_k}$, find an accurate orthogonal basis for the corresponding invariant subspace of T and thereby of $(A + E)$.

(ii) Given a bound ϵ for $\|E\|$ and a tolerance η find the optimum grouping of the eigenvalues of T so that the corresponding invariant subspaces of $(A + E)$ differ from those of A by less than η. Naturally we want the groups to be as small as possible so that we shall determine an individual eigenvector whenever we can guarantee that an eigenvector of $A + E$ differs from the corresponding eigenvector of A by less than η. (We purposely leave vague the definition of the 'difference' between two orthogonal bases).

Of these the solution of (i) is comparatively satisfactory. Notice first that if we group together the first k eigenvalues of T and write

$$T = \begin{bmatrix} T_{11} & T_{12} \\ 0 & T_{22} \end{bmatrix} \tag{5.2}$$

where T_{11} is a square matrix of order k then

$$T I_k = I_k T_{11} \tag{5.3}$$

where I_k is the $n \times k$ matrix given by the first k columns of the identity matrix. Hence I_k gives an orthogonal basis for the relevant invariant subspace of T and $Q_k = Q I_k$ (i.e. the first k columns of Q) gives the relevant orthogonal basis of the invariant subspace of $A + E$.

For any other group of k eigenvalues of T the problem is solved if we can find an orthogonal transformation Q_1 such that

$$Q_1^H T Q_1 = \begin{bmatrix} T_{11}^{(k)} & T_{12}^{(k)} \\ 0 & T_{22}^{(k)} \end{bmatrix} = T^{(k)} \tag{5.4}$$

where $T^{(k)}$ is still upper triangular but the required group of eigenvalues are now the diagonal elements of $T_{11}^{(k)}$ the leading principal minor of order k of $T^{(k)}$. Since

$$Q_1^H Q^H (A+E) Q Q_1 = Q_1^H T Q_1 = T^{(k)} \tag{5.5}$$

we see that the relevant orthogonal basis is $(Q Q_1) I_k$, i.e. the first k columns of $Q Q_1$.

Ruhe has described a convenient method for determining Q_1 as the product of plane rotations. It is based on the following observation. If

$$T = \begin{bmatrix} p & q \\ 0 & r \end{bmatrix} \tag{5.6}$$

is a (2×2) upper triangular matrix then there is a plane rotation R such that

$$RTR^H = \begin{bmatrix} r & \bar{q} \\ 0 & p \end{bmatrix} \tag{5.7}$$

This plane rotation gives an interchange of r and p. In fact if we write

$$R = \begin{bmatrix} \bar{c} & \bar{s} \\ -s & c \end{bmatrix} \quad \text{where} \quad |c|^2 + |s|^2 = 1 \tag{5.8}$$

then we require

$$\begin{bmatrix} \bar{c} & \bar{s} \\ -s & c \end{bmatrix} \begin{bmatrix} 0 & q \\ 0 & r-p \end{bmatrix} \begin{bmatrix} c & -\bar{s} \\ s & \bar{c} \end{bmatrix} = \begin{bmatrix} r-p & \bar{q} \\ 0 & 0 \end{bmatrix}$$

or

$$\begin{bmatrix} 0 & \bar{c}q + \bar{s}(r-p) \\ 0 & -sq + c(r-p) \end{bmatrix} \begin{bmatrix} c & -\bar{s} \\ s & \bar{c} \end{bmatrix} = \begin{bmatrix} r-p & \bar{q} \\ 0 & 0 \end{bmatrix} \tag{5.9}$$

giving

$$-sq + c(r-p) = 0 \quad \text{or} \quad s = (r-p)/x \qquad c = q/x \tag{5.10}$$

where $x = \{|q|^2 + |r-p|^2\}^{1/2}$. It can readily be verified that this gives the required transformation. When p, q and r are real the plane rotation is real. By a sequence of similarity transformations using plane rotations any ordering may be induced in the diagonal elements. Hence Q_1 may be found as the product of plane rotations.

In practice we are often concerned with finding the eigensystem of a real matrix. The QR algorithm gives an orthogonal reduction to a quasi-triangular real matrix having real 2×2 blocks on the diagonal corresponding to each complex conjugate pair of eigenvalues. It is easy to modify Ruhe's technique so as to order the diagonal elements and 2×2 blocks as required, using real elementary orthogonal similarities. Once the grouping has been decided on we can obtain an orthogonal basis for each invariant subspace by the above technique. If we denote these orthogonal bases by $S_1, S_2, \ldots, S_r$, then even when the effect of rounding errors are taken into account we have

$$A[S_1 S_2 \cdots S_r] = [S_1 S_2 \cdots S_r] \begin{bmatrix} T_1 & 0 & \\ 0 & T_2 & \\ & & \ddots \\ & & & T_r \end{bmatrix} + F \tag{5.11}$$

where we have taken r groups, each T_i is a quasi-triangular real matrix, and F is small and directly related to the machine precision. Each S_i has orthogonal columns but the columns of S_i will not in general be orthogonal to those of S.

It should be emphasized that the size of F is not at all dependent on how well' we have performed the grouping. Indeed even if A is defective, the computed T will usually not be when rounding errors are involved; in this case even if we find individual eigenvectors, so that $r = n$ and each S_i has one column, the residual F will be no larger than that corresponding to the best grouping. The emergence of a small generalized residual F merely verifies that the computer has performed correctly and the algorithm has been correctly programmed. A small generalized residual F is guaranteed by an *a priori* error analysis.

Writing $S = S_1 S_2 \cdots S_r$ equation (5.11) gives

$$S^{-1}[A + FS^{-1}]\,S = \begin{bmatrix} T_1 & & & \\ & T_2 & \cdot & \\ & & \cdot & \\ & & & T_r \end{bmatrix} \tag{5.12}$$

so that the S_i are exact invariant subspaces for $A + FS^{-1}$. If the grouping has been done well, $\|S^{-1}\|$ must not be large (i.e. S must be well-conditioned with respect to inversion) otherwise FS^{-1} will not, in general, be small.

Although this condition is necessary, it is certainly not sufficient to guarantee that the S_i should be accurate (in the absolute sense) for A itself. Indeed if A were a real symmetric matrix the T given by the QR algorithm would be diagonal and if we grouped the eigenvalues separately we have $S = Q$ (no further Q_1 transformation being necessary). Hence the Q_i would each have one column and each Q_i would be orthogonal to all other Q_j. Nevertheless we could not guarantee that the absolute error in the computed invariant subspaces (i.e. the individual eigenvectors) would be small. It would still be necessary to group eigenvalues together. *A well-conditioned S matrix merely guarantees that we have dealt successfully with small s_i.*

At the moment it would appear that a reasonably economical algorithm for solving problem (ii) has not yet been devised.

VI. NUMERICAL EXAMPLE

A good illustration of the problem of determining accurate invariant subspaces is provided by the Frank matrices F_n the general form of which is illustrated by

$$F_4 = \begin{bmatrix} 4 & 3 & 2 & 1 \\ 3 & 3 & 2 & 1 \\ & 2 & 2 & 1 \\ & & 1 & 1 \end{bmatrix} \tag{6.1}$$

As n increases the eigenproblem of F_n becomes increasingly ill-conditioned; in fact F_{16} is already very ill-conditioned.

In Table I we give the nine smallest computed eigenvalues together with the true values. The computation was carried out on the computer KDF9 which has a 39 binary digit mantissa, roughly the equivalent of 12 decimals, but for convenience we give only 4 significant decimals.

Table I

Computed Eigenvalues	True Eigenvalues
$\lambda_{16} = -0\cdot02710 + i(0\cdot04506)$	$\lambda_{16} = 0\cdot02176$
$\lambda_{15} = -0\cdot02710 - i(0\cdot04506)$	$\lambda_{15} = 0\cdot03133$
$\lambda_{14} = 0\cdot06121 + i(0\cdot09907)$	$\lambda_{14} = 0\cdot4517$
$\lambda_{13} = 0\cdot06121 - i(0\cdot09907)$	$\lambda_{13} = 0\cdot06712$
$\lambda_{12} = 0\cdot1882 + i(0\cdot06248)$	$\lambda_{12} = 0\cdot1051$
$\lambda_{11} = 0\cdot1882 - i(0\cdot06248)$	$\lambda_{11} = 0\cdot1775$
$\lambda_{10} = 0\cdot3342$	$\lambda_{10} = 0\cdot3307$
$\lambda_9 = 0\cdot6809$	$\lambda_9 = 0\cdot6809$
$\lambda_8 = 1\cdot469$	$\lambda_8 = 1\cdot469$

These smaller eigenvalues are so ill-conditioned that changes of the order of $10^{-12}\|A\|$ lead, in general, to a matrix having three complex pairs of eigenvalues although all eigenvalues of all F_n are real.

The invariant subspaces of dimensions 2, 4, 6, 7, 8, 9 were computed, where, in general, that of dimension r was derived from the r smallest eigenvalues. (Since we compute real orthogonal bases, we did not determine the subspaces of dimensions 1, 3 and 5). In Table II we give the angles between the computed invariant subspaces and the corresponding exact invariant subspaces.

Table II

Dimension	Angle between computed and true subspace
2	$3\cdot05 \times 10^{-2}$
4	$1\cdot73 \times 10^{-2}$
6	$6\cdot23 \times 10^{-4}$
7	$1\cdot74 \times 10^{-6}$
8	$1\cdot73 \times 10^{-8}$
9	$2\cdot67 \times 10^{-10}$

We observe that the invariant subspace of order 2 is very inaccurate, as also is that of order 4 but by the time we group the 6 smallest eigenvalues together the subspace is correct "to about three decimals" even though it is derived from eigenvalues which have no correct figures. When we group the 9 smallest eigenvalues together the subspace is beginning to approach the full working accuracy.

We may describe this in an alternative way. If we regard the given matrix F_{16} as having relative errors of order 10^{-12} in its elements then the invariant subspace corresponding to its two smallest eigenvalues is scarcely determined at all while the invariant subspace corresponding to its 9 smallest eigenvalues is determined highly accurately.

VII. ATTAINABLE ACCURACY

It cannot be too strongly emphasized that when rounding errors are involved the accuracy attainable with a prescribed precision of computation is strictly limited by the sensitivity of the eigensystem. The Householder-QR combination and the Jacobi method for Hermitian matrices and the Householder-QR combination for real non-normal matrices give exact eigenvalues of some $A + E$, where for computation with t binary digits in the base b

$$\|E\| \leqslant b^{-t}\|A\|f(n) \tag{7.1}$$

where $f(n)$ is a fairly innocuous function of n (for the Householder-QR combination $f(n)$ is usually less than n in practice), are in our opinion 'best possible' algorithms. Although more efficient algorithms may be developed it is almost inconceivable that they will be appreciably more accurate.

Some error analysts disapprove of this use of the expression 'best possible' on the grounds that it is likely to discourage research into better methods. In our opinion quite the reverse is true; anyone who has not studied error analysis sufficiently carefully to appreciate the meaning of this expression in unlikely to develop improved stable methods. A realisation of its truth serves to direct research in the most hopeful directions.

In order to improve on the accuracy of the 'best possible' methods it is generally necessary to intervene in the rounding procedure itself or to perform at least some part of the computation to higher precision. Most commonly one finds that this can be achieved by performing a vital part of the computation exactly, even though ostensibly working only to finite precision. We may illustrate this in a simple context by considering the calculation of the zeros of

$$x^3 - 3 \cdot 0000012x^2 + 3 \cdot 0000017x - 1 \cdot 0000021 \tag{7.2}$$

This is an ill-conditioned polynomial and the accuracy attainable by iteration on an 8-digit decimal computer is strictly limited. However it is evident that all three roots are close to unity. Suppose we use the subtitution $y = x - 1$. There is a standard algorithm for forming the transformed polynomial; effectively the coefficients are obtained by dividing repeatedly by $x - 1$. Suppose the algorithm for this reduction is performed in standard single-precision arithmetic on our 8-digit computer. Unless it has quite unreasonable rounding procedures we will obtain the polynomial

$$y^3 - 0 \cdot 0000012 y^2 - 0 \cdot 0000007 y - 0 \cdot 0000016 \qquad (7.3)$$

which is well conditioned; the roots of the original polynomial may now be found much more accurately via the modified polynomial. However we think it is reasonable to describe iteration in the original polynomial as a 'best possible' method. Although the reduction appears to have been performed 'in single-precision arithmetic' it has in fact been done *exactly* i.e. in infinite precision. This is precisely the type of stratagem one must employ in order to achieve greater accuracy.

As a more sophisticated example we may consider the solution of a system of positive definite equations $Ax = b$ when an approximate Cholesky factorisation $A = LL^T$ has been derived using single-precision arithmetic. Provided A is not so ill-conditioned that the first computed solution has no accuracy it will usually be possible to obtain indefinite accuracy without ever using 'more than single-precision' computation by the following stratagem. An iterative cycle is used in which the computed solution is successively updated. At each stage the exact residual $r_s = b - Ax_s$ is determined. A single precision version $\bar{r}_s$ of r_s is then used to obtain a correction δ_s via the solution of

$$LL^T \delta_s = \bar{r}_s \qquad (7.4)$$

using the computed L. We then have

$$r_{s+1} = r_s - A\delta_s \qquad (7.5)$$

where in this updating of r_{s+1} we must compute $A\delta_s$ exactly. If the elements of A itself may all be represented exactly with the same exponent using single-precision mantissae, $A\delta_s$ will usually be computable exactly merely using single-precision multiplication with accumulation of innerproducts. Again the result is achievable because we can perform a key part of the computation exactly while ostensibly using single-precision computation. As an exercise this technique was used to find the inverse of ill-conditioned submatrices of the Hilbert inverse correct to 1536 binary places on the computer ACE.

The question arises whether there is an analogous technique for finding accurate eigensystems while ostensibly working to limited accuracy. There does not appear to be any technique which is quite as convenient as the one which has just been described for solving equations but the following has been used on ACE to give eigensystems of high-accuracy.

Suppose we have a system of approximate left-hand and right-hand eigenvectors $\bar{y}_i$ and $\bar{x}_i$ and approximate eigenvalues $\bar{\lambda}_i$ determined by a 'best-possible' single precision algorithm. Unless the eigensystem is too ill-conditioned, the following cycle may be used to improve, say, λ_1 and x_1 while working always with the original $\bar{y}_i, \bar{x}_i, \bar{\lambda}_i$ ($i = 2, \ldots, n$). The cycle is as follows. Suppose the current approximations to λ_1 and x_1 are $\lambda_1^{(s)}$ and $x_1^{(s)}$. Then if

$$x_1^{(s)} = x_1 + \sum_2^n \alpha_i x_i \qquad (7.6)$$

$$r_s = Ax_1^{(s)} - \lambda_1^{(s)}x_1^{(s)} = (\lambda_1 - \lambda_1^{(s)})x_1 + \sum_2^n \alpha_1(\lambda_i - \lambda_1^{(s)})x_i \qquad (7.7)$$

Provided only r_s is computed exactly we may find corrections to $\lambda_1^{(s)}$ and $x_1^{(s)}$ via the equations

$$\delta\lambda_1^{(s)} = (\bar{y}_1^T r_s)/(\bar{y}_1^T \bar{x}_1)$$

$$\alpha_i = (\bar{y}_i^T r_s)/[(\bar{\lambda}_i - \bar{\lambda}_1)(\bar{y}_i^T \bar{x}_i)] \qquad (7.8)$$

$$\delta x_1^{(s)} = \text{rounded version of } \sum_{i=2}^n \alpha_i \bar{x}_i$$

The most difficult part of the computation is the accurate updating of r_s. We have

$$r_{s+1} = r_s + A\delta x_1^{(s)} - \lambda_1^{(s)}\delta x_1^{(s)} - \delta\lambda_1^{(s)}x_1^{(s)} - \delta\lambda_1^{(s)}\delta x_1^{(s)} \qquad (7.9)$$

The difficult terms in the correction are $\lambda_1^{(s)}\delta x_1^{(s)}$ and $\delta\lambda_1^{(s)}x_1^{(s)}$ but even these involve only the multiplication of multi-precision numbers by a 'single-precision' number. Nowhere do we have the multiplication of two multi-precision numbers which would be intolerably time consuming.

The above algorithm is merely a *tour-de-force*, but a simplified version in which all computation is performed in single precision (except that at each stage r_s is recomputed from $Ax_1^{(s)} - \lambda_1^{(s)}x_1^{(s)}$ using accumulation of inner-products) may be used to produce eigensystems which will be correct to single-precision. In order to show what can be done with an eigensystem which is mainly of low accuracy we have considered the refinement of only one eigenvalue and eigenvector. In practice one can of course improve all eigenvalues and eigenvectors in each cycle, always using the latest values. When the process is terminated the final correction to each eigenvalue and eigenvector will usually be such that if it is added on using double precision one will obtain an eigensystem which is correct to more than single-precision. When A is well-conditioned the final eigensystem obtained in this way will be correct almost to double precision.

VIII. LOW RELATIVE ERRORS IN SMALL EIGENVALUES

The results obtained by error analysis show that, in general, computed eigenvalues are at best correct for some $A + E$ where $\|E\|$ is small compared with A. Such results can never guarantee low relative errors in eigenvalues which are small with respect to $\|A\|$. Indeed unless A can be regarded as exact *and is held exactly in the computer* there can, in general, be little point in computing small eigenvalues to high accuracy since end figure changes in the elements of A may even change their signs.

However consider a matrix of the type

$$\begin{bmatrix} a_1 & a_2 & a_3 \\ b_1\epsilon & b_2\epsilon & b_3\epsilon \\ c_1\epsilon^2 & c_2\epsilon^2 & c_3\epsilon^2 \end{bmatrix} \tag{8.1}$$

where ϵ is very small and the a_i, b_i, c_i are of order unity. Such a matrix will have one eigenvalue of order ϵ^2, one of order ϵ and one of order unity unless the a_i, b_i and c_i happen to be very specially related; moreover, small relative changes in individual elements will make small relative changes even in the very small eigenvalue. For matrices of this kind it may be meaningful and important to determine the small eigenvalues accurately.

Not a great deal has been written about problems having the above characteristics. Indeed it is not easy to frame theorems of sufficient generality to justify publication.

The following result however, is quite general and is important in practice. It has been well known to the authors for a number of years but does not appear to have been published. In a fundamental paper Givens[8] gave an error analysis of the algorithm for finding the eigenvalues of a symmetric tridiagonal matrix by the method of bisection, using the Sturm sequence property of leading principal minors. This was modified for floating-point computation by Wilkinson, and the appropriate result is that the error in any computed eigenvalue λ_i is bounded by $k2^{-t}|\lambda_1|$ where k is a constant (less than 10 with ordinary rounding procedure), the computation is in floating point binary with a t-digit mantissa and λ_1 is the eigenvalue of maximum modulus.

The bound is independent of n and for most problems is extremely satisfactory. However tri-diagonal matrices frequently arise in practice having elements which vary enormously in order of magnitude. As a rather exaggerated example we may consider the matrix with elements

$$\alpha_i = a_{ii} = 10^{n-i}, \quad \beta_i = a_{i,i+1} = a_{i+1,i} = 1, \quad a_{ij} = 0 \quad \text{otherwise} \quad (8.2)$$

The largest eigenvalue of this matrix is approximately 10^{n-1} and the smallest is of order unity. Suppose $n = 20$ and we are working with a 40 binary digit mantissa. The bound for the error in any computed eigenvalue is approximately

$$k.2^{-40} \times 10^{19} \doteq 10^7 \qquad (8.3)$$

which is quite useless as a bound for the smaller eigenvalues.

We now show that the Sturm-sequence property plus bisection gives an error in each computed eigenvalue which is no greater than can be accounted for by relative changes *in each element of the tridiagonal matrix* of a few parts in 2^t. Hence if an eigenvalue of A is well-determined by the data it will be obtained accurately even if it is very small relative to λ_1.

We assume for convenience that the rounding procedures are such that

$$fl(x_1 \pm x_2) = (x_1 \pm x_2)(1+\epsilon); \quad fl(x_1 x_2) = x_1 x_2 (1+\epsilon) \text{ with } |\epsilon| \leqslant 2^{-t} \quad (8.4)$$

In the Givens algorithm we calculate the leading principal minors p_r of $A - \mu I$ from the relations

$$p_0 = 1, \quad p_1 = \alpha_1 - \mu, \quad p_r = (\alpha_r - \mu)p_{r-1} - \beta_r^2 p_{r-2} \quad (r = 2, \ldots, n) \quad (8.5)$$

The *computed* p_r therefore satisfy *exactly* relations of the form

$$p_r = [(\alpha_r - \mu)(1+\epsilon_{1r})p_{r-1}(1+\epsilon_{2r}) - \beta_r^2(1+\epsilon_{3r})p_{r-2}(1+\epsilon_{4r})](1+\epsilon_{5r})$$

$$(8.6)$$

where $|\epsilon_{ir}| \leqslant 2^{-t}$. We may write (8.6) in the form

$$p_r = [\alpha_r(1+F_r) - \mu(1+F_r)]p_{r-1} - \beta_r^2(1+G_r)p_{r-2}$$

where

$$\begin{aligned} 1+F_r &= (1+\epsilon_{1r})(1+\epsilon_{2r})(1+\epsilon_{5r}) \\ 1+G_r &= (1+\epsilon_{3r})(1+\epsilon_{4r})(1+\epsilon_{5r}) \end{aligned} \qquad (8.7)$$

It would be nice at this stage to be able to claim that we have an exact sequence for the matrix with elements $\alpha_r(1+F_r), \beta_r(1+G_r)^{1/2}$ for some slightly perturbed value μ'. Unfortunately we cannot do this. The term $\mu(1+F_r)$ occurring in (8.7) is different for each value of r. All we can say is that the sequence of computed values is exact for the given value of μ for a matrix with elements

$$\alpha_r' = \alpha_r(1+F_r) - \mu F_r \quad \text{and} \quad \beta_r' = \beta_r(1+G_r)^{1/2} = \beta_r(1+H_r) \quad (8.8)$$

From equations (8.7) we deduce certainly that

$$|F_r| < (3.1)2^{-t}, \quad |H_r| < (1.6)2^{-t} \tag{8.9}$$

for any reasonable value of t. The Sturm sequence is exact for $A + \delta A$ where δA consists of two parts, a tridiagonal matrix δA_1 with elements $\alpha_r F_r$ and $\beta_r H_r$ and a diagonal matrix δA_2 with elements μF_r.

Let us ignore δA_2 for the moment and consider the set of all δA_1 with F_r and H_r satisfying equation (8.9). If the eigenvalues of A are λ_i (in monotonic decreasing order) the set of all λ_i' corresponding to all permissible perturbations δA_1 lie in intervals containing the λ_i as illustrated in Figure 1 by the firm lines

FIGURE 1

For convenience in Figure 1 we have shown disjoint intervals. The size of each interval depends on the sensitivity of the corresponding λ_i but notice that the end points of the λ_i interval represent the extreme range covered by the ith eigenvalue of all matrices obtained from A by varying the last figure of each α_i by up to 3.1 units and the last figure of β_i by up to 1.6 units. These intervals are independent of μ though the precise set of λ_i' corresponding to a δA_1 are of course a function of μ.

Now consider the further effect of δA_2. If the eigenvalues of $A + \delta A_1 + \delta A_2$ are λ_i'' we have

$$|\lambda_i' - \lambda_i''| \leq (3.1)2^{-t}|\mu| \tag{8.10}$$

Hence for any particular μ the intervals above must be extended by $(3.1)2^{-t}\mu$ on either side as shown in Figure 1 by the broken lines.

Now we are interested in the number of agreements in sign of the sequence of computed p_i, this number being the number of λ_i greater than μ; in practice we find the exact sequence for some $A + \delta A_1 + \delta A_2$. Suppose now we are trying to find λ_r by bisection. If we take any value of μ, then if μ is to the left of the extended interval based on λ_r, the relevant number of λ_i'' is less than r and hence from the computed p_r we will correctly diagnose that λ_r is to the right of that μ. Similarly if μ is to the right of the extended interval based on λ_r we will correctly diagnose that λ_r is to the left of that μ. In other words we make a correct diagnosis if

$$\left. \begin{aligned} \mu &< \lambda_r - I_r - (3.1)2^{-t}|\mu| \\ \text{or} \\ \mu &> \lambda_r + I_r + (3.1)2^{-t}|\mu| \end{aligned} \right\} \tag{8.11}$$

This analysis is not changed if intervals overlap; we have only to remark that we do not require a correct Sturm sequence count; we need only know the correct answer to the question "Is the number of agreements in sign less than r or not?". Hence if $I_r = \theta_r \lambda_r$ we achieve the correct diagnosis when

$$\mu < \lambda_r (1 - \theta_r)/(1 \pm (3.1)2^{-t})$$
$$\mu > \lambda_r (1 + \theta_r)/(1 \pm (3.1)2^{-t})$$

$$(8.12)$$

where we take the upper or lower signs according as λ_r is positive or negative. The result now follows immediately and we see that λ_r may be obtained with a relative error which is bounded by $(\theta_r + (3.1)2^{-t})/(1 - (3.1)2^{-t})$. The effect of δA_2 is therefore wholly unimportant and provided θ_r is small, i.e. λ_r is little affected by end figure changes in α_r and β_r, it will be well determined by our algorithm.

REFERENCES

1. Hoffman, A. J., and Wielandt, H. W., 1953, The variation of the spectrum of a normal matrix. *Duke Math. J.* **20**, 37–39.
2. Bauer, F. L. and Fike, C. T., 1960, Norms and exclusion theorems. *Num. Math.* **2**, 137–141.
3. Wilkinson, J. H., 1965, *The Algebraic Eigenvalue Problem*. Oxford University Press, London.
4. Osborne, E. E., 1960, On pre-conditioning of matrices. *J. Ass. Comp. Mach.* **7**, 338–348.
5. Parlett, B. N. and Reinsch, C., 1969, Balancing a matrix for calculation of eigenvalues and eigenvectors. *Num. Math* **13**, 293–304.
6. Ruhe, A., 1970, Properties of a matrix with a very ill-conditioned eigenproblem. *Num. Math.* **15**, 57–60.
7. Wilkinson, J. H., 1972, Note of matrices with a very ill-conditioned eigenproblem. *Num. Math.* **19**, 176–178 (1972).
8. Givens, J. W., 1954, Numerical computation of the characteristic values of a real symmetric matrix. Oak Ridge National Laboratory, ORNL–1574.
9. Wilkinson, J. H., 1963, Rounding Errors in Algebraic Processes. *Notes on Applied Science No. 32*, Her Majesty's Stationary Office, London; Prentice Hall, New-Jersey.
10. Varah, J. M., Invariant subspace perturbations for a non-normal matrix. Presented at IFIP Congress 71.
11. Stewart, G. W., Error bounds for approximate subspaces of closed linear operators in Hilbert space. (To appear).
12. Wilkinson, J. H. and Reinsch, C., 1971, *Handbook for Automatic Computation, Vol. II, Linear Algebra*. Springer-Verlag, Berlin, Heidelburg, New York.

Buckling of a Beam under Axial Compression with Elastic Support

A. P. GALLAGHER

Department of Engineering Mathematics,
The Queen's University of Belfast, Belfast, Northern Ireland

I. INTRODUCTION

The problem considered here is that of the buckling due to axial compression of a uniform beam, hinged at both ends which is in contact with a semi-infinite elastic medium acting as a foundation, (Fig. 1). This problem is of some interest in engineering practice with particular reference to the reinforcement of beams using steel bars under compression, the buckling of piles and the diffusion of a load from a stiffener into a sheet.

The usual method of solution uses the Bernoulli-Euler law for the beam and the Winkler theory for the medium[1]. This theory treats the latter as a continuously distributed set of springs, the stiffness of which is defined by "the modulus of the foundation" k. It is defined as the ratio of the force w of the reaction per unit area to the corresponding deflection y of the beam, i.e. $k = w/y$. In practice k is assumed constant for a given medium although various empirical formulae which include the beam parameters are also used. Using this approach the differential equation and end conditions are easily found to be

$$E_B I y^{(\mathrm{iv})}(x) + P y''(x) + k y(x) = 0, \quad y(\pm l) = y''(\pm l) = 0 \tag{1}$$

where E_B is the Young's modulus of the beam material,

I is the second moment of the area of cross-section of the beam per unit width,

P is the axial compression force per unit width,

y is the deflection of the neutral axis,

y'', $y^{(\mathrm{iv})}$ are the second and fourth derivatives of y respectively,

$2l$ is the length of the beam.

The end conditions express the fact that the beam is hinged at both ends.

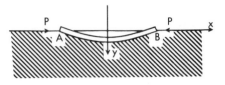

The differential system (1) is an eigenvalue problem for the critical values of P for which buckling is possible. These are easily found to be

$$P = E_B I \left(\frac{n^2 \pi^2}{4 l^2} + \frac{k 4 l^2}{n^2 E_B I \pi^2} \right) \tag{2}$$

where n denotes the buckling mode. The graphs of $\alpha = P/E_B I$ against $k/E_B I$ with $l = 1$ are shown in Fig. 2 for $n = 1,2,3,4$. The corresponding deflection is $y = A \sin n\pi(x+l)/2l$. Since the Winkler theory does not take into account the fact that the reaction at any one point depends not only on the deflection at that point but also at all other points of the beam-medium interface, a more realistic theory should be obtained if the medium is assumed to be in a state of plane strain. In order to simplify the

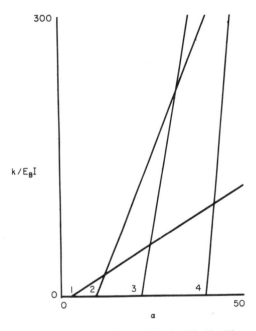

FIGURE 2. Critical values of α, k for the Winkler Theory.

problem mathematically another important assumption is made, namely that the shear stresses at the beam-medium interface are zero. This assumption is justifiable if the beam is in fractionless contact with the supporting medium. In this case, however, since the medium cannot exert a reaction on the beam which is positive, the range of relevant solutions obtained is rather limited. If, however, the beam is in bonded contact with the medium such a reaction is possible but the assumption of zero shear stress is not justifiable. Nevertheless it is felt that the solutions obtained provide insight into this physical problem.

The theory of plane strain gives the reaction as a singular integral. This, coupled with the Bernoulli-Euler theory for the beam, leads to a fourth-order integro-differential equation, which, by expanding the slope of the deflection in a series of Chebyshev polynomials gives the eigenvalues as the latent roots of an infinite matrix with the coefficients of the series as the corresponding latent column vectors. It is found that the graphs of the critical values of P are similar to those of the Winkler theory (cf. Figs. 2 and 3) insofar as the form of (2) is found to be valid for certain ranges of the parameter λ (related to the stiffness of the medium) but k is found to be proportional to $1/l$ and depends also on the mode of buckling. An advantage of the present theory is that k (difficult to measure in practice) is expressed in terms of Young's modulus E and Poisson's ratio σ of the medium. Another interesting feature is that, in contrast to the Winkler theory, non-elementary modes of buckling are found.

II. FORMULATION OF THE INTEGRO-DIFFERENTIAL EQUATION

Let $y = y(x), -l \leq x \leq l$, denote the deflection of the neutral axis of a uniform beam, hinged at both ends, under an axial compression force P per unit width and load $w(x)$. Then the Bernoulli-Euler law gives the deflection as the solution of the differential equation

$$E_B I y^{(\mathrm{iv})}(x) + P y''(x) = w(x), \quad y(\pm l) = y''(\pm l) = 0 \qquad (3)$$

The following assumptions are made:
  (i) that the supporting medium is in a state of plane strain;
  (ii) that the slope of beam-medium interface is the same as that of the neutral axis;
  (iii) that the shear stresses at the interface are zero;
  (iv) that the ends of the beam (A,B in Fig. 1) are just touching the medium surface.

Under these conditions the reaction of the medium may be taken as

$$w(x) = \frac{E\sqrt{l^2 - x^2}}{2\pi(1 - \sigma^2)} \int_{-l}^{l} \frac{y'(t)\,dt}{\sqrt{l^2 - t^2}(t - x)} \tag{4}$$

where E and σ are the Young's modulus and Poisson's ratio respectively of the medium [2, p489].

The problem is made non-dimensional by taking l as a representative length. To this end we let

$$\xi = x/l, \quad \eta = y/l, \quad \tau = t/l, \quad \alpha = \frac{Pl^2}{E_B I}, \quad \lambda = \frac{El^3}{(1 - \sigma^2)E_B I}. \tag{5}$$

Then (3), (4) and (5) give

$$\phi'''(\xi) + \alpha\phi'(\xi) = \omega(\xi), \quad \eta(\pm 1) = \phi'(\pm 1) = 0, \tag{6}$$

where for convenience we let

$$\phi = \eta' \quad \text{and} \quad \omega(\xi) = \frac{\lambda}{2\pi}\sqrt{1 - \xi^2} \int_{-1}^{1} \frac{\phi(\tau)\,d\tau}{\sqrt{1 - \tau^2}(\tau - \xi)}. \tag{7}$$

III. ANALYSIS OF THE EVEN MODES OF BUCKLING

In the following let $T_r(\xi)$ and $U_r(\xi)$ denote the Chebyshev polynomials of the first and second kind respectively. Assuming an expansion of the form

$$\phi(\xi) = \sum_{n=1}^{\infty} a_{2n-1} T_{2n-1}(\xi) \tag{8}$$

and utilizing

$$\int_{-1}^{1} \frac{T_r(\tau)\,d\tau}{\sqrt{1 - \tau^2}(\tau - \xi)} = \pi U_{r-1}(\xi), \quad r \geq 1, \quad -1 \leq \xi \leq 1$$

[3, p833], then formal substitution of (8) into (6) gives

$$\sum_{n=1}^{\infty} a_{2n-1} T'''_{2n-1}(\xi) + \alpha \sum_{n=1}^{\infty} a_{2n-1} T'_{2n-1}(\xi) = \frac{\lambda}{2}\sqrt{1 - \xi^2} \sum_{n=1}^{\infty} a_{2n-1} U_{2n-2}(\xi) \tag{9a}$$

and

$$\sum_{n=1}^{\infty} (2n - 1)^2 a_{2n-1} = 0. \tag{9b}$$

Using the orthogonality of the Chebyshev polynomials, multiplication of (9a) by $4U_{2m-2}(\xi)/\pi$ and formal integration term by term of the series over $(-1, 1)$ yields

$$\sum_{n=1}^{\infty} a_{2n-1}(I_{mn} + \alpha J_{mn}) = \lambda a_{2m-1}, \quad m \geq 1, \tag{10}$$

where
$$I_{mn} = \frac{4}{\pi} \int_{-1}^{1} U_{2m-2} T'''_{2n-1} \, d\xi,$$

$$= I_{mn}^{(1)} \text{ for } m \text{ or } n = 1,$$

$$= I_{mn}^{(1)} + I_{mn}^{(2)} \text{ for } m, n \geqslant 2,$$

$$I_{mn}^{(1)} = \frac{8}{3\pi} (2m-1)(2n-1)^2 [(2n-1)^2 - 1],$$

$$I_{mn}^{(2)} = \frac{512}{\pi} (2n-1) \sum_{r=1}^{m-1} \sum_{s=1}^{n-1} \sum_{t=1}^{s} \frac{r^2 s}{(2t-1)^2 - 4r^2},$$

(11)

and
$$J_{mn} = \frac{4}{\pi} \int_{-1}^{1} U_{2m-2} T'_{2n-1} \, d\xi,$$

$$= \frac{8}{\pi} \text{ for } m = 1, \quad n \geqslant 1,$$

$$= \frac{8}{\pi} + \frac{16}{\pi} (2n-1)^2 \sum_{r=1}^{m-1} \frac{1}{(2n-1)^2 - 4r^2}, \quad m \geqslant 2, \quad n \geqslant 1.$$

(12)

For the evaluation of these integrals the reader is referred to the Appendix. Applying condition (9b) to (10) and (11) the value for $I_{mn}^{(1)}$ becomes

$$I_{mn}^{(1)} = \frac{8}{3\pi} (2m-1)(2n-1)^4. \tag{13}$$

If α is treated as a parameter the corresponding critical values for λ are the latent roots, and $\{a_{2n-1}\}$ the corresponding latent column vectors of the infinite matrix

$$A = [I_{mn} + \alpha J_{mn}],$$

where these integrals are given by (11), (12) and (13). The expressions for the functions occurring in (7) and for $\eta''(\xi)$ are

$$\eta(\xi) = \int^{\xi} \phi(\tau) \, d\tau = \frac{1}{2} \left[a_1 \xi^2 + \sum_{n=2}^{\infty} a_{2n-1} \left\{ \frac{T_{2n}(\xi)}{2n} - \frac{T_{2n-2}(\xi)}{2n-2} \right\} \right] + C, \tag{14a}$$

$$\phi(\xi) = \sum_{n=1}^{\infty} a_{2n-1} T_{2n-1}(\xi), \tag{14b}$$

$$\omega(\xi) = \frac{\lambda}{2} \sqrt{1-\xi^2} \sum_{n=1}^{\infty} a_{2n-1} U_{2n-2}(\xi), \tag{14c}$$

$$\eta''(\xi) = \phi'(\xi) = \sum_{n=1}^{\infty} (2n-1) a_{2n-1} U_{2n-2}(\xi), \tag{14d}$$

using the properties of the Chebyshev polynomials (see appendix). The condition $\eta(\pm 1) = 0$ is used to find the constant of integration C in (14a). It should be remembered that these functions are indeterminate to the extent of a multiplicative constant.

IV. ANALYSIS OF THE ODD MODES OF BUCKLING

Proceeding as in the previous section we assume an expansion of the form

$$\phi(\xi) = \sum_{n=0}^{\infty} a_{2n} T_{2n}(\xi) \tag{15}$$

which when inserted into (6) formally yields

$$\sum_{n=1}^{\infty} a_{2n} T_{2n}'''(\xi) + \alpha \sum_{n=1}^{\infty} a_{2n} T_{2n}'(\xi) = \frac{\lambda}{2} \sqrt{1 - \xi^2} \sum_{n=1}^{\infty} a_{2n} U_{2n-1}(\xi) \tag{16a}$$

and

$$\sum_{n=1}^{\infty} n^2 a_{2n} = 0. \tag{16b}$$

Multiplication of (16a) by $4 U_{2m-1}(\xi)/\pi$ and formal term by term integration gives

$$\sum_{n=1}^{\infty} a_{2n}(K_{mn} + \alpha L_{mn}) = \lambda a_{2m}, \quad m = 1, 2, 3, \cdots, \tag{17}$$

where

$$
\left.
\begin{aligned}
K_{mn} &= \frac{4}{\pi} \int_{-1}^{1} U_{2m-1} T_{2n}''' \, d\xi, \\
&= K_{mn}^{(1)} \quad \text{for } m \text{ or } n = 1, \\
&= K_{mn}^{(1)} + K_{mn}^{(2)} \quad \text{for } m, n = 2, 3, \ldots, \\
K_{mn}^{(1)} &= \frac{64}{3\pi} \left[m(4n^4 - 3n^3 - n^2) + 3(n^3 - n^2) \right], \\
K_{mn}^{(2)} &= \frac{128}{\pi} n \sum_{r=1}^{m-1} \sum_{s=1}^{n-1} \sum_{t=1}^{s} \frac{(2r+1)^2(2s+1)}{4t^2 - (2r+1)^2},
\end{aligned}
\right\} \tag{18}
$$

and

$$L_{mn} = \frac{4}{\pi} \int_{-1}^{1} U_{2m-1} T_{2n}' \, d\xi = \frac{64}{\pi} n^2 \sum_{r=1}^{m} \frac{1}{4n^2 - (2r-1)^2}. \tag{19}$$

The reader is referred to the Appendix for the evaluation of these integrals. Applying condition (16b) to (16a) and (18) the value for $K_{mn}^{(1)}$ now becomes

$$K_{mn}^{(1)} = \frac{64}{3\pi} [m(4n^4 - 3n^3) + 3n^3]. \tag{20}$$

The critical values of λ and the coefficients a_{2n} $(n = 1,2,3,\ldots)$ are the latent roots and latent column vectors, respectively, of the infinite matrix

$$B = [K_{mn} + \alpha L_{mn}],$$

where the values of these integrals are given by (18), (19) and (20).

The expansions for the functions in (7) and the function $\eta''(\xi)$ for bending moment are found to be

$$\eta(\xi) = \int_0^\xi \phi(t)\,dt = a_0\xi + \tfrac{1}{2}\left[\sum_{n=1}^\infty a_{2n}\left\{\frac{T_{2n+1}(\xi)}{2n+1} - \frac{T_{2n-1}(\xi)}{2n-1}\right\}\right],$$

$$\phi(\xi) = \sum_{n=0}^\infty a_{2n} T_{2n}(\xi),$$

$$\omega(\xi) = \frac{\lambda}{2}\sqrt{1-\xi^2}\sum_{n=1}^\infty a_{2n} U_{2n-1}(\xi),$$

$$\eta''(\xi) = \phi'(\xi) = 2\sum_{n=1}^\infty n a_{2n} U_{2n-1}(\xi),$$

where the coefficient a_0 is determined from the condition that $\eta(\pm 1) = 0$, to the extent of an arbitrary multiplicative constant.

V. NUMERICAL RESULTS

Successive approximations to the infinite matrices A, B were made by taking the first N rows and columns and computing the latent roots and column vectors of the truncated matrices. Convergence to the eigenvalues λ and the functions $\eta(\xi)$, $\omega(\xi)$, $\eta''(\xi)$ was found to be very rapid. Thus for instance in the case $\alpha = 50$, it was found that $N = 6$, 12, 18 gave the solutions for the first four modes to 2,3,4 significant figures, respectively. The rapidity of the convergence of the series expansions for these functions was due to the fact that the coefficients a_r tended to zero very rapidly for all the modes examined: for instance, the coefficients a_{35} (for even modes), a_{36} (for odd modes), were numerically less than 10^{-6} for all the values of α examined ($\alpha \leqslant 200$).

It will be noted from (1) and (6) that $\lambda = k = 0$ corresponds to the buckling of an unsupported beam for which the critical values of α are $n^2\pi^2/4$ from (2). The corresponding solution for the deflection is $\sin n\pi(\xi+1)/2$ where the arbitrary constant and the half-length l have been normalized to unity. On substituting these values for α into the matrices A and B a zero eigenvalue λ was found (to four significant figures with $N = 18$). The

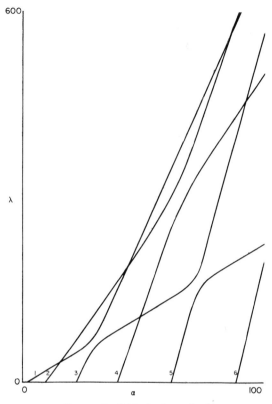

FIGURE 3. Critical values of α, λ.

corresponding solution for the deflection was given to an even higher degree of accuracy.

The graphs of the first six eigenvalues λ versus α are shown in Fig. 3. Those curves have been labelled $1, 2, 3, \ldots$ which correspond to the deflection $\sin n\pi(\xi + 1)/2$ for $\alpha = n^2\pi^2/4$ and $\lambda = 0$, $n = 1, 2, 3, \ldots$. If these are compared with those of the Winkler theory (Fig. 2) it will be seen that there is a remarkable similarity between the two sets of curves in that the present theory also predicts a linear relationship between λ and α but only for certain ranges of λ, depending on the mode. Unlike the Winkler theory, however, the curves of the even modes (labelled 1,3,5) never intersect, the same being true of the odd mode curves (labelled 2,4,6). This pattern was found to continue for higher modes and larger values of α than those shown.

Thus it can be said that for $\lambda \le 40$ the two theories are in close agreement insofar as the relationship between λ and α is of the form

$$\alpha = \frac{n^2\pi^2}{4} + C_n\lambda, \quad n = 1,2,3,\ldots, \tag{21}$$

where C_n is a constant depending on the mode only. If (2) and (21) are compared k is found to be $(n^2\pi^2/4)C_nE/l(1-\sigma^2)$ and so is a function of $1/l$, and the mode of buckling. The values of $(n^2\pi^2/4)C_n$ for the first six modes are listed in Table I.

Table I

n	1	2	3	4	5	6
$\dfrac{n^2\pi^2}{4}C_n$	1·05	1·77	2·66	3·39	4·23	4·97

The graphs of the dimensionless deflection and reaction, $\eta(\xi)$ and $\omega(\xi)$ respectively, for curve 1 of Fig. 3 are shown in Figs. 4, 5; those for curve 3 in Figs. 6, 7. They have all been normalized to unity for ease of plotting. The maximum values for $|\omega(\xi)|$ and for the dimensionless bending moment $|\eta''(\xi)|$ for max $|\eta| = 1$ are given in Table II. The most significant feature of the deflection curves $\eta(\xi)$ is that in contrast to the Winkler theory, non-elementary modes of buckling are present. It is interesting to observe the change of the mode of buckling as values of α and λ are chosen from curve 1 of Fig. 3. Initially at $\alpha = \pi^2/4$, $\lambda = 0$ and

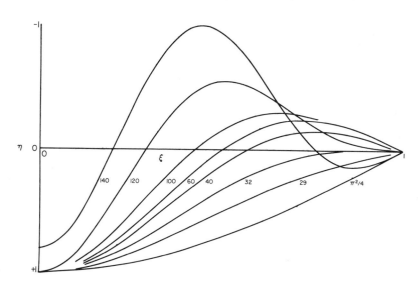

FIGURE 4. Normalized deflection curves for various values of α on curve 1 of Fig. 3.

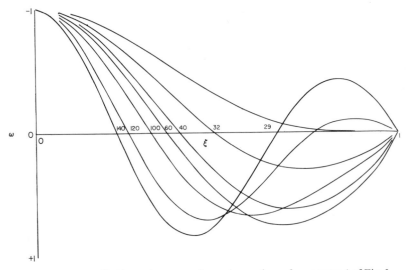

FIGURE 5. Normalized reaction curves for various values of α on curve 1 of Fig. 3.

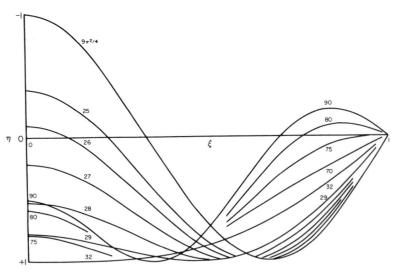

FIGURE 6. Normalized deflection curves for various values of α on curve 3 of Fig. 3.

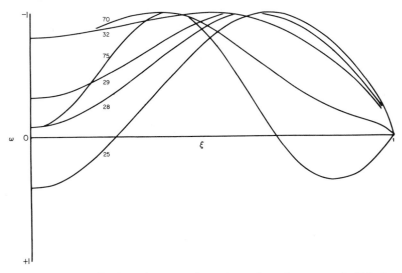

FIGURE 7. Normalized reaction curves for various values of α on curve 3 of Fig. 3.

Table II

Maximum values of the dimensionless reaction and bending moment for critical values of α, λ on curves 1 and 3, Fig. 3 (max $|\eta| = 1$).

α	25	29	32	40	60	65	75	100	140
max $|\omega_1|$	56·9	78·2	111	213	480	550	698	1201	4029
max $|\omega_3|$	59·6	84·8	72·5	83·1	129	144	291	1086	1524
max $|\eta_1''|$	3·40	5·29	7·73	10·6	12·6	12·9	13·6	17·4	47·2
max $|\eta_3''|$	18·0	8·38	4·09	2·56	2·72	3·14	13·3	25·3	15·2

for points (α, λ) on the first linear section, the deflection (Fig. 4) is of a general mode 1 type of buckling. On the second linear section of curve 1 (e.g. $\alpha = 60$) the corresponding deflection has merged into a third mode form of buckling. This second linear section has in fact the same slope as the first linear section of curve 3, Fig. 3. As α is increased further (e.g. $\alpha = 140$) so that (α, λ) lies on the third linear section of curve 1 (not shown in Fig. 3) the mode of buckling has the general form of the fifth mode, the slope of this third linear section being the same as that of the first linear section of curve 5.

This same pattern of modal change is found to occur also as (α, λ) traverses curve 3 starting with a mode 3 type of buckling at $\alpha = 9\pi^2/4$

(Fig. 6), merging into a mode 1 at $\alpha = 32$ and eventually into a mode 5 at about $\alpha = 90$. It should be pointed out that the form of deflection is only a perfect cosine curve at $\alpha = n^2\pi^2/4$ with $\lambda = 0$. The terms "mode 1,3,5 etc." are only intended to express the general shape.

The reaction curves $\omega(\xi)$ show an interesting feature, namely that the reaction can have the same sign as the deflection whereas the basis of the Winkler theory is that the reaction always has the opposite sign to the deflection. As noted in the introduction, if the medium is in frictionless contact with the beam then it cannot sustain a reaction which is positive. Hence only those solutions for which the reaction is negative along the entire length of the beam are applicable to this problem for instance, $\alpha \leq 29$ and $28 \leq \alpha \leq 70$ approximately, for values of α, λ on curves 1 and 3 respectively. If the beam is in bonded contact the medium can sustain such reactions so that all solutions are applicable but since the shear stress was assumed zero at the interface they present a less accurate picture of the physical problem.

APPENDIX

For the convenience of the reader we list the properties of the Chebyshev polynomials used and outline the method of evaluation of the integrals occurring in Section III and Section IV.

Elementary Properties of the Chebyshev Polynomials

By definition

$$T_n(\xi) = \cos n\theta, \quad U_n(\xi) = \frac{\sin (n+1)\theta}{\sin \theta}, \quad \xi = \cos \theta.$$

Hence

$$T_n(1) = 1, \quad T_n(-1) = (-1)^n, \quad U_n(1) = n+1, \quad U_n(-1) = (-1)^n(n+1),$$

and

$$T_n'(1) = n^2, \quad U_n'(1) = \frac{n(n+1)(n+2)}{3}, \quad T_n''(1) = \frac{n^2(n^2-1)}{3}.$$

The relations

$$T_n' = nU_{n-1}, \quad T_n = \frac{U_n - U_{n-2}}{2}, \quad U_n' - U_{n-2}' = 2nU_{n-1}$$

are easily proved.

From the first of these two equations it follows that

$$\int T_n \, d\xi = \tfrac{1}{2}\left[\frac{T_{n+1}}{n+1} - \frac{T_{n-1}}{n-1}\right], \quad n > 1$$

while from the last, one obtains

$$U'_{2n} = 4 \sum_{r=1}^{n} r U_{2r-1}, \quad U'_{2n-1} = 2 \sum_{s=0}^{n-1} (2s+1) U_{2s}.$$

For details of some of these formulae see [4].

Evaluation of Integrals

The following integrals are easily evaluated directly for the special cases m or $n = 1$. We consider $m,n \geqslant 2$.

(1) $\displaystyle\int_{-1}^{1} U_{2m-2} T'''_{2n-1} \, d\xi = (2n-1) \int_{-1}^{1} U_{2m-2} U''_{2n-2} \, d\xi$

$$= (2n-1) \left[2 U_{2m-2}(1) U'_{2n-2}(1) - \int_{-1}^{1} U'_{2m-2} U'_{2n-2} \, d\xi \right]$$

and

$$\int_{-1}^{1} U'_{2m-2} U'_{2n-2} \, d\xi = 16 \sum_{r=1}^{m-1} \sum_{s=1}^{n-1} rs \int_{-1}^{1} U_{2r-1} U_{2s-1} \, d\xi.$$

Now

$$\int_{-1}^{1} U_{2r-1} U_{2s-1} \, d\xi = \int_{0}^{\pi} \frac{\sin 2r\theta \sin 2s\theta}{\sin \theta} \, d\theta$$

which is evaluated using integration by parts and the formula

$$\int \frac{\sin 2s\theta}{\sin \theta} \, d\theta = 2 \sum_{t=1}^{s} \frac{\sin (2t-1)\theta}{2t-1}.$$

(2) $\displaystyle\int U_{2m-2} T'_{2n-1} \, d\xi = (2n-1) \int_{-1}^{1} U_{2m-2} U_{2n-2} \, d\xi$

$$= (2n-1) \int_{0}^{\pi} \frac{\sin (2m-1)\theta \sin (2n-1)\theta}{\sin \theta} \, d\theta.$$

This last integral is evaluated using the formula

$$\int \frac{\sin (2m-1)\theta}{\sin \theta} \, d\theta = \sum_{r=1}^{m-1} \frac{\sin 2r\theta}{r} + \theta \quad \text{and integration by parts.}$$

(3) $\displaystyle\int_{-1}^{1} U_{2m-1} T'''_{2n} \, d\xi = 2n \int_{-1}^{1} U_{2m-1} U''_{2n-1} \, d\xi$

$$= 2n \left[2 U_{2m-1}(1) U'_{2n-1}(1) - \int_{-1}^{1} U'_{2m-1} U'_{2n-1} \, d\xi \right].$$

Now

$$\int_{-1}^{1} U'_{2m-1} U'_{2n-1} \, d\xi$$

$$= 4 \int_{-1}^{1} \left[1 + \sum_{r=1}^{m-1} (2r+1) U_{2r} + \sum_{s=1}^{n-1} (2s+1) U_{2s} + \sum_{r=1}^{m-1} \sum_{s=1}^{n-1} U_{2r} U_{2s} \right] d\xi.$$

The last integral has already occurred in integral (2) and

$$\int_{-1}^{+1} U_{2r} \, d\xi = \frac{2T_{2r+1}}{2r+1} = \frac{2}{2r+1}$$

(4) $$\int_{-1}^{1} U_{2m-1} T'_{2n} \, d\xi = 2n \int_{-1}^{1} U_{2m-1} U_{2n-1} \, d\xi = 2n \int_{0}^{\pi} \frac{\sin 2m\theta \sin 2n\theta}{\sin \theta} \, d\theta$$

which has been dealt with in integral (1).

REFERENCES

1. Hetényi, M.: *Beams on Elastic Foundations*. University of Michigan Press, 1946.
2. Muskhelishvili, N. I.: *Some Basic Problems of the Mathematical Theory of Elasticity*. Noordhoff, Groningen, 1963.
3. Gradshteyn, I. S. and Ryzhik, I. M.: *Table of Integrals, Series and Products*. Academic Press, New York and London, 4th Edition, 1965.
4. Fox, L. and Parker, I. B.: *Chebyshev Polynomials in Numerical Analysis*. Oxford University Press, London, 1968.

The Frequency Approach
to Numerical Analysis

R. W. HAMMING

Bell Telephone Laboratories, Incorporated
Murray Hill, New Jersey, U.S.A.

I. INTRODUCTION

One of Cornelius Lanczos' most important contributions is the systematic use of the frequency approach in numerical analysis, (see *Applied Analysis*, Prentice Hall 1956, and *Linear Differential Operators*, Van Nostrand 1961). Put simply, the frequency approach uses approximations made in terms of sines and cosines as the basis functions, while classical numerical analysis uses powers of x (polynomials). Abstractly speaking, any complete set of linearly independent functions could be used as a basis for representation, but the three sets:

1. powers of x (polynomials),
2. sines and cosines (Fourier series and integrals), and
3. exponentials (exponential sums and Laplace transforms)

have a preferential role, (along with combinations of them); they are the *only* sets which are independent of coordinate translations so that the choice of the origin will not affect the answer in any fundamental way.

The instinctive reaction of the classical numerical analyst to the frequency approach is that the computation of the sines and cosines will take much more time than does the evaluation of the powers of x. But, as we shall see, this observation is irrelevant because most formulas deal directly with weighted sums of the samples of the functions; the basis functions enter in the design of the formula and in the interpretation of the results, but not directly in the computation. Furthermore, the development of the Fast Fourier Transform (Cooley and Tukey, 1965) has reduced the computational labor of finding the coefficients of a Fourier expansion from the order of N^2 operations to the order of $N \ln N$ operations. Thus this objection to the use of trigonometric functions as a basis for representation is also not valid.

151

Since the appearance of Lanczos' *Applied Analysis* only one generally available book on computing has emphasized the frequency approach,† although the approach is widely used by electrical engineers and many papers have been published by them in various engineering journals. It is difficult to understand the continued neglect of this powerful method, and it is the purpose of this chapter to make the frequency approach a bit more palatable to the mathematically inclined.

It appears that in spite of his efforts to explain what he was doing, the mathematically inclined regard Lanczos more as a mathemagician than a mathematician. It has been said of Sir Isaac Newton that if he had waited for the rigor he would never have discovered the calculus; so too it is likely that had Lanczos waited to make everything mathematically rigorous he would have done much less. We shall try to avoid the appeal to physical intuition as a basis for proof, but shall use it to motivate what is going on; thus we will run some of the same risks that Lanczos ran, but the subject matter seems to require outside "explanations" to motivate what is going on in the formal mathematics.

We assume that the reader is reasonably well acquainted with classical numerical analysis, has a passing acquaintence with Fourier series, and a smattering of knowledge about the Fourier integral, but we shall avoid all of the deeper parts as not proper in a first approach to the field. We also assume familiarity with equivalence of the sines and cosines to the complex exponentials

$$e^{i\omega x} \quad \text{and} \quad e^{-i\omega x}.$$

We shall use the latter for ease of notation and clarity of thought, though, of course, the ultimate computing must be done in the domain of real numbers. Contrary to first impressions, the complex notation does not lead to complicated computations or conceal great quantities of arithmetic.

II. THE EFFECTS OF SAMPLING – ALIASING

Much of numerical analysis is based on taking samples of the function and then using these samples for estimating the results of applying some linear operator to the function. The linear operators are often infinite operators in the sense that they involve taking a limit – for example integration. It is reasonable, therefore, to examine carefully the effects of sampling.

† Hamming, R. W. *Numerical Methods for Scientists and Engineers*, second edition, McGraw Hill, New York, 1973.

The classical polynomial interpolation theory evades this question, and states that the error is expressible in terms of some high-order derivative of the function. We shall, therefore, give a direct examination of the effects of sampling which will parallel the corresponding result for the frequency approach.

Central to the polynomial theory of sampling based on n points is the polynomial of degree n

$$\pi(x) = (x - x_1)(x - x_2) \cdots (x - x_n),$$

where the x_k are assumed, for convenience, to be distinct. Now given any polynomial $P_m(x)$ of degree m (we can think of m as being much larger than n) we divide the polynomial $P_m(x)$ by $\pi(x)$ to get the quotient $Q(x)$ and the remainder $R(x)$, with the remainder of degree less than n,

$$P_m(x) = Q(x)\pi(x) + R(x).$$

This expression provides the obvious generalization of the remainder theorem, namely *at the sample points*

$$P_m(x_k) = R(x_k), \qquad (k = 1, 2, \cdots, n)$$

and the functions cannot be distinguished from each other *if* all we have are the values of the function at the sample points.

If we imagine doing this for each x^m, $(m = 1, 2, \cdots)$ we have the correspondence

$$x^m \to R_m(x).$$

Now for a Taylor's series expansion of any function $f(x)$

$$f(x) = \sum_{m=0}^{\infty} a_m x^m \to \sum_{m=0}^{\infty} a_m R_m(x)$$

and at the sample points

$$f(x_k) = \sum_{m=0}^{\infty} a_m R_m(x_k).$$

The summation is *the* interpolating polynomial of degree less than n which we would get if we passed the interpolating polynomial through the samples of the function (neglecting roundoff, of course).

We now turn to the effects of sampling sines and cosines *at a set of equally spaced points*, which is the case of overwhelming practical importance. For convenience let the sample points be at

$$x = 0, \quad \pm 1, \quad \pm 2, \cdots,$$

or

$$x_k = k, \quad (k = 0, \quad \pm 1, \quad \pm 2, \cdots).$$

Now consider the trigonometric identity

$$\cos \pi (1+\epsilon)x - \cos \pi (1-\epsilon)x = -2 \sin \pi x \sin \pi \epsilon x.$$

We see immediately that at the sample points $x_k = k$ the $\sin \pi k = 0$, and the higher frequency $(1+\epsilon)$ cosine is identical with the lower frequency $(1-\epsilon)$ cosine *at these points*. Thus the higher frequencies are "aliased" (go under different names) into the lower frequencies due to the process of sampling at equally spaced points.

III. THE FINITE FOURIER SERIES

In this Section we examine the fact that the Fourier series functions are orthogonal not only over a continuous interval but also over suitable sets of equally spaced points. This remarkable fact has many consequences. Among others it allows us to look at the same calculation in either of two ways:

1. We can view the finite sampling of a function as an approximation to the continuous function at the point where the various integrals must be computed from the samples of the function, or else
2. As a finite approximation at the start, after which no further approximations are made (except for roundoff).

The continuous Fourier series over the interval $(0 \leqslant x \leqslant 2N)$ is given by

$$f(x) = \frac{a_0}{2} + \sum_{k=0}^{\infty} \left[a_k \cos \left(k \frac{\pi}{N} x \right) + b_k \sin \left(k \frac{\pi}{N} x \right) \right]$$

where the coefficients are computed from the formulas

$$a_k = \frac{1}{N} \int_0^{2N} f(x) \cos \left(k \frac{\pi}{N} x \right) dx$$

$$b_k = \frac{1}{N} \int_0^{2N} f(x) \sin \left(k \frac{\pi}{N} x \right) dx$$

These formulas follow directly from the orthogonality relations

$$\int_0^{2N} \cos \left(p \frac{\pi}{N} x \right) \cos \left(q \frac{\pi}{N} x \right) dx = \begin{cases} 0, & p \neq q \\ N, & p = q \neq 0 \\ 2N, & p = q = 0 \end{cases}$$

$$\int_0^{2N} \cos \left(p \frac{\pi}{N} x \right) \sin \left(q \frac{\pi}{N} x \right) dx = 0$$

$$\int_0^{2N} \sin \left(p \frac{\pi}{N} x \right) \sin \left(q \frac{\pi}{N} x \right) dx = \begin{cases} 0, & p \neq q \\ N, & p = q \neq 0 \\ 0, & p = q = 0 \end{cases}$$

For the corresponding theory of $2N$ samples of the function, we begin by proving the orthogonality of the $2N$ complex functions

$$e^{ip(\pi/N)k} \qquad (p = -N+1, \cdots, 0, 1, \cdots, N)$$
$$(k = 0, 1, \cdots, 2N-1)$$

As a lemma we first prove

$$S(m) = \sum_{k=0}^{2N-1} e^{im(\pi/N)k} = \begin{cases} \dfrac{1 - e^{2\pi im}}{1 - e^{im(\pi/N)}} = 0, & e^{im(\pi/N)} \neq 1 \\ \\ 2N, & e^{im(\pi/N)} = 1 \end{cases}$$

which follows from the fact that the series is a geometric progression. The condition

$$e^{im(\pi/N)} = 1$$

means that $m = 2N\nu$ for some integer ν.

To prove the orthogonality of the p-th function of the set with respect to the q-th function we use (as is customary in dealing with complex numbers) the complex conjugate of the second function in the sum of the products

$$\sum_{k=0}^{2N-1} e^{ip(\pi/N)k}\, e^{-iq(\pi/N)k} = \sum_{k=0}^{2N-1} e^{i(p-q)(\pi/N)k} = S(p-q)$$

By the lemma this is zero except when

$$p - q = 2N\nu \qquad (\nu \text{ is an integer})$$

For the $2N$ functions we are using this happens *only* if $p = q$, and we have the desired result.

Since the Fourier functions are periodic over the continuous interval $(0 \leq x \leq 2N)$ it is reasonable to suppose that shifting the sample points an amount α would preserve the orthogonality. To check this, note that in the formula for $S(m)$ each term will have an extra factor

$$S(m) = 2Ne^{im(\pi/N)\alpha} = 2Ne^{(2\pi i\alpha)\nu}.$$

In particular, for the midpoint of the intervals, which is the main case of interest

$$\alpha = \tfrac{1}{2},$$

and we have

$$S(m) = 2Ne^{\pi i\nu} = (-1)^{\nu}2N.$$

Expressed in terms of the "real" trigonometric functions we have

$$S(m) = \sum_{k=0}^{2N-1} [\cos m(\pi/N)(k+\alpha) + i \sin m(\pi/N)(k+\alpha)] = 2Ne^{2\pi i\alpha\nu}$$

or taking the real and imaginary parts

$$\sum_{k=0}^{2N-1} \sin m \frac{\pi}{N} (k+\alpha) = \begin{cases} 0, & m \neq 2N\nu \\ 2N \cos 2\pi\alpha\nu, & m = 2N\nu \end{cases}$$

$$\sum_{k=0}^{2N-1} \sin m \frac{\pi}{N} (k+\alpha) = \begin{cases} 0, & m \neq 2N\nu \\ 2N \sin 2\pi\alpha\nu, & m = 2N\nu. \end{cases}$$

From the addition formulas of trigonometry we can now get the orthogonality of the "real" functions as follows

$$\sum_{k=0}^{2N-1} \cos p \frac{\pi}{N} (k+\alpha) \cos q \frac{\pi}{N} (k+\alpha)$$

$$= \frac{1}{2} \sum_{k=0}^{2N-1} \left\{ \cos\left[(p+q) \frac{\pi}{N} (k+\alpha) \right] + \cos\left[(p-q) \frac{\pi}{N} (k+\alpha) \right] \right\}$$

Each time $p+q$ or $p-q$ is a multiple of $2N$ (including the zero multiple) we get a contribution to the sum. Similar identities work for the other two products needed in the orthogonality relations.

In the case of $2N$ equally spaced sample points and $\alpha = 0$ it is customary to use the $2N$ functions

$$1, \quad \cos \frac{\pi}{N} x, \quad \cos 2 \frac{\pi}{N} x, \cdots, \quad \cos (N-1) \frac{\pi}{N} x, \quad \cos \pi x$$

$$\sin \frac{\pi}{N} x, \quad \sin 2 \frac{\pi}{N} x, \cdots, \quad \sin (N-1) \frac{\pi}{N} x$$

If $\alpha = \frac{1}{2}$ (the midpoint of the intervals) then $\cos \pi x = 0$ at all the sample points and is replaced by $\sin \pi x$. Other values of α are not important so that the linear relation between the two functions $\cos \pi x$ and $\sin \pi x$ need not be discussed here.

For the finite Fourier series with $\alpha = 0$ we have the representation of an arbitrary function $f(x)$ as

$$f(x) = \frac{A_0}{2} + \sum_{m=1}^{N-1} \left[A_m \cos m \frac{\pi}{N} x + B_m \sin m \frac{\pi}{N} x \right] + \frac{A_N}{2} \cos \pi x$$

The coefficients (we have used capital letters for the finite case to distinguish it from the continuous case where we used lower case letters) are given by

$$A_m = \frac{1}{N} \sum_{k=0}^{2N-1} f(k) \cos m \frac{\pi}{N} k \quad (m = 0, 1, \cdots, N)$$

$$B_m = \frac{1}{N} \sum_{k=0}^{2N-1} f(k) \sin m \frac{\pi}{N} k \quad (m = 1, 2, \cdots, N-1).$$

It is natural to ask how the coefficients A_m and B_m of the finite Fourier series (based on $2N$ equally spaced points) are related to the a_m and b_m of the continuous expansion. To find out we merely multiply the continuous expansion by the proper cosine or sine and sum over the discrete points. On the left hand side we will get the corresponding coefficient of the finite expansion while on the right we will get the sum of all the terms that survive the summation process, one term for each value of ν. The results are

$$A_m = a_m + \sum_{\nu=1}^{\infty} (a_{2N\nu-m} + a_{2N\nu+m})$$

$$B_m = b_m + \sum_{\nu=1}^{\infty} (-b_{2N\nu-m} + b_{2N\nu+m}).$$

This shows how the individual high frequency terms are aliased into the frequencies of the finite Fourier series. Figure 1 shows graphically how the various frequencies spread along a line (imagined to go toward the right) are folded back as if on a ribbon, and how each frequency appears above the one it is aliased into.

This aliasing is a familiar phenomenon. It can be seen as a simple trigonometric identity; it can be seen by looking at the graphs of various sinusoids and noting how they coincide at the sample points; it can be seen on television and movie screens when the increasingly rapid rotations of the stage coach wheels make the rapid rotations appear (due to the sampling of the picture process) look like slower, or even backward, rotations; and it can be seen when using a stroboscopic light. Thus it is nothing new to the reader.

But let us note an important fact. For the Fourier set of sines and cosines the sampling process aliases a single function to a single function,

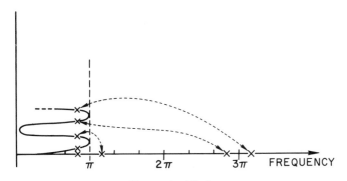

Figure 1. Aliasing.

while for the powers x^n a single power can come down to $2N$ lower powers, $1, x, x^2 \cdots, x^{2N-1}$. Thus in a very real sense the Fourier functions, sines and cosines, are the proper, the characteristic, the eigenfunctions for equally spaced sampling.

In the complex notation we have corresponding to the continuous variable $(0 \leqslant x \leqslant 2N)$

$$f(x) = \sum_{m=-\infty}^{\infty} c_m e^{\pi i (m/N) x}$$

where

$$c_m = \int_0^{2N} f(x) e^{-\pi i (m/N) x} \, dx$$

and for the discrete case

$$f(x) = \sum_{m=-N+1}^{N} C_m e^{\pi i (m/N) x}$$

where

$$C_m = \sum_{k=0}^{2N-1} f(k) e^{-\pi i (m/N) k}$$

and finally

$$C_m = \sum_{\nu=-\infty}^{\infty} c_{m+2N\nu}.$$

If one stops and thinks a bit then it becomes clear that the exponentials (real or complex) are *also* the proper, the eigenfunctions for the linear operations of differentiation and integration. Thus for three purposes, coordinate translation, equally spaced sampling, and the calculus, the exponentials are the natural, eigenfunctions, while the polynomials only have one of these properties. It follows, therefore, that the exponentials will in many cases have a simpler theory than will the classic polynomial theory. In the following material we will give a number of examples illustrating this point.

IV. THE NYQUIST INTERVAL AND THE SAMPLING THEOREM

The Fourier series is used to represent functions in a finite interval. This is equivalent to representing periodic functions with a finite period. The Fourier integral

$$f(x) = \int_{-\infty}^{\infty} F(\sigma) e^{2\pi i \sigma x} \, d\sigma$$

is used to represent functions over the whole real axis $(-\infty \leqslant x \leqslant \infty)$. It uses a continuum of frequencies σ rather than a discrete set as does the Fourier series.

The aliasing discussed in Section II that results from sampling happens not only for the discrete frequencies that occur in the Fourier series but it also happens for all the frequencies that appear in the Fourier integral. The highest frequency that is not aliased down to a lower frequency is called the "folding frequency," or the Nyquist frequency. This occurs at the point in the frequency domain for which there are exactly two samples for that frequency. If there are less than two samples in a cycle then it will appear equivalent to a frequency that has more than two samples in its period. This aliasing was portrayed in Figure 1 and is the same for the discrete and continuous cases. The same physical arguments also apply.

We now turn to the reconstruction of the signal from the samples. For polynomials the Lagrange interpolation formula used the functions

$$\frac{\pi_i(x)}{\pi_i(x_i)}$$

where

$$\pi_i(x) = (x-x_1) \cdots (x-x_{i-1})(x-x_{i+1}) \cdots (x-x_n),$$

the i-th factor being omitted. This function has the property that it takes on the value 1 at the i-th point and zero at all other sample points. For the infinite set of equally spaced sample points $x = k$ we have the corresponding sampling function

$$\frac{\sin \pi(x-k)}{\pi(x-k)}$$

which takes on the value 1 at $x = k$ and is zero at all other integers. Thus in the same formal way as for Lagrange interpolation the function

$$f(x) = \sum_{k=-\infty}^{\infty} f(k) \left[\frac{\sin \pi(x-k)}{\pi(x-k)} \right]$$

is the interpolating function.

If there are frequency components in $f(x)$ above the Nyquist frequency we cannot be expected to reconstruct them from the samples alone. For this reason the frequency approach to numerical analysis constantly uses the expression "band limited" functions, meaning functions all of whose frequencies are confined to a band, usually the Nyquist band.

Given a suitable band limited function our interpolating function has a reasonable chance of reconstructing it. But is the reconstructed function band limited? To see that it is consider the integral for which $F(\sigma) = 1$ for $(-\frac{1}{2} \leqslant \sigma \leqslant \frac{1}{2})$ and zero elsewhere. We have by the definition of the

Fourier integral

$$\int_{-\infty}^{\infty} F(\sigma)e^{2\pi i\sigma x}\,d\sigma = \int_{-1/2}^{1/2} e^{2\pi i\sigma x}\,d\sigma$$

$$= \frac{e^{2\pi i\sigma x}}{2\pi ix}\bigg|_{-1/2}^{1/2}$$

$$= \frac{e^{\pi ix} - e^{-\pi ix}}{2\pi ix}$$

$$= \frac{\sin \pi x}{\pi x}$$

Since, reversing the two sides,

$$\frac{\sin \pi x}{\pi x} = \int_{-\infty}^{\infty} F(\sigma)e^{2\pi i\sigma x}\,d\sigma$$

it follows that $\dfrac{\sin \pi x}{\pi x}$ is band limited to the proper interval.

This is the essence of the famous sampling theorem; a band limited function can be reconstructed from its samples *provided* there are at least two samples for the highest frequency present. We have of course, not proved this important theorem, we have only made it plausible—a rigorous proof would be out of place in a simple introduction.

V. THE DIFFERENCE TABLE AND NOISE

Central to most of the finite difference calculus is the difference table. In the classic polynomial theory there are three results that bear on the matter. First is the theorem that the $(k+1)$-th differences of a polynomial of degree k are zero. Second, an isolated error in a difference table (Figure 2) propagates rapidly with binomial coefficients and alternating signs. Third, the successive differences in the k-th column of differences can be shown to have a negative correlation coefficient of

$$\frac{-k}{(k+1)}$$

These three results are combined in the following way. First, it is hoped that the function has been tabulated sufficiently closely so that the ideal higher differences will be small. Second, it is believed that the random roundoff errors each give rise to a rapidly growing error as the difference order increases. Last, it is believed that the alternating signs in some

x	y	Δy	Δ²y	Δ³y	Δ⁴y
.					
.					
.					
−3	0				
		0		0	
−2	0		0		ε
		0		−ε	
−1	0		ε		−4ε
		−ε		3ε	
0	ε		−2ε		6ε
		ε		−3ε	
1	0		ε		−4ε
		0		ε	
2	0		0		ε
		0			
3	0				
		0			
4	0				

FIGURE 2. A Difference Table.

difference column mark where the information about the function has been completely swamped by the roundoff noise. From this column of differences an estimate can be made of the probable roundoff in the original data using the formula

$$\text{original noise variance} = \frac{(k!)^2}{(2k)!} \text{ [mean square of } k\text{-th differences]}$$

Thus the difference table is an amplifier of the noise, and hence makes the noise more readily detectable and measurable.

On the other hand, the frequency approach to the difference table proceeds as follows. We suppose not that the function is written as a polynomial (with roundoff noise added) but as a complex frequency

$$y_k = e^{2\pi i \sigma k} = e^{i\omega k} \quad \text{(where } 2\pi\sigma = \omega)$$

σ being the rotational frequency and ω the angular frequency of the complex sinusoid. The linear operation of differencing leads to

$$\Delta y_n = e^{i\omega(n+1)} - e^{i\omega n} = ie^{i\omega/2}\left[2 \sin \frac{\omega}{2}\right] y_n$$

Thus we see that differencing the function produces (1) a factor $e^{i\omega/2}$ that affects the phase but not the amplitude, (2) a factor $2 \sin \omega/2$, that affects the amplitude but not the phase, and (3) the original function y_n.

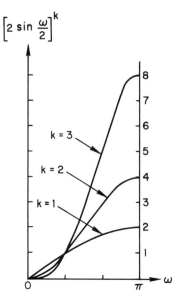

FIGURE 3. Effect of differences.

Thus the k-th difference will be simply

$$\Delta^k y_n = i^k e^{i\omega k/2} \left[2 \sin \frac{\omega}{2} \right]^k y_n.$$

From Figure 3 we see that the process of differencing amplifies the upper two thirds of the Nyquist interval and attenuates the lower one third. If we believe that the difference table did in fact uncover the noise then we are led to identify noise with the upper two thirds of the Nyquist interval and the signal with the lower third.

As we shall later see, we can easily design linear operations (such as the example of differencing) which will amplify any fraction of the interval we please. In this sense we can generalize the difference operator so that we can separate signal and noise wherever we think the dividing line in the frequency domain lies. In practice the dividing line is often much lower than the one-third point, but sometimes it is well above it.

VI. A QUICK LOOK AT INTEGRATION

For periodic functions, or equivalently functions defined over a finite range, the Fourier series gives a valid representation, and the a_0 coefficient gives effectively the integral of the function. The formula for the finite

Fourier series has the coefficient A_0 computed by using an equally spaced, equally weighted sum of samples of the integrand.

Looking at this formula more closely, we see that first it is the equivalent of the Newton-Cotes formula since it uses equally spaced samples and is exactly true for the functions 1, sin x, cos x, etc. for at least as many functions as sample points. But it is also the Gauss quadrature formula since in fact the formula is exact all the way up through cos Nx, i.e. in all for $2N$ functions. Lastly, the formula is the Chebyshev formula since it uses equal weights on the samples. Thus in the frequency approach this single formula combines all the features of three rather unrelated formulas in the classical polynomial approach to numerical analysis. Again we see the essential unity of the frequency approach.

The formula we gave for the aliasing of the zero frequency

$$A_0 = a_0 + 2 \sum_{\nu=1}^{\infty} a_{2N\nu}$$

gives a readily understandable formula for the error, since A_0 is the computed answer and a_0 is the true answer. We see immediately that it is only certain higher harmonics (frequencies) that matter. Often we can, using outside information from the source of the problem, make realistic estimates of these frequencies in the sense of knowing bounds on them, so that we can make *a priori* error estimates.

Experience and special test cases bear out these observations that the error in the integration can be directly connected with the problem that gave rise to the integration.

VII. SMOOTHING

Although the problem of smoothing data is obviously important, classical numerical analysis tends to avoid the topic. Least squares and Chebyshev fitting both imply smoothing, and we shall now examine them from the frequency point of view to see more clearly what they do and how they compare.

We take the specific case of five equally spaced sample points and first fit a least squares quadratic to the five points. The value of this quadratic at the midpoint is taken as the smoothed value (which is the same as if we had used a cubic). For convenience suppose that the origin is at the middle point of the five. We have to minimize

$$f(a,b,c) = \sum_{k=-2}^{2} [a + bk + ck^2 - y_k]^2$$

The normal equations are:

$$\text{multipliers}$$

$$5a + 0b + 10c = \sum y_k \qquad 17$$

$$0a + 10b + 0c = \sum x_k y_k \qquad 0$$

$$10a + 0b + 34c = \sum x_k^2 y_k \qquad -5$$

The smoothed value we want is when $x = 0$, so we need to find only a,

$$\bar{y}_0 = a = -\frac{3y_{-2} + 12y_{-1} + 17y_0 + 12y_1 - 3y_2}{35}$$

$$= y_0 - \frac{3}{35}\Delta^4 y_{-2}$$

This is the formula for smoothing equally spaced data using the least squares principle.

To find out what it does to the various frequencies that are present in the original signal, we suppose

$$y(t) = e^{i\omega t} \quad (\omega = 2\pi\sigma)$$

and get

$$\bar{y} = \left[-\frac{3e^{-2i\omega} + 12e^{-i\omega} + 17 + 12e^{i\omega} - 3e^{2i\omega}}{35} \right] e^{i\omega t}$$

Since the formula is linear we get as the multiplier of the original function

$$H(\omega) = \frac{17 + 24\cos\omega - 6\cos 2\omega}{35}$$

The graph of the function, Figure 4, shows the curve, and we see that this process of smoothing by fitting a least squares quadratic attenuates the higher frequencies and "passes" the lower frequencies, with the constant (zero frequency) exactly correct.

If we were to do more drastic smoothing and fit a straight line (it will be the same as fitting a constant) we will get the average value of the five as the smoothed value, that is

$$\bar{y} = \frac{y_{-2} + y_{-1} + y_0 + y_1 + y_2}{5}$$

Again substituting

$$y(t) = e^{i\omega t}$$

we get

$$H(\omega) = \frac{1 + 2\cos\omega + 2\cos 2\omega}{5}$$

$$= \frac{\sin 5\omega/2}{\sin \omega/2}$$

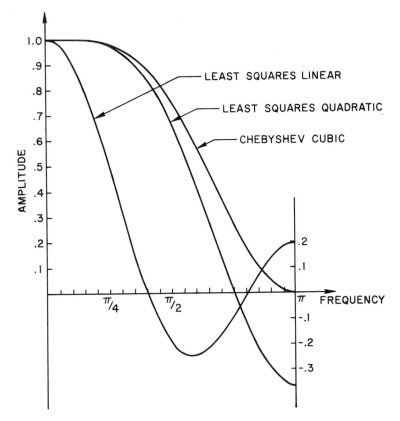

FIGURE 4. Smoothing Formulas.

Figure 4 shows that this is far more severe than the quadratic in attenuating the higher frequencies, while still giving the correct value when the data is a constant ($\omega = 0$).

Lastly, if we want to fit a Chebyshev curve through the five points and use it as the smoothed middle value, then we begin by fitting a quartic to the five points. Next we find an equal ripple quartic on the five points. This is easily seen to be

$$T(x) = \frac{2x^4 - 8x^2 + 3}{3}$$

so that

$$T(k) = (-1)^k$$

Subtracting the proper multiple of this from the original quartic we can eliminate the fourth degree term. The result turns out to be

$$\bar{y} = \frac{-y_{-2} + 4y_{-1} + 10y_0 + 4y_1 - y_2}{16}$$

$$= y_0 - \frac{\Delta^4 y_{-2}}{16}$$

Once more we set

$$y = e^{i\omega t}$$

and find out what happens to the various frequencies. We get the factor

$$\tfrac{1}{8} (5 + 4 \cos \omega - \cos 2\omega)$$

which is also shown in Figure 4. We see immediately that this Chebyshev smoothing is less severe than is the corresponding least squares smoothing in the sense that it passes more of the lower frequencies. It also rejects the higher frequencies a bit better than does least squares.

Thus on a single chart we have a comparison of three different smoothing formulas and can state how they compare with each other. Having earlier shown that the higher differences in a difference table amplify the upper two-thirds of the Nyquist interval of frequencies, and argued that classically it is the difference table that is used to detect noise in a table, we can see how the formulas treat the upper two thirds of the interval. We need not, however, believe that always the noise is in the upper two-thirds — often it may be the upper nine-tenths or more, and hence severe attenuation in the upper part is necessary to eliminate as much noise as possible. On the other hand sometimes the signal covers at least nine-tenths of the interval.

VIII. THE GIBBS PHENOMENON

A. A. Michelson found when using his harmonic analyzer that the sum of a finite number of terms from an infinite Fourier expansion gave a sum that overshot and undershot the true function near a point of discontinuity. After some time Gibbs showed that this was to be expected and gave an explicit formula for the effect.

A simple explanation of this can be based on the rectangular wave function (Figure 5)

$$f(x) = \begin{cases} 0, & -\pi \leqslant x \leqslant 0 \\ 1, & 0 < x < \pi \end{cases}$$

Since the function is the sum of $\frac{1}{2}$ plus an odd function there are no cosine terms (except for $k = 0$) in the Fourier expansion, and we have from the

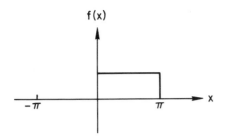

FIGURE 5. Rectangular wave.

formula for the coefficients (and a little integration), the sum of a finite number of terms

$$f_{2n}(x) = \frac{1}{2} + \frac{2}{\pi} \sum_{k=1}^{n} \frac{1}{2k-1} \sin (2k-1)x$$

Writing this in the form

$$f_{2n}(x) = \frac{1}{2} + \frac{2}{\pi} \sum_{k=1}^{n} \int_{0}^{x} \cos (2k-1)t \, dt$$

$$= \frac{1}{2} + \frac{2}{\pi} \int_{0}^{x} \sum_{k=1}^{n} \cos (2k-1)t \, dt$$

and using

$$\sum_{k=1}^{n} \cos (2k-1)t = \frac{\sin 2nt}{2 \sin t}$$

we get

$$f_{2n}(x) = \frac{1}{2} + \frac{1}{\pi} \int_{0}^{x} \frac{\sin 2nt}{\sin t} \, dt$$

To find the maximum and minimum of $f_{2n}(x)$ we have only to differentiate and set equal to zero;

$$\frac{d f_{2n}(x)}{dx} = \frac{1}{\pi} \frac{\sin 2nx}{\sin x} = 0$$

Thus for

$$x = \frac{\pi m}{2n}$$

we will have extremes which depend only slightly on n.

Lanczos suggested that since these oscillations could be expected to arise near any discontinuity, and since the period of the oscillation is known and arose from the neglected terms of the infinite expansion, then smoothing the truncated Fourier series by averaging over one period of the oscillation will tend to remove this oscillation effect due to the truncation process. Thus he suggested that for the general Fourier series

$$f_n(x) = \frac{a_0}{2} + \sum_{k=1}^{n} [a_k \cos kt + b_k \sin kt]$$

we replace $f_n(x)$ by the integral over a period of the oscillation.

$$\bar{f}_n(x) = \frac{n}{2\pi} \int_{x-\pi/n}^{x+\pi/n} f(t)\, dt$$

$$= \frac{n}{2\pi} \left[\frac{a_0}{2} \frac{2\pi}{n} + \sum_{k=1}^{n} \left(a_k \frac{\sin kt}{k} - b_k \frac{\cos kt}{k} \right) \right]\Bigg|_{x-\pi/n}^{x+\pi/n}$$

$$= \frac{a_0}{2} + \frac{n}{2\pi} \sum_{k=1}^{n} \left(\frac{a_k}{k} [\sin k(x+\pi/n) - \sin k(x-\pi/n)] \right.$$

$$\left. - \frac{b_k}{k} [\cos k(x+\pi/n) - \cos k(x-\pi/n)] \right)$$

$$= \frac{a_0}{2} + \frac{n}{2\pi} \sum_{k=1}^{n} \left(\frac{2a_k}{k} \sin \frac{k\pi}{n} \cos kx + \frac{2b_k}{k} \sin \frac{k\pi}{n} \sin kx \right)$$

$$= \frac{a_0}{2} + \sum_{k=1}^{n} \frac{\sin(k\pi/n)}{(k\pi/n)} (a_k \cos kx + b_k \sin kx)$$

Set

$$\frac{\sin(k\pi/n)}{(k\pi/n)} = \sigma_k(n) = \sigma_k$$

Therefore the smoothing Lanczos proposed amounts to taking the original Fourier coefficients and multiplying them by the appropriate sigma factors. The n arises from the number of terms kept and the k relates to the term the factor is applied to. Note that $\sigma_n(n)$ is zero and is the first term that is neglected in the expansion in practice.

The often used simple truncation of the Fourier series can be modified to eliminate the last sine term and take one half of the last cosine term as in the finite Fourier series. This does a little bit of smoothing of the Gibbs phenomena. Lanczos sigma factors reduces the ripples by about a factor of 10. The mathematician's favorite, Cesaro smoothing is somewhat more drastic, see Figure 6. Obvious other smoothing of the Fourier coefficients would give other effects in the smoothed function, and we will look again at this matter.

These same sigma factors arise in numerical differentiation if we space

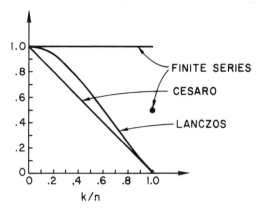

FIGURE 6. Smoothing weights.

the two values a full period apart. Thus we use the approximation

$$\frac{df(x)}{dx} \sim \frac{f(x+\pi/n)-f(x-\pi/n)}{2\pi/n}$$

we are led directly to the sigma factors.

IX. ELEMENTARY FILTER THEORY

Many of the formulas in numerical analysis are linear expressions in the function values at the sample points. Thus the substitution of the function

$$y(t) = e^{i\omega t} \quad (\omega = 2\pi\sigma)$$

into the formula will give an expression of the form

$$H(\omega)e^{i\omega t}$$

The function $H(\omega)$ is called the "transfer function" of the formula—it shows how to "transfer" from the input $y(t)$ to the output $H(\omega)y(t)$ by simply multiplying the input by $H(\omega)$. The sequence of k such linear operations will result in the product

$$H_1(\omega)H_2(\omega) \cdots H_k(\omega)e^{i\omega t}$$

For the frequency approach to numerical analysis the process of evaluating a given formula is now clear, and we therefore turn to the elementary theory of designing formulas rather than merely evaluating them.

Suppose we wish to remove (filter out) the upper half of the Nyquist band of frequencies using a filter of the form

$$\bar{y}_0 = A_{-2}y_{-2} + B_{-1}y_{-1} + C_0y_0 + B_1y_1 + A_2y_2$$

If we wish to avoid the phase shift (factors depending on i) we need to put $A_2 = A_{-2}$ and $B_2 = B_{-2}$. As a result the transfer function is (dropping subscripts on the coefficients)

$$H(\omega) = Ae^{-2i\omega} + Be^{-i\omega} + C + Be^{i\omega} + Ae^{2i\omega}$$

$$= C + 2B \cos \omega + 2A \cos 2\omega$$

It is reasonable to require that the formula be exactly true for the constant function $y(t) = c$, and this means that for $\omega = 0$ we must have

$$1 = C + 2B + 2A.$$

As a result we automatically have

$$\left. \frac{dH(\omega)}{d\omega} \right|_{\omega=0} = 0$$

Suppose, to further limit the class of filters we require, that

$$H(\pi) = 0$$

which is equivalent to

$$C - 2B + 2A = 0$$

From these two equations we get the one parameter family of filters

$$C = \tfrac{1}{2} - 2A$$

$$B = \tfrac{1}{4}$$

$$H(\omega) = (\tfrac{1}{2} - 2A) + \tfrac{1}{2}\cos \omega + 2A \cos 2\omega$$

Figure 7 shows nine cases corresponding to the transfer function $H(\omega)$ for

$$A = -\tfrac{4}{16}, \ -\tfrac{3}{16}, \cdots, \ \tfrac{4}{16}$$

and it is easy (using the mid value) to pick out a simple filter to fit various smoothing needs.

Of course we could, and often do, use other conditions other than $H(\pi) = 0$. The simple approach we used should make it clear that we first assume the form of the filter we want to use, and then impose as many conditions as there are free parameters. As a result of some algebra we have the desired filter.

Low pass filters (which let through low frequencies and block out high frequencies) can be converted to high pass filters by simply subtracting the low pass filter from the allpass filter $H(\omega) = 1$.

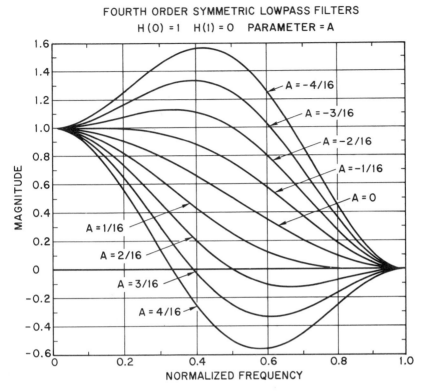

FIGURE 7. Fourth order symmetric lowpass filters $H(0) = 1$, $H(1) = 0$, parameter $= A$.

We have examined smoothing filters. If we wanted a filter to differentiate as well as smooth some data we would want an approximation in the pass part of the filter response.

$$H(\omega) \sim i\omega$$

since

$$\frac{de^{i\omega t}}{dt} = i\omega e^{i\omega t}$$

For such filters we want odd symmetry rather than even symmetry so that we will get sine terms in $H(\omega)$ rather than cosine terms, and we will also get the proper phase factor i.

Similarly for integration filters we want odd symmetry. We will not go further into this matter of designing simple filters because the process is fairly obvious and not worth further space in this tutorial.

X. MORE ADVANCED FILTER THEORY

The elementary approach to designing filters is adequate for fairly crude ones, but many times carefully designed filters are necessary, and these unfortunately often involve many terms. The simple approach of the last section leads, therefore, to many free parameters to be determined, and this is rarely a successful process. Instead we need a method for getting a very good first approximation, which if necessary can be subjected to a final refinement using any of a number of known ways.

Before plunging into the details let us review the elements of Fourier transforms. For this purpose we start with the simplest case, a pure sinusoid signal

$$f(t) = \sin \omega t = \frac{1}{2i} \left(e^{i\omega t} - e^{-i\omega t} \right)$$

In the frequency domain this is a pair of spikes, Figure 8a. This is the Fourier transform correspondence.

Suppose now we chop out a finite piece of the infinitely long sinusoid signal, that is we multiply the time function $f(t)$ by a rectangular pulse $p(t)$, Figure 8b. The transform of this pulse is

$$P(\omega) = \int_{-\infty}^{\infty} p(t) e^{-i\omega t} dt$$

$$= \int_{-T/2}^{T/2} e^{-i\omega t} dt = -\frac{1}{i\omega} \left[e^{-i\omega T/2} - e^{i\omega T/2} \right]$$

$$= \frac{\sin (\omega T/2)}{\sin (\omega/2)}$$

The larger T the narrower the main lobe of the sin x/x curve.

The product of the signal times the pulse is, by the convolution theorem, the convolution of the spikes times the sin x/x function, Figure 8c.

Now if the true signal were the sum of a finite number of pure sinusoids then the chopped off signal would be the corresponding sum of the convolutions of the spikes with the sin x/x function — the pure lines of the spectrum would be smeared out. Going to the limit of a continuous distribution of sinusoids we arrive at the corresponding convolution integral in the frequency space.

Returning to the filter design problem, we start with the shape of the desired transfer function in the frequency space Figure 9a. For pure smoothing it will be an even function of the frequency, while for differentiation and integration it will be an odd function.

We expand this transfer function into a Fourier series. If we wish to use the sample points we search for an expansion in the integral multiples of

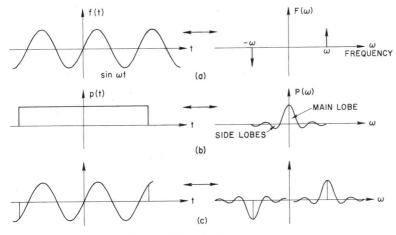

FIGURE 8. Fourier Transforms.

the frequency, while if we wish to obtain the answers at the midpoints of the intervals of measurement then we want an expansion in terms of the odd half multiples of the frequency.

The coefficients of the Fourier expansion are pictured in the time domain on the left of line (a) of Figure 9.

Next we truncate the infinite series to a finite number of terms (to make the computation practical). If we merely chop off the extra terms then this

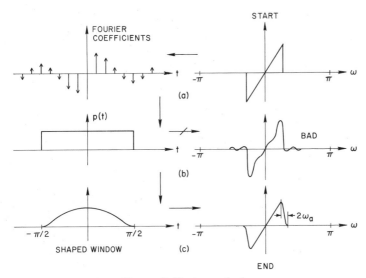

FIGURE 9. Design method.

is the same as multiplying by the familiar rectangular pulse which in the frequency domain has the sin x/x behavior and leads to what is shown on line b of Figure 9 (the convolution carried out).

What is wrong with this approach is simply the Gibbs phenomenon discussed in Section 8. Because of the ripples we simply do not like the filter that results.

What shape pulse (window) could we use (perhaps the previous one used by Lanczos) so as to keep the ripples down? We want a shape in the time domain (line c of Figure 9) such that outside the interval $(-T/2 \leq t \leq T/2)$ it is zero but which also has a corresponding shape in the frequency domain that is "concentrated" without the large side ripples.

Perhaps the first window used in this connection is the one with weights in the frequency domain

$$\cdots, 0, 0, 0, \tfrac{1}{4}, \tfrac{1}{2}, \tfrac{1}{4}, 0, 0, \cdots$$

The corresponding function in the time domain is

$$\frac{1 + \cos t}{2}$$

A slightly more elegant window is

$$\cdots 0, 0. .23, .54, .23, 0. 0. \cdots$$

which gives

$$.54 + .46 \cos t$$

and has more in the main lobe and smaller side lobes than has the first one.

A method given by Kaiser[†] for the design of a smoothing filter is based on the idea that the prolate spheroidal functions are optimal in a certain sense of keeping the tails of the distributions down in both domains at the same time. He proposed using the approximation

$$w(t) = \begin{cases} \dfrac{I_0[\omega_a \sqrt{(T/2)^2 - t^2}]}{I_0(\omega_a T/2)}, & |t| < T/2 \\[2mm] 0, & |t| > T/2 \end{cases}$$

$$W(\omega) = \frac{2}{I_0(\omega_a T/2)} \frac{\sinh\left[(T/2)(\sqrt{\omega_a^2 - \omega^2})\right]}{\sqrt{\omega_a^2 - \omega^2}}$$

where ω_a is approximately the half width of the main lobe and the I_0 is the modified Bessel function of the first kind and zeroth order. The quantity $(\omega_a T/2)$ is the parameter of the window. Note that when $|\omega| > \omega_a$ then

[†] Kaiser, J. F. *System Analysis by Digital Computer* pp. 228–243. Wiley, New York, 1966.

the sinh has sinusoidal behavior, but when $|\omega| < \omega_c$ it has the normal sinh behavior.

Thus we have shown how the Lanczos and other methods of smoothing of Fourier series fit into a more general theory of smoothing windows. Clearly each specific problem requires its own appropriate window to achieve best results for the particular problem, and no universal windows can be given for all cases.

XI. SUMMARY

In summary, we began the introduction to the frequency approach to numerical analysis by considering it as an alternate to the classical polynomial approach, and showed how the classical polynomial approach based on differences, especially the difference table, leads to the idea of the separation of the frequency content of the signal and the noise. We also looked at integration from the frequency approach and found that three different methods, Newton–Cotes, Gauss, and Chebyshev are all the same when the basis functions are sines and cosines.

We then turned to new material such as smoothing, and showed how this topic can be given a uniform framework that unites least-squares and Chebyshev smoothing into the more general concept of a transfer function.

Finally we gave a brief introduction to the design of filters in a more general sense, and indicated both an elementary approach and a slight view of a more sophisticated theory.

What we did not do was to indicate many of the other aspects of the theory and its applications, nor did we indicate the application of the frequency approach to the topic of algorithms. The excuse is the shortness of space available.

Detouring around Computational Roadblocks — A Tale of Two Integrals

FORMAN S. ACTON

School of Engineering, Princeton University, U.S.A.

I address this article to two groups: to the engineers who must cope with instabilities and the other awkwardnesses of computational algorithms, and to those mathematicians who are kind enough to be interested in helping. We have here a two-culture communication gap—widely deplored, though not often bridged since it appeared in the 19th century. Professor Lanczos has given much of his professional life to closing this gap, and the engineers of America owe him a great debt for his sympathetic efforts to make his mathematical insights available in familiar terms. His *Applied Analysis* (1956) remains useful today though it was written before the electronic computer became commonplace and much of its analysis is motivated by a desire to help the man with the desk calculator. A book that survives fifteen years of this rapidly changing computer world may be considered a monument indeed. It was my own good furtune to know Professor Lanczos for three years during my first professional employment at the Institute for Numerical Analysis (U.S. National Bureau of Standards) at UCLA. His example, imperfectly followed, has motivated much of my own work, and I hope that this article, dedicated to Niels as a teacher, will perhaps be useful to the two groups he has served so well.

Persons without computational experience tend to assume that because the answers came from a computer they must be correct. This is, of course, merely an up-to-date version of "I know it's true because I read it in the newspaper". But even quite sophisticated scientists still hold the belief that if they possess a correct mathematical expression the computer will evaluate it "correctly", thereby producing the answers they seek. Anyone who computes knows it does not. The problems arise in part from the finiteness of computer arithmetic, which combines with such

177

things as slowly converging alternating series to cancel out most or all of the significant digits, leaving one with an "answer" made up largely of rounding errors. This kind of problem is easily overlooked until one has been burnt a few times. The familiar mathematical theorems, after all, worry mainly about whether a series converges rather than how quickly or with what requirements for intermediate computational precision.

Other difficulties arise because definite integrals often exist even though their integrands become infinite somewhere on the range of integration. It seems to escape the classically trained mind that numerical quadrature techniques are unconcerned about the integrals' existence, but are extremely sensitive to the zero in the denominator. That singularity *must* be removed, and not by the computer. These are quite "obvious" points, but anyone sitting near a university computing center soon realizes that our students are not learning about them, in their calculus courses or anywhere else. Apparently it takes a painful succession of wrong numerical results before the facts about instability, loss of significance, and slowness of convergence begin to assume their proper place in the technical human consciousness. With the greatly expanded use of computers in science and engineering, these topics must find their way into the undergraduate curriculum. Otherwise we shall merely continue to get wrong answers faster.

Faced with evaluating a mathematical function an engineer is concerned first with finding a method that works and, secondly, a method that works efficiently. He does not seek a proof that the solution exists unless, perchance, the proof is constructive and provides an algorithm. His concern for efficiency also extends to his own efforts. After he has found two or three different techniques that give identical numerical results he will choose the best one and use it. Thus it frequently happens that a practical problem gets solved by inelegant techniques. Not because the engineer cannot be bothered searching for elegant solutions — far from it — but simply because practical problems often have to be solved in a hurry. Still, like mathematicians, good engineers are artists, with a reluctance to abandon a subject half understood. Occasionally they can indulge in the luxury of returning to a problem after it has ceased to be of practical concern, perhaps thereby discovering how they "should" have carried out their computations. In this paper I set forth the history of a rather typical engineering calculation. It was intended to give undergraduates realistic experience in computation, complete with instabilities and singularities. It also had the essential engineering characteristic that the correct results were not known *a priori* — not even to the instructor!

The computational problem

Physically, the original problem was to evaluate the potential field at arbitrary points around the two-dimensional finite parallel-plate condenser by several methods in order to find which of them gives the most efficient computational algorithm. Since the details of the "best" method, the Rosser iterative mapping, have been reported elsewhere[2] we shall ignore that technique and here concentrate on two integrals that arise in an integral equation method that was second best. They are

$$J_1(x, n, b) = \int_{-1}^{1} \frac{\xi^{2n}}{\sqrt{1 - \xi^2}} \ln \left[b^2 + (x - \xi)^2 \right] d\xi$$

$$C_1(x, n, b) = \int_{-1}^{1} \frac{T_{2n}(\xi)}{\sqrt{1 - \xi^2}} \ln \left[b^2 + (x - \xi)^2 \right] d\xi$$

(1)

We also do not discuss the exact conformal map[4] for the problem — (which ran a poor third in our investigations) — although it produces some interesting computational difficulties of its own.

Computationally, our paper deals with the various difficulties that arose while trying to evaluate these two integrals to 10 significant digits. The narrative includes discussion of methods that ultimately worked but were inefficient, as well as some that were beautifully efficient — whenever they were not unstable. (At one point we even managed to lose 12 significant digits from a single number!) We offer this case history in the belief that the experiences are applicable to a variety of engineering computations. It should guide those with similar problems and, perhaps as important, illustrate for the purer mathematicians some of the computational interests of the engineer together with the constraints under which he works.

Mathematically, we were operating at an undergraduate level. The first integral (J_1) was given as a computational project to third and fourth year engineering undergraduates with varying backgrounds. In particular, we could not expect proficiency in complex variable methods, although they were not excluded. Thus the students were encouraged to apply their underclass calculus to an integral containing unpleasant singularities in its integrand. They were also encouraged to explore such compendia as Bierens De Haan and the *Handbook of Mathematical Functions* (AMS55) for possibly applicable formulae, even if they did not initially understand them. Their approach was thus partly experimental — and the project contained the essential engineering element that the "right answers" were not available for checking the correctness of their work. For this reason, *duplicate computation by two distinctly different methods was mandatory.* In addition to possible mathematical errors, there are just too many factors of 2 or π and too many minus signs that can get lost during derivations, transcriptions into FORTRAN, and typing into the computer to permit any less stringent form of verification.

The parallel-plate condenser — an integral equation strategy

If we take a finite two-dimensional parallel-plate condenser, normalizing the plate length to 2 and calling the plate separation 2h, we apply the obvious rectangular coordinate system of Figure 1. (The plates may be considered to extend indefinitely in the direction perpendicular to the page.) If the upper plate is charged to the constant potential of + 100 volts and the lower to − 100 volts the problem has an even symmetry in x and an odd symmetry in y. The even symmetry in x is important in the sequel.

The potential $P(x,y)$ at an arbitrary point (x,y) can be expressed in terms of the (unknown) charge $g(\xi)$ along the plates* by the expression

$$P(x,y) = C_p \int_{-1}^{1} \frac{g(\xi)}{\sqrt{1-\xi^2}} \ln\frac{1}{r_1} d\xi - C_p \int_{-1}^{1} \frac{g(\xi)}{\sqrt{1-\xi^2}} \ln\frac{1}{r_2} d\xi \qquad (2)$$

where the term C_p is a constant of proportionality that depends on the units. It will cancel out so we can set it to unity immediately. The distances r_1 and r_2 are shown in Figure 1 and are given by

$$r_1{}^2 = (x-\xi)^2 + (h-y)^2 \qquad r_2{}^2 = (x-\xi)^2 + (h+y)^2$$

For a given condenser geometry, h, and a given point (x,y) both integrals

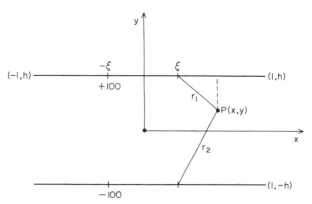

FIGURE 1. A Finite Parallel-Plate Condenser.

---

* The charge is actually $g(\xi)/\sqrt{1-\xi^2}$, carrying a singularity of this form at the ends of the plate. The function $g(\xi)$ is merely the well behaved part.

have the same form—a form that depends only on the charge function. Since (2) represents the potential *on* the plate as well as elsewhere, we may let (x,y) lie anywhere on the top plate to give the integral equation

$$100 = \tfrac{1}{2} \int_{-1}^{1} \frac{g(\xi)}{\sqrt{1-\xi^2}} \ln \left[(h-y)^2 + (x-\xi)^2 \right] d\xi$$

$$- \tfrac{1}{2} \int_{-1}^{1} \frac{g(\xi)}{\sqrt{1-\xi^2}} \ln \left[(h+y)^2 + (x-\xi)^2 \right] d\xi \tag{3}$$

for the unknown $g(\xi)$.

From physical considerations we have good reason to believe $g(\xi)$ to be a slowly varying symmetric function of ξ. An immediately appealing procedure is to represent it by a series, hopefully short, in ξ^2

$$g(\xi) = A_0 + A_1 \xi^2 + A_2 \xi^4 + \cdots + A_n \xi^{2n} \tag{4}$$

thereby turning equation (3) into

$$100 = \frac{A_0}{2} \left[\int_{-1}^{1} \frac{1}{\sqrt{1-\xi^2}} \ln \left[(x-\xi)^2 \right] d\xi \right.$$

$$- \int_{-1}^{1} \frac{1}{\sqrt{1-\xi^2}} \ln \left[4h^2 + (x-\xi)^2 \right] d\xi \right]$$

$$+ \frac{A_1}{2} \left[\int_{-1}^{1} \frac{\xi^2}{\sqrt{1-\xi^2}} \ln \left[(x-\xi)^2 \right] d\xi - \cdots \right]$$

a linear combination of $2n$ integrals, each of the form J_1. When any definite value of x $(0 \leqslant x \leqslant 1)$ is chosen, each of these integrals has a definite computable value, giving us one equation, linear in the $n+1$ coefficients A_n. If we pick $n+1$ distinct points on the right half of the top condenser plate, we obtain $n+1$ such equations that are linearly independent and may therefore be solved.

Once we possess the A_n we can return to equation (2), replacing $g(\xi)$ there by (4) to give

$$P(x,y) = \frac{A_0}{2} \left[\int_{-1}^{1} \frac{1}{\sqrt{1-\xi^2}} \ln \left[b_1^2 + (x-\xi)^2 \right] d\xi \right.$$

$$- \int_{-1}^{1} \frac{1}{\sqrt{1-\xi^2}} \ln \left[b_2^2 + (x-\xi)^2 \right] d\xi \right]$$

$$+ \frac{A_1}{2} \left[\int_{-1}^{1} \frac{\xi^2}{\sqrt{1-\xi^2}} \ln \left[b_1^2 + (x-\xi)^2 \right] d\xi - \cdots \right]$$

from which the evaluation of $P(x,y)$ at any interesting point (x,y) only requires the evaluation of $2n$ more integrals of type J_1 which are then combined linearly with the weights $A_n/2$. An efficient algorithm to evaluate these integrals thus lies at the heart of our integral equation strategy.

The integral equation symmetric in x

Although we know our physical problem is symmetric in x, the integral J_1 appears to be a function of x rather than of x^2. But if we split it into two integrals, one on the range $(-1,0)$ and one on the range $(0,1)$ then change the sign of the dummy variable of the first integral, we may recombine the two parts into the single integral over the reduced range $(0,1)$ to obtain

$$J_2(x,n,b) = \int_0^1 \frac{\xi^{2n}}{\sqrt{1-\xi^2}} \ln \{[b^2 + (x-\xi)^2][b^2 + (x+\xi)^2]\} \, d\xi$$

$$= \int_0^1 \frac{\xi^{2n}}{\sqrt{1-\xi^2}} \ln [(b^2+x^2)^2 + 2(b^2-x^2)\xi^2 + \xi^4] \, d\xi$$

in which we can see that only x^2 and b^2 occur. In seeking computational algorithms we can begin from either form of our integral: J_1 symmetric in its range of integration, or J_2 symmetric in its arguments. Most derivational methods will produce algorithms from both forms. Your author had an initial prejudice in favor of J_2 on the intuitive grounds that it had two singularities (instead of three) and that its range of integration was half that of J_1 — an advantage in numerical quadrature.

Numerical quadrature — singularity removal

Since the integral J_1 exists for all positive x and b, some numerical quadrature technique is a natural first thought. The singularities from the factor $\sqrt{1-\xi^2}$ are easily removed by the standard substitution of $\cos\theta$ for ξ, after which a straightforward use of (say) Simpson's rule will give an accurate evaluation of J_1 for most values of the parameters. It is only when we worry about *efficiency* that we become unhappy with numerical quadratures. The error term for Simpson's rule is proportional to h^5 from which we can see that on the order of 100 evaluations of the integrand will be needed to restrict our error to approximately 10^{-10}. Since series expansions and recurrence relations typically succeed with less than 1/10 that amount of computation, we should be reluctant to use numerical quadratures as our standard technique whenever the functional subroutine is to be used frequently. Still it will be useful as an alternate computational method for checking any more efficient algorithm we may derive and, should our more efficient techniques fail to work for some small set of parameter values, we just might be forced to fall back upon numerical quadratures there. We have already commented upon the necessity for validation by alternate computation, and the relative simplicity of numerical quadratures makes them attractive for that purpose. They may be expensive ways to get answers but you can be rather sure the answers they give are right.

When b is zero and x lies on $(0, 1)$ — i.e., when our point P lies *on* the condenser plate — the argument of the logarithm in J_1 becomes zero when ξ is equal to x and so the integrand there is infinite. We must somehow remove this singularity if our numerical quadrature technique is to be usable when b is zero. Moreover, we need to remove this singularity *even if b is small* because the standard numerical quadrature methods assume the geometry of the integrand is well-fitted by a polynomial. The presence of a singularity just outside the region of integration gives a distinctly nonpolynomic shape to the integrand that requires many further subdivisions before the polynomial approximation can yield the desired accuracy. Thus efficiency requires the removal of the near-singularity even if it is not computationally necessary for correct numerical results.

Given the usual human tendency to ignore unpleasant difficulties whenever possible, it is not surprising that most engineers will prefer to forget our logarithmic problems for small b — dealing only with the unavoidable case of b equal to zero which can more easily be handled by a special derivation. It is, however, possible to give a treatment that succeeds both for b small and b zero. While mathematically simple, it seems not to be commonly known. Since it also points the way to other more efficient algorithms, we give it here.

The basic tool for dealing with a logarithmic singularity in an integral is to introduce a factor that goes to zero at the same place. Usually this can be done by integration by parts with a *judicious choice for the constant of integration*. In elementary calculus courses we are so thoroughly conditioned by the standard treatment of ln x, singular at the origin, by

$$\int_0 \ln x \, dx = [x \ln x]_0 - \int_0 x \frac{1}{x} dx = x \, ln \, x - x$$

that we seldom think to deal with

$$\int_a \ln (x - a) \, dx$$

by the same device

$$\int_a \ln (x - a) dx = [(x - a) \ln (x - a)]_a - \int_a (x - a) \frac{1}{x - a} dx$$

yet for many practical integrals, this is the useful trick. We take our integral in the form

$$J_3 = \int_0^{\pi/2} \cos^{2n-1}\theta \ln\,[BQ]\cos\theta\,d\theta$$

where

$$BQ = (b^2+x^2)^2 + 2(b^2-x^2)\cos^2\theta + \cos^4\theta$$
$$= [\,b^2 + (x-\cos\theta^2)\,][\,b^2 + (x+\cos\theta)^2\,]$$

and $\cos\theta_x = x$
hence $\sin\theta_x = \sqrt{1-x^2}$
in the sequel.

We integrate by parts to obtain

$$\{(\sin\theta - \sin\theta_x)\cos^{2n-1}\theta\ln\,[BQ]\}_0^{\pi/2}$$

$$+ \int_0^{\pi/2} (\sin\theta - \sin\theta_x)(\sin\theta)\left\{(2n-1)\cos^{2n-2}\theta\ln[BQ]\right.$$

$$\left.+\cos^{2n-1}\theta\frac{[4(b^2-x^2)+4\cos^2\theta]\cos\theta}{BQ}\right\}d\theta$$

where we have chosen our factor $(\sin\theta - \sin\theta_x)$ to become zero exactly where the biquadratic argument, BQ, of the logarithm goes to zero if b^2 is zero. Simplifying, we obtain

$$J(x,n,b) = \sqrt{1-x^2}\ln\{[\,b^2 + (x-1)^2\,][\,b^2 + (x+1)^2\,]\}$$

$$+ (2n-1)\int_0^{\pi/2}\cos^{2n-2}\theta\sin\theta(\sin\theta - \sin\theta_x)\ln\,[BQ]\,d\theta$$

$$+ 4\int_0^{\pi/2}\cos^{2n-1}\theta\sin\theta\frac{(\cos^2\theta - x^2 + b^2)(\sin\theta - \sin\theta_x)}{[\,b^2 + (x-\cos\theta)^2\,][\,b^2 + (x+\cos\theta)^2\,]}\,d\theta$$

and now we can employ numerical quadratures on the two integrals. Both integrands maintain human-sized values throughout the range of integration, although both become indeterminate when θ equals θ_x and b is zero. But that set of special parameters can always be tested for and the limiting values supplied by the subroutine that does the evaluation. For the first integral, the critical value is zero. Although it takes a bit of work to demonstrate it, the other integral gives $-0{\cdot}5\cos^{2n-1}\theta_x$ when b is zero and θ approaches θ_x.

Some series that work for distant points

Whenever the argument of the logarithm is sufficiently removed from zero, a series replacement for that function permits explicit analytic integration to give an attractive computational algorithm. Several variations of this technique are possible, differing in their computational efficiency by factors of 2 to 4. Conceptually simplest, we rewrite J_3 as

$$J_4(x, n, b) = \ln (b^2 + x^2)^2 \int_0^{\pi/2} \cos^{2n}\theta \, d\theta + \int_0^{\pi/2} \cos^{2n}\theta$$

$$\times \ln \left[1 + \frac{2(b^2 - x^2)\cos^2\theta + \cos^4\theta}{(b^2 + x^2)^2} \right] d\theta \qquad (5)$$

and then use the series

$$\ln (1 + \epsilon) = \epsilon - \epsilon^2/2 + \epsilon^3/3 \cdots$$

to generate a series in $\cos^2\theta$ for the logarithm of the second integral. The resultant integrals, all of the form

$$I_{2k} = \int_0^{\pi/2} \cos^{2k}\theta \, d\theta = \frac{2k-1}{2k} I_{2k-2} = \frac{2k-1}{2k} \frac{2k-3}{2k-2} \cdots \frac{1}{2} \frac{\pi}{2}$$

do not contain the parameters x^2 and b^2. They all can be generated at the beginning and stored for efficient use. The distressing aspects of this algorithm are its lack of convergence whenever the denominator $b^2 + x^2$ becomes less than unity and its rapidly growing complexities as higher terms of the logarithmic series are taken — ϵ^n requiring a binomial expansion.

This same device may be used on the trigonometric version of J_1 to produce a series that converges slightly more slowly — the practical difference being a need to evaluate about two more terms for comparable accuracy.

A (slightly) more efficient series

We observe that ξ^{2n} on $(0,1)$ is very small over most of the range if n is at all large. Thus most of the contribution to our integral J_2 must occur in the vicinity of ξ equal to unity, which corresponds to θ equal to zero. But we have expanded our logarithm in even powers of the *cosine* — an expansion around $\pi/2$. We might expect a somewhat better representation (which implies fewer terms of the series for a given precision) if we expand in even powers of the sine. The change is easily made. We rewrite the biquadratic argument BQ as

$$BQ = E^2 + 2G(1 - \sin^2\theta) + (1 - \sin^2\theta)^2$$

$$BQ = (E^2 + 2G + 1) - 2(G + 1)\sin^2\theta + \sin^4\theta$$

whence

$$\ln BQ = \ln{(E^2+2G+1)} + \ln\left[1 - \frac{2(G+1)\sin^2\theta}{E^2+2G+1} + \left(\frac{\sin^2\theta}{E^2+2G+1}\right)^2\right]$$

(6)

and the further manipulations follow as before except that the integrals are now of the type

$$I_{n,k} = \int_0^{\pi/2} \cos^{2n}\theta \sin^{2k}\theta\,d\theta$$

A trial of this algorithm confirmed that it did, indeed, converge faster than (5) — about 1 or 2 terms faster for the precision we require. Typically it required 14 terms instead of 15 — scarcely worth it.

A more efficient series

In order to reduce the length of our series we heeded Professor Lanczos' frequent advice to seek an orthogonal expansion. We wanted a represent-ation of our logarithm in terms of Chebyshev polynomials* of b^2 and x^2 rather than powers of these parameters. Some searching in AMS55[1] revealed (22.9.8) a useful expansion. We get

$$\ln{[1 - 2u\zeta + \zeta^2]} = -2\sum_{k=1}^\infty \frac{T_k(u)}{k}\zeta^k$$

We can write J_2 as

$$2\ln{(b^2+x^2)}\int_0^1 \frac{\xi^{2n}}{\sqrt{1-\xi^2}}\,d\xi + \int_0^1 \frac{\xi^{2n}}{\sqrt{1-\xi^2}} \ln\left\{1 - 2\frac{(x^2-b^2)}{R}\frac{\xi^2}{R} + \left(\frac{\xi^2}{R}\right)^2\right\}d\xi$$

with

$$R^2 = b^2 + x^2$$

and then identify $(x^2-b^2)/R$ with u and ξ^2/R with ζ to give

$$J_5(x,n,b) = 2I_{2n}\ln{(b^2+x^2)} - 2\sum_{k=1}^\infty \frac{T_k(u)}{k}\frac{I_{2k}}{R^k}$$

(7)

Computationally and aesthetically this series is much more satisfying than the one for J_4. Not only is it simpler in its derivation and structure, but its use is extremely efficient because of the standard three-term recur-rence scheme for evaluating any finite series of Chebyshev polynomials.

---

* Lanczos has written about these polynomials so much that one student confused his name with still another spelling of Chebyshev.

Thus, given a point (x,b) we may compute the coefficients I_{2k}/kR^k until they become sufficiently small, then evaluate u, and finally the finite series we have just generated — without ever evaluating any of the $2k$ Chebyshev polynomials themselves. The device is "well known" — meaning that most people who need it have never heard about it. We give it here. To evaluate the finite Chebyshev series

$$S = C_0 T_0(u) + C_1 T_1(u) + \cdots + C_8 T_8(u)$$

in which the $\{C_n\}$ and u have known values we calculate a sequence of numbers $\{b_n\}$ via the recurrence

$$b_n = -b_{n+2} + (2u)b_{n+1} + C_n \qquad n = 8,7, \cdots, 0$$

starting with

$$b_9 = b_{10} = 0$$

Then

$$S = b_0 - b_1 u$$

This algorithm requires $3k + 6$ arithmetic operations where k is the degree of the highest Chebyshev polynomial. It is, of course, a specific instance of a general algorithm that evaluates any finite series of functions having a three-term recurrence relation.

A second Chebyshev series

We can expect a slightly more efficient Chebyshev series if we first express the algorithm in terms of $\sin^2(\theta)$ as in (6). Here we obtain

$$\ln \left\{ 1 - \frac{2(G+1)\eta^2}{E^2 + 2G + 1} + \frac{\eta^4}{E^2 + 2G + 1} \right\} \equiv \ln [1 - 2u\zeta + \zeta^2]$$

and we identify ζ with $\eta^2/(E^2 + 2G + 1)^{1/2}$ and u with $(G+1)/(E^2 + 2G + 1)^{1/2}$. The final expansion contains the integrals

$$I_{n,k} = \int_0^{\pi/2} \cos^{2n}\theta \, \sin^{2k}\theta \, d\theta$$

which can be precomputed or generated recursively as needed. The result is

$$J_6(x,n,b) = I_{n,0} \ln (E^2 + 2G + 1) - 2 \sum_{k=1}^{\infty} \frac{T_k(u)}{k} \frac{I_{n,k}}{(E^2 + 2G + 1)^{k/2}} \qquad (8)$$

where

$$E = b^2 + x^2$$
$$G = b^2 - x^2$$

as before.

Regions of convergence

The two series that involve the Chebyshev expansion of the logarithm are useful in somewhat different regions of the (x, b) plane. The series (7) requires $b^2 + x^2$ to be greater than unity, where (8) requires $(b^2 + x^2)^2 + 2(b^2 - x^2) + 1$ to be greater than unity. The regions where these two expansions fail are shown in Figure 2, from which we see that the latter expansion has a noticeably larger utility, permitting its use even inside a considerable part of the unit circle. This fact is helpful later when we come to check on the validity of our recurrence relations. They often diverge outside the unit circle and the region of overlap is distinctly desirable. (Of course we always have numerical quadratures for purposes of validation, but for higher precision they are decidedly expensive of computational labor. Agreement between two relatively cheap practical schemes is much more comforting.)

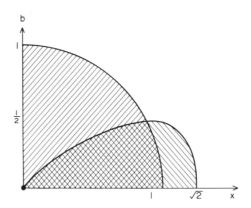

FIGURE 2. Region of Divergence for the Two Chebyshev Series.

Recurrence relations (a first attempt)

Since our integral simplifies considerably when n is zero, we sought to compute J for higher values of n via recurrence relations. Our integration by parts to tame the logarithmic singularity has already pointed one path to such relations. For values of (b, x) away from the condenser plate we have no logarithmic singularity in the integrand and hence can proceed rather more simply. We write J_2 as

$$J_{2n} = J_2(x, n, b) = \int_0^{\pi/2} \cos^{2n-1}\theta \ln [BQ] \cos \theta \, d\theta$$

$$= \left[\sin\theta \cos^{2n-1}\theta \ln[BQ] \right]_0^{\pi/2} + \int_0^{\pi/2} \sin^2\theta \Big\{ (2n-1) \cos^{2n-2}\theta \ln[BQ]$$

$$+ \frac{[4(b^2-x^2)+4\cos^2\theta]\cos^{2n}\theta}{BQ} \Big\} d\theta$$

The first term disappears at both limits for all positive n, so we obtain

$$J_{2n} = (2n-1)(J_{2n-2}-J_{2n}) + 4\int_0^{\pi/2} \cos^{2n}\theta \, \frac{(G+\cos^2\theta)(1-\cos^2\theta)}{E^2+2G\cos^2\theta+\cos^4\theta} \, d\theta$$

hence

$$J_{2n} = \left(1-\frac{1}{2n}\right) J_{2n-2} + \frac{2}{n}\int_0^{\pi/2} \cos^{2n}\theta\left[-1+\frac{(G+1)\cos^2\theta+(E^2+G)}{BQ}\right] d\theta$$

so

$$J_{2n} = \left(1-\frac{1}{2n}\right) J_{2n-2} + \frac{2}{n}\left[-I_{2n}+(G+1)H_{2n+2}+(E^2+G)H_{2n}\right] \qquad (9)$$

where

$$H_{2n} = \int_0^{\pi/2} \frac{\cos^{2n}\theta \, d\theta}{[b^2+(x-\cos\theta)^2][b^2+(x+\cos\theta)^2]} \quad \text{and} \quad I_{2n} = \int_0^{\pi/2} \cos^{2n}\theta \, d\theta.$$

Rewriting the fraction in H_{2n} as

$$\cos^{2n-4}\theta - \frac{2G\cos^{2n-2}\theta + E^2\cos^{2n-4}\theta}{\cos^4\theta+2G\cos^2\theta+E^2}$$

we immediately have the recurrence for H_{2n}

$$H_{2n} = I_{2n-4} - 2G\,H_{2n-2} - E^2\,H_{2n-4}. \qquad (10)$$

The recurrences (9) and (10) require almost no computational labor after starting values are found. We shall spare the reader the details of finding H_0 and H_2. Suffice it here to remark that both of them factor into two integrals, each with a quadratic in $\cos\theta$ for its denominator. The standard substitution of $\tan(\theta/2)$ for the dummy variable converts each of these into biquadratic denominators, which factor, and a further separation of each into two integrals by partial fractions leaves us with a total of four quite similar integrals that have the elementary forms

$$\int_0^1 \frac{a \pm bu}{1 \pm cu + u^2} \, du$$

that can be evaluated explicitly.

J_0 is different. We have

$$J(x, 0, b) = \int_0^{\pi/2} \ln\{[b^2 + (x - \cos\theta)^2][b^2 + (x + \cos\theta)^2]\}\, d\theta$$

$$= \int_{-1}^1 \frac{\ln[b^2 + (x - \xi)^2]}{\sqrt{1 - \xi^2}}\, d\xi = 2\pi\, Re\left\{\ln\left[\frac{x + ib + \sqrt{(x + ib)^2 - 1}}{2}\right]\right\}$$

from Bierens de Haan. Note that for the special case of b equal to zero we obtain two forms, according to whether x lies on the plate ($|x| \leq 1$) or on its extension ($x > 1$).

$$J(x, 0, 0) = \begin{cases} -2\pi \ln 2 & |x| \leq 1 \\ 2\pi \ln \dfrac{x + x^2 - 1}{2} & |x| > 1 \end{cases}$$

The value of $J(x, 0, 0)$ for points on the plate is *independent of* x — a result that surprised your author both mathematically and physically.

A disadvantage of this whole recurrence scheme is its breakdown when x is zero because of the forms taken and by H_0 and H_2. An analogous and simpler recurrence is easily derived for this special case but the need for it rankles. Similarly, when b is zero we need *two* more recurrences, one for $|x|$ less than unity, (a point on the condenser plate) and another for $|x|$ greater than unity — but here the trouble arises because of the appearance of the logarithmic singularity in the original integral rather than an artifact of a formal solution.

Stability of the recurrence

The recurrence (10) for H_{2n} has coefficients that depend on b and x so it should be examined to find those regions of the (b,x) plane where it is stable. The functions H_{2n} certainly *decrease* monotonically with increasing n (for all b and x) since their numerators contain the factor $(\cos\theta)^{2n}$ while the denominator is independent of n and never negative.

The error in H_{2n} satisfies the homogeneous difference equation

$$\epsilon_{2n} + 2G\epsilon_{2n-2} + E^2\epsilon_{2n-4} = 0$$

whose characteristic equation is

$$p^2 + 2Gp + E^2 = 0 \tag{11}$$

If the error is not to grow exponentially as n increases (and thereby swamp our decreasing function H_{2n}) both roots of (11) must lie inside or at least on the unit circle. We have

$$p = -G \pm \sqrt{G^2 - E^2} = x^2 - b^2 \pm \sqrt{-4b^2x^2}$$

so

$$p = x^2 - b^2 \pm i2bx$$
$$|p|^2 = (b^2 + x^2)^2$$

Thus we see that the interior of the unit circle in the (b,x) plane is the stable region for this recurrence.

To see how rapidly the instability might hurt us, we ran a set of values, all for x at 0·4 but with b at 0·8, 1·2, 1·6, and 2·5. The number of correct significant figures at several values of n are shown in Table I. We note that even outside the region of strict stability, the recurrence is usable for a while provided we begin with several extra (guarding) digits. Thus, with two series adequate for larger values of the b and x, and a set of recurrences that work inside and even somewhat outside the first quadrant of the unit circle, with good agreement in a region of overlap, and with some spot checks via numerical quadratures, we felt reasonably confident about our ability to evaluate the integral J_1.

Table I
The number of significant figures given by the recurrence when x is 0·4.

n	0·8	1·2	1·6	2·5
2	12	12	13	12
4	12	12	12	11
6	12	12	10	9
8	12	12	10	8
10	12	11	9	6

The equation system

Having the integrals J_1 now well in hand, we returned to the original problem and set up the linear equations for the coefficients A_k of the charge series. We used three terms and got respectable results for such a small number of fitted parameters. Our test was a point off the end of the condenser plate in a region where the potential is rapidly varying. The correct value was 60 volts and we obtained 59·999939. The matrix of that equation was composed of human-sized numbers and its determinant was 127·6.

Next we tried seven terms of the series (4). The determinant of our equation system shrank alarmingly to 10^{-8}, but the potential improved to 60·000000498. Finally we tried eleven terms. The matrix was still composed of numbers that generally had magnitudes between 0·1 and 2·0 but the determinant was now 10^{-28}. And the potential at our test point was no longer respectable ($-634\cdot7$).

The smallness of the determinant is a symptom of near linear dependence between the equations of their system. This arises because the functions of our series expansion, ξ^{2n} for $n = 1,2, \cdots$, are nearly linearly dependent on the range $(-1,1)$ — famous theorems to the contrary notwithstanding. The theorems, of course, are correct. — the $\{\xi^{2n}\}$ are *not* *exactly* linearly dependent, but the discrepancies between ξ^{20} and the best approximation to ξ^{20} by the linear combination

$$\sum_{k=0}^{9} \beta_k \xi^{2k}$$

are so small that on a 14 decimal machine they might as well be zero — i.e., ξ^{20} *is*, for most computational purposes, *linearly dependent upon the lower even powers of* ξ.

This phenomenon is, of course, well-known. It merely is not usually heeded until it intrudes, unpleasantly and decisively, into the computational process. Like many before him, your author had gambled that he could obtain satisfactory results with few enough terms so that the effect would not become controlling. He lost.

The cure is also well-known. One expands the unknown charge, $g(\xi)$, in a series of orthogonal functions. Here the obvious expansion uses Chebyshev polynomials. Having discovered algorithms for computing J_1 it is a simple matter to derive analogous methods for evaluating integrals of the form

$$\int_0^{\pi/2} \cos 2n\theta \ln [BQ] \, d\theta$$

that have $\cos 2n\theta$ where before we had $\cos^{2n} \theta$.

But before going on to discuss the two interesting developments, we jump ahead to report on the results of the equation system that we obtained when we expanded the charge in the orthogonal functions. The matrix looked much the same as before — the numbers were quite reasonably scaled. The determinant, however, of the equation system was drastically different. For 3, 7, and 11 terms in the Chebyshev expansion, the determinants of the corresponding equation system were 204, 914, and 136 respectively. And the voltages at our test point converged nicely to 60 as we fitted more terms.

The Other Integral

After demonstrating the ultimate futility of expanding the charge function $g(\xi)$ in even powers of ξ, we turned, somewhat reluctantly, to representing it as a series of Chebyshev polynomials. Our reluctance was based on the realization that it would be very difficult to evaluate

$$C_2(x, n, b) = \int_0^1 \frac{T_{2n}(\xi)}{\sqrt{1-\xi^2}} \ln\left[b^2 + (x-\xi)^2\right] d\xi$$

by numerical quadratures if n were at all large. The rapid oscillations of the Chebyshev polynomials would cause large arithmetic cancellations and would require a large number of ordinates. For practical purposes, we would have to verify our computational algorithms without the help of Simpson's rule. Fortunately by this time we knew how to find series expansions and recurrence relations, and there was a realistic chance that agreement within their joint regions of validity would provide the verification necessary to any computer programs. The derivations differ only slightly from the corresponding ones for the first integral, so we will not give them here — except for one that produced an interesting computational problem. It was an unstable recurrence that could be stabilized.

The reluctant recurrence

Our fundamental integral containing the Chebyshev polynomial $T_{2n}(\xi)$ of degree $2n$ is

$$C_1(x, n, b) = \int_{-1}^1 \frac{T_{2n}(\xi)}{\sqrt{1-\xi^2}} \ln\left[b^2 + (x-\xi)^2\right] d\xi$$

and it has the corresponding biquadratic form

$$C_2(x, n, b) = \int_0^1 \frac{T_{2n}(\xi)}{\sqrt{1-\xi^2}} \ln\left\{\left[b^2 + (x-\xi)^2\right]\left[b^2 + (x+\xi)^2\right]\right\} d\xi$$

as well as both trigonometric forms that remove the singularities at the ends of the range of integration.

$$C_3(x, n, b) = \int_0^\pi \cos 2n\theta \ln\left[b^2 + (x-\cos\theta)^2\right] d\theta$$

$$C_4(x, n, b) = \int_0^{\pi/2} \cos 2n\theta \ln\left[E^2 + 2G\cos^2\theta + \cos^4\theta\right] d\theta$$

$$\left\{ \begin{aligned} E &= b^2 + x^2 \\ G &= b^2 - x^2 \end{aligned} \right\}$$

Integrating C_4 by parts we have

$$C_4 = \left[\frac{\sin 2n\theta}{2n}\ln[BQ]\right]_0^{\pi/2} + \frac{1}{2n}\int_0^{\pi/2}\sin 2n\theta\,\frac{(4G\cos\theta+4\cos^3\theta)\sin\theta}{BQ}\,d\theta$$

$$= \frac{1}{2n}\int_0^{\pi/2}\frac{\sin 2n\theta}{BQ}2\sin 2\theta\,(G+\cos^2\theta)\,d\theta$$

$$= \frac{1}{2n}\int_0^{\pi/2}\frac{[\cos(2n-2)\theta-\cos(2n+2)\theta]}{BQ}(G+\cos^2\theta)\,d\theta$$

Hence

$$C_4 = \frac{1}{2n}\left[(H_{2n-2}-H_{2n+2})+G(K_{2n-2}-K_{2n+2})\right]$$

where

$$K_{2n} = \int_0^{\pi/2}\frac{\cos 2n\theta\,d\theta}{E^2+2G\cos^2\theta+\cos^4\theta}$$

$$H_{2n} = \int_0^{\pi/2}\frac{\cos 2n\theta\cos^2\theta\,d\theta}{E^2+2G\cos^2\theta+\cos^4\theta}$$

so that we have now to find recurrences for H_{2n} and K_{2n}.

If we start with the recurrence relation for the even cosines

$$(4\cos^2\theta-2)\cos 2n\theta = \cos(2n+2)\theta+\cos(2n-2)\theta \qquad (12)$$

then divide by our biquadratic factor and integrate, we immediately have

$$4H_{2n} = K_{2n+2}+2K_{2n}+K_{2n-2} \qquad (13)$$

Multiplying (12) by $\cos^2\theta$, dividing by (BQ), and integrating gives

$$H_{2n+2}+H_{2n-2} = \int_0^{\pi/2}\cos 2n\theta\,\frac{4\cos^4\theta-2\cos^2\theta}{\cos^4\theta+2G\cos^2\theta+E^2}\,d\theta$$

$$= \int_0^{\pi/2}\cos 2n\theta\left[4-\frac{(8G+2)\cos^2\theta+4E^2}{BQ}\right]d\theta$$

$$= -(8G+2)H_{2n}-4E^2K_{2n}$$

and we obtain

$$-4E^2K_{2n} = H_{2n+2}+(8G+2)H_{2n}+H_{2n-2} \qquad (14)$$

which together with (13) forms an interlocking pair of recurrences for H_{2n} and K_{2n}.

Unfortunately, direct use of this pair of recurrences in the direction of increasing n is unstable — violently so if E or G is large. For example, when b is $2\cdot5$ and x is $0\cdot8$ we have $G=5\cdot61$ and $E^2=47\cdot4721$. Then the K_{2n} computed from (13) and (14) are the numbers shown in Table II where the correct values are given just underneath for comparison. Although the

Table II
Values of K_{2n} via the {unstable/stable}
Recurrence

$2n$	K_{2n}	
0	0·29577 47389 59	-1
0	0·29577 47389 59	-1
2	$-0·16918\ 86709\ 06$	-2
	$-0·16918\ 86709\ 06$	-2
4	0·62198 28329 96	-4
	0·62198 28247 05	-4
6	$-0·15774\ 92068\ 99$	-5
	$-0·15774\ 51539\ 06$	-5
8	0·18652 52279 19	-7
	0·17380 55683 98	-7
10	$-0·26693\ 43750\ 57$	-7
	0·85498 09568 53	-9
12	0·26054 78396 79	-6
	$-0·69351\ 23142\ 82$	-10

starting values for the recurrences are correct to at least 10 significant figures, the first application of the recurrence reduces the number of correct figures to 8, and the next to 5. At the third (K_8) there are about 2 figures correct. At K_{10} not only are there *no* significant figures left, even the exponent is wrong! Indeed, the errant behavior of the recurrence was spotted because of the exponents. Both H_{2n} and K_{2n} should decrease monotonically with n (as is evident from the strongly oscillatory behavior of their numerators and the lack of dependence on n of their denominators). Our numbers decrease through n equal to 4 and thereafter increase. The correct values, as always in engineering computations, were not available at that time. And in the heat of battle for computer time, theory was temporarily forgotten. The numbers were, however, uncompromising—the recurrence was unstable, and it was unstable even for $b^2 + x^2$ *within* the unit circle.

Stabilizing the recurrence

Since both K_{2n} and H_{2n} should decrease in magnitude and since our recurrence relations are linear, our first instinct might be to try to run them backwards, starting with large n. The interlocking, however, causes problems with this strategy, since it is necessary to assume values for both H_{2n} and K_{2n} that are wrong by unknown and presumably different scale factors. To avoid this difficulty we might try to use a recurrence for K_{2n} alone. We therefore multiply equation (14) by 4 and substitute for $4H_{2n}$ throughout. We get

$$K_{2n+4} + (4 + 8G)K_{2n+2} + (6 + 16G + 16E^2)K_{2n} + (4 + 8G)K_{2n-2} + K_{2n-4} = 0$$

(15)

—which does not look overly promising as we now need 4 starting values.

At this point, however, a remarkable fact comes to light*—the characteristic polynomial of this difference equation (15) is a *reciprocal polynomial* of 4th degree—i.e., it has two pairs of roots that are reciprocals of one another. This property follows immediately from the symmetry of the coefficients about the middle term. The general solution for (15) is thus

$$K_{2n} = A_1(r_1)^{2n} + A_2(r_2)^{2n} + A_3(1/r_1)^{2n} + A_4(1/r_2)^{2n}$$

To get *decreasing* values of K when $|r_1|$, $|r_2| > 1$ it is necessary that the coefficients A_1 and A_2 be zero. Rounding error guarantees in practice that they are not, and hence these terms assume greater and greater prominence as our recurrence computations progress—soon completely swamping the decreasing terms that are the correct solution. For proper behavior we want the two roots *inside* the unit circle and we emphatically do not want the others. If we find these two roots, we may construct the *quadratic* equation having these roots and then we can immediately write a three term recurrence for K_{2n} that should be stable. Indeed, we could write down the general solution but the recurrence is apt to be the more efficient computational algorithm. The derivation of the stable 3-term recurrence is given in the Appendix. The numerical values from this recurrence are the stable values, shown in Table II, good to at least 12 significant figures.

If we derive the 5-term recurrence for H_{2n} we find it identical to that for K_{2n}—only the starting values are different. Thus the stable 3-term recurrence for H is also identical to that for K.

How the integrals should have been evaluated

Some time after the algorithms given above were implemented and the physical problem solved, your author found simpler algorithms for J_1 and C_1. An important formal simplification of the problem occurs if we

---

*(Or at least it did to H. F. Trotter, to whom I showed this form of the recurrence!)

make use of the observation that

$$2 \operatorname{Re}\{\ln (u+iv)\} = \ln (u^2 + v^2)$$

to express our integrals as

$$J_1(z,n) = 2 \operatorname{Re} \int_{-1}^{1} \frac{\xi^{2n}}{\sqrt{1-\xi^2}} \ln (z-\xi) \, d\xi \, \Bigg|$$

$$z = x + ib$$

$$C_1(z,n) = 2 \operatorname{Re} \int_{-1}^{1} \frac{T_{2n}(\xi)}{\sqrt{1-\xi^2}} \ln (z-\xi) \, d\xi \, \Bigg|$$

Note that the integrations are still real. The complex variable z is only a convenient packaging of our parameters x and b—to be unpacked at the end. From these forms it is possible to derive another complete set of algorithms, both series expansions for large parameter values and recurrences for smaller ones. These are simpler because, loosely speaking, expressions that used to be quartic are now quadratic and previous quadratic expressions are now linear. The recurrences that were unstable before are still unstable, still have a symmetric polynomial for a (now quadratic) characteristic equation, and are still reducible to a stable (now one term!) recurrence. It was this last discovery that finally led us to seek the "correct" evaluation technique, since the one-term recurrence for C_1 showed it to be proportional to

$$\frac{1}{n} \operatorname{Re} (\rho^2)^n$$

where ρ was the smaller root of the quadratic

$$\rho^2 - 2z\rho + 1 = 0$$

Surely such a simple result ought to be derivable more or less directly. It is! The crucial tool has been used earlier: the expansion of the logarithm in a series of Chebyshev polynomials.

We write

$$C_2 = 2 \operatorname{Re} \int_{-1}^{1} \frac{T_{2n}(\xi)}{\sqrt{1-\xi^2}} \ln \left[\frac{1 - 2\xi p + p^2}{2p} \right] d\xi$$

and identify

$$\frac{1}{2p} + \frac{p}{2} - \xi \equiv z - \xi$$

so that

$$p^2 - 2zp + 1 = 0 \tag{16}$$

Then we may write

$$C_2 = -2 \operatorname{Re} \ln (2p) \int_{-1}^{1} \frac{T_{2n}(\xi)}{\sqrt{1-\xi^2}} \, d\xi$$

$$-4 \sum_{k=1}^{\infty} \operatorname{Re} \left(\frac{p^k}{k}\right) \int_{-1}^{1} \frac{T_{2n}(\xi)}{\sqrt{1-\xi^2}} T_k(\xi) \, d\xi$$

where the orthogonality relations between the Chebyshev polynomials guarantee that only one term will survive. Thus we finally obtain

$$C_2(x,n,b) = \begin{cases} -\pi [\ln 4 + \ln (p_{\text{real}}^2 + p_{\text{imag}}^2)], & n = 0 \\ -\dfrac{\pi}{n} \operatorname{Re} (p^2)^n & , & n \neq 0 \end{cases} \qquad (18)$$

In this form C_2 is stably evaluable provided we take care:
(i) to evaluate p from (16) sensibly, and then
(ii) to evaluate $\operatorname{Re} (p^2)^n$ carefully.
the object being to avoid subtractions of nearly equal quantities. For the quadratic this is easy. We merely write

$$p = \frac{1}{z + \sqrt{z^2 - 1}} \qquad (19)$$

(Why is the quadratic formula

$$x = \frac{-2a}{b \pm \sqrt{b^2 - 4ac}}$$

never taught in school? It is the only useful form when an apparent quadratic

$$ax^2 + bx + c = 0$$

has a small value for a—it is really a *perturbed linear equation*, a form that very frequently occurs in physical problems.)

The evaluation of the real part of p^{2n} requires that it be cast into the polar form

$$\operatorname{Re}(p^{2n}) = r^{2n} \cos 2n\theta \qquad (20)$$

while computational efficiency dictates that we start with r^2 and also that we evaluate the successive $\cos n(2\theta)$ *via* their recurrence. With these *caveats* the computational process (19), (20), (18) is completely stable for all (b, x) in the first quadrant. It also suffers no unpleasantness at the points previously crucial: the end of the plate $(0, 1)$ and its center $(0, 0)$.

Finally we clearly can evaluate our original integral J_1 by these same methods. The only additional step is the replacement of ξ^{2n} by its representation as a finite series of Chebyshev polynomials. Then we have a double summation in (17) from which a single *finite* series emerges each term of which has the form (18). These terms are combined linearly with some binomial coefficients from the Chebyshev expansion of ξ^{2n}. It's all very simple, afterwards!

APPENDIX

Factorization of a reciprocal polynomial

The symmetric polynomial

$$r^4 + Ar^3 + Br^2 + Ar + 1 = 0$$

is to be factored into two quadratics of the form

$$(r^2 + Tr + S)(r^2 + T/S + 1/S) = 0$$

The original polynomial has two pairs of reciprocal roots, as may be seen via the substitution $x = 1/r$ to produce precisely the same polynomial. Our strategy is to factor this quartic into two quadratics with mutually reciprocal roots—i.e., one having both roots inside the unit circle and the other having the roots that are outside. Such quadratics have identical coefficients but in the opposite order. On multiplying out and comparing coefficients, we have

$$T + T/S = A \quad \text{or} \quad T(1 + 1/S) = A \tag{21}$$

and

$$1/S + S + T^2/S = B = 1/S + S + \frac{A^2}{S(1 + 1/S)^2}$$

hence

$$B = (1/S + S) + \frac{A^2}{(S + 1/S) + 2}$$

If we now let

$$y = S + 1/S \tag{22}$$

then

$$B = y + \frac{A^2}{y + 2} \tag{23}$$

and we may solve first for y from (23) then for S from (22) and finally get T from (21). The required recurrence in our problem is then

$$K_{2n+2} + TK_{2n} + SK_{2n-2} = 0$$

provided the solutions of the several quadratics have been chosen to obtain the smaller roots. To guarantee the smaller root takes a little care. We rewrite (22) as

$$S^2 - yS + 1 = 0$$

hence

$$S = y/2 \pm \sqrt{(y/2)^2 - 1} = R \pm \sqrt{R^2 - 1}$$

To make S the smaller root, choose the $-$ sign, but compute S *via* the relation

$$S = \frac{1}{y/2 + \sqrt{(y/2)^2 - 1}}$$

to avoid the subtraction.

REFERENCES

1. Abramowitz, M. and Stegun, I. A., *Handbook of Mathematical Functions* (AMS 55), U.S. National Bureau of Standards, Washington, 1964.
2. Acton, F. S., and J. Barkley Rosser, An Iterative Algorithm for Calculating Potentials Near Small Groups of Finite Charged Plates, Mathematical Research Center Report 1120, University of Wisconsin, December 1970.
3. Lanczos, C., *Applied Analysis*, Prentice-Hall, Englewood Cliffs, 1956.
4. Love, A. E. M., Some Electrostatic Distributions in Two Dimensions, *Proc. London Math. Soc.*, Ser. 2, **22** (1923), 337–369.

The Hypercircle Method

J. L. SYNGE

Dublin Institute for Advanced Studies, Dublin, Ireland

I. SIMPLIFICATION

Imagine a flat piece of country. Imagine in it two straight narrow roads which intersect at right angles. Imagine two blind men set down by helicopter, one on each road. Provide them with a long measuring tape, stretching from the one man to the other man. Let the tape be embossed, so that they can read it, even though blind. Without moving, let them tighten the tape and read it. Suppose it reads 100 yards. Thus, the men know that they are 100 yards apart.

But that is not the question we are to put to them. We are to ask each of them how far he is from the cross-roads. These men, though blind, are very intelligent, and they know that the hypotenuse of a right-angled triangle is greater than either of the other two sides. So each of the men says that his distance from the cross-roads is less than (or possibly equal to) 100 yards.

But we have a second question: relative to the pair of you, where might the cross-roads lie? These men are also aware that the angle in a semi-circle is a right angle, and so they answer at once: *the cross-roads must lie on the circle having the stretched tape for diameter* (Fig. 1).

That circle is the prototype of the hypercircle. To pass from this too-simple model to the general case, we are to make certain generalisations:

1. We are to replace the flat piece of country (a Euclidean 2-space) by a Hilbert space or function-space. We need only a *real* Hilbert space, which is in fact a Euclidean space of infinite dimensions.

2. We are to replace the two straight roads by linear subspaces of the Hilbert space, say L' and L''. These two subspaces are to have only one point in common (the cross-roads) and they are to be *completely*

201

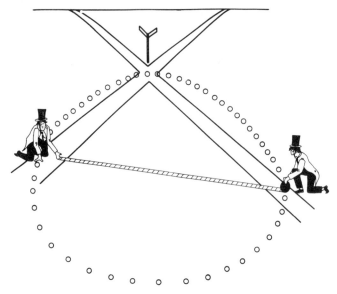

FIGURE 1. Two blind men and their "hypercircle".

orthogonal. That means that any vector lying in L' is orthogonal (in the Hilbertian sense) to any vector lying in L''.†

3. We are to replace the two blind men by two points, one in L' and the other in L'', say S' and S''. The measuring tape measures the distance from S to S'' in the sense appropriate to Hilbert space.

4. If we now ask how far S' and S'' lie from the (unique) intersection of L' and L'', the answer is that these two distances are less than the distance from S' to S''; a right-angled triangle in Hilbert space is precisely the same as any familiar right-angled triangle.

5. If we ask where the intersection lies in Hilbert space, the answer is that it lies on a hypersphere having for diameter the straight line joining S' and S''. Note that I say "hypersphere" at this point; the hypercircle will come later.

---

† The two roads satisfy this condition, but it is not altogether easy to see what happens when we increase the dimensionality to three. True, we can satisfy the condition of complete orthogonality by taking our linear subspaces to be straight lines in 3-space intersecting at right angles. But planes will not work. Two planes cannot intersect in a single point; they are either parallel or they intersect in a line; and if they are orthogonal in the ordinary sense, they are not *completely* orthogonal, the line common to them not being at right angles to itself. The geometry of Hilbert space demands some intuitions rather different from what we are accustomed to in elementary geometry.

II. AN EXAMPLE: THE DIRICHLET PROBLEM

Let me illustrate the method by applying it to the problem of Dirichlet in a plane. Let B be a simple closed curve (Fig. 2). We seek a function harmonic in the domain V bounded by B and taking prescribed values on B.

We have to distinguish clearly between two geometries. First, there is the geometry of the plane in which the problem is formulated, and I shall call that plane *P*-space (*P* for *physical*, since in contemplating the Dirichlet problem we probably have some physical situation in mind). Secondly, there is a Hilbert space or function-space, not yet introduced; I shall call it *F-space* (*F* for *function*). Thus we shall have *P*-points and *F*-points, *P*-vectors and *F*-vectors.

The Dirichlet problem is stated as follows:

$$\Delta u = 0, \quad (u)_B = f. \tag{2.1}$$

We have no intention of solving that problem, i.e. obtaining some exact expression for the function u. We are interested rather in finding out some things about the solution, which we know exists under suitable conditions of smoothness imposed on the curve B and the function f.

You might think that the proper procedure would be to let a vector in *F*-space correspond to a function u in *P*-space. But that would not do at all. We must be more subtle.

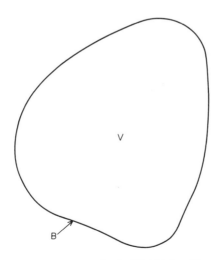

FIGURE 2. *P*-space for the Dirichlet problem.

Let p_i (subscripts taking the values 1,2)† be any vector field in P-space. We are to regard this vector field as a point or vector **S** in F-space, and we indicate this correspondence by writing

$$\mathbf{S} \leftrightarrow p_i.$$

If we add another vector field p_i', we get in F-space the vector

$$\mathbf{S} + \mathbf{S}' \leftrightarrow p_i + p_i'. \tag{2.2}$$

If we multiply p_i by a real number a, we get

$$a\mathbf{S} \leftrightarrow ap_i. \tag{2.3}$$

Here we invoke only the properties of a linear vector F-space. It becomes a Hilbert space only when we have defined a scalar product satisfying certain algebraic conditions. Thus, given two vector fields, p_i and p_i', corresponding to the F-vectors **S** and **S'**, how are we to define the scalar product of **S** and **S'**?

It transpires that the definition suitable to the Dirichlet problem is this:

$$\mathbf{S}.\mathbf{S}' = \int p_i p_i' \, dV, \tag{2.4}$$

with integration over the domain V of P-space.‡ Note that the integrand is itself the usual scalar product (in P-space) of two vectors, summation (Einstein's convention) being understood when a suffix is repeated. It is important to note that

$$\mathbf{S}^2 = \mathbf{S}.\mathbf{S} = \int p_i p_i \, dV, \tag{2.5}$$

which is positive-definite, vanishing only if $p_i = 0$. This positive-definiteness is an essential property of Hilbert space.

To get back to the Dirichlet problem. We have to *split* it in such a way as to recognise that the solution is the unique point of intersection of two completely orthogonal linear subspaces. This is done as follows. It is a matter of *relaxation*.

Consider the following formulae:

$$L': \quad \mathbf{S}' \leftrightarrow p_i', \quad p_i' = u_{,i}', \quad (u')_B = f. \tag{2.6}$$

$$L'': \quad \mathbf{S}'' \leftrightarrow p_i'', \quad p_{i,i}'' = 0. \tag{2.7}$$

---

† For simplicity, I take the Dirichlet problem in a plane; if we use indicial notation, it would not complicate matters at all to take in space of three or more dimensions. Heavy type is reserved for F-points or F-vectors.

‡ I find the notation **S.S'** briefer and more suggestive than the usual notation of Hilbert space, viz. (**S, S'**).

The formulae (2.6) invite us to choose any function u' satisfying the boundary condition as in (2.1), and form a vector field p'_i by taking its gradient (the comma indicates partial differentiation with respect to the coordinates); finally we are to consider F-vectors $\mathbf{S}'$ corresponding to p'_i. On the other hand (2.7) invites us to take any vector field p''_i with vanishing divergence and consider the corresponding F-vector $\mathbf{S}''$. The symbols L' and L'' indicate that we have before us two subspaces of F-space.

There are two important things about these subspaces. First, they have an F-point in common, and only one. To test this, put

$$u' = u, \quad p'_i = p''_i = p_i, \quad \text{say}. \tag{2.8}$$

Then it is clear that u satisfies (2.1), and we know that the solution of the Dirichlet problem is unique.

But there is more to it than that: the subspaces L' and L'' are linear subspaces. But what is a linear subspace? We need to delve a little into the geometry of F-space.

Just as in ordinary geometry, it is necessary to have clear ideas about points and vectors, and to recognise how useful it is to talk of free vectors and bound vectors. If we have two F-vectors, $\mathbf{S}'$ and $\mathbf{S}''$, we may think of them as the position vectors of two points, relative to the origin. What then is the straight line joining these points? It is, of course, the totality of points

$$\mathbf{S} = a'\mathbf{S}' + a''\mathbf{S}'', \quad a' + a'' = 1, \tag{2.9}$$

a' and a'' being two real numbers with unit sum. Here we merely take over into F-space a familiar formula of ordinary space.

A *linear subspace* is such that all points of the straight line joining any two points in it are in the subspace. That is a definition. Let us see whether L' and L'', as in (2.6) and (2.7), are linear subspaces.

To test L', consider two points in it:

$$\mathbf{S}' \leftrightarrow p'_i, \quad p'_i = u'_{,i}, \quad (u')_B = f,$$
$$\hat{\mathbf{S}}' \leftrightarrow \hat{p}'_i, \quad \hat{p}'_i = \hat{u}'_{,i}, \quad (\hat{u}')_B = f. \tag{2.10}$$

Let a', $\hat{a}'$ be any two real numbers. Then

$$a'\mathbf{S}' + \hat{a}'\hat{\mathbf{S}}' \leftrightarrow a'p'_i + \hat{a}'\hat{p}'_i = p_i, \quad \text{say}. \tag{2.11}$$

Define

$$u = a'u' + \hat{a}'\hat{u}'. \tag{2.12}$$

Then, if $a' + \hat{a}' = 1$, as demanded by (2.9), we have

$$a'\mathbf{S}' + \hat{a}'\hat{\mathbf{S}}' \leftrightarrow p_i = u_{,i}, \quad (u)_B = f \tag{2.13}$$

and this states that the point $a'\mathbf{S}' + \hat{a}'\hat{\mathbf{S}}'$ is in L'. The result is proved: L' is a linear subspace. Likewise (or more simply) L'' is a linear subspace.

(Note that the numbers with which we multiply F-vectors, as in (2.9) or (2.11), are always *constants*.)

We have now found that L' and L'' are linear subspaces with a unique point of intersection. But are they completely orthogonal, in the sense that any vector lying in L' is orthogonal to any vector lying in L''?

Of course this does not mean anything until we define "orthogonal" and "lying in." The first is very simple: two vectors $\mathbf{S}$ and $\mathbf{S}'$ are orthogonal if their scalar product vanishes:

$$\mathbf{S}.\mathbf{S}' = 0. \tag{2.14}$$

As for 'lying in', let $\mathbf{S}$ and $\mathbf{S}'$ be two *points* in a linear subspace, these being their position vectors relative to the origin. Then $\mathbf{S} - \mathbf{S}'$ is an F-vector *lying in* the linear subspace, provided we regard it as a bound vector emanating from $\mathbf{S}$ or $\mathbf{S}'$. (This merely carries over into F-space a commonplace of ordinary space.)

Referring to (2.10), we recognise

$$\mathbf{T}' = \mathbf{S}' - \hat{\mathbf{S}}' \tag{2.15}$$

as an arbitrary vector lying in L'. It is clear that

$$\mathbf{T}' \leftrightarrow p_i' - \hat{p}_i', \quad p_i' - \hat{p}_i' = (u' - \hat{u}')_{,i}, \quad (u' - \hat{u}')_B = 0. \tag{2.16}$$

In words, the F-vector $\mathbf{T}'$ which lies in L' corresponds to a P-vector field which is the gradient of a scalar field which vanishes on the boundary B. We may express (2.16) in the form

$$\mathbf{T}' \leftrightarrow q_i', \quad q_i' = v_{,i}', \quad (v')_B = 0. \tag{2.17}$$

Referring to the definition of L'' in (2.7), we see that any F-vector lying in L'' may be written

$$\mathbf{T}'' \leftrightarrow q_i'', \quad q_{i,i}'' = 0; \tag{2.18}$$

in words, it corresponds to a P-vector field with vanishing divergence.

Is $\mathbf{T}'$ as in (2.17) orthogonal to $\mathbf{T}''$ as in (2.18)? Let us try: the scalar product being always of the form (2.4), we have

$$\mathbf{T}'.\mathbf{T}'' = \int q_i' q_i'' \, dV = \int v_{,i}' q_i'' \, dV = 0, \tag{2.19}$$

on integrating by parts and using (2.17) and (2.18).

We have thus *split the problem. The solution* $\mathbf{S}$ *is the unique intersection of two completely orthogonal linear subspaces, L' and L'', defined as in (2.6) and (2.7).*

The origin of F-space corresponds to a zero P-vector field. Note then that L'' contains the origin, but L' does not (since $f \neq 0$). This is a rather accidental property of this particular problem.

III. THE PAY-OFF IN THE DIRICHLET PROBLEM

What has been done above might interest geometers, but so far there is no indication how, by thus splitting the Dirichlet problem, we are in any way informed about the solution. But let us transcribe the splitting formulae (2.6), (2.7):

$$L': S' \leftrightarrow p'_i, \quad p'_i = u'_{,i}, \quad (u')_B = f. \tag{3.1}$$

$$L'': S'' \leftrightarrow p''_i, \quad p''_{i,i} = 0.$$

It is very easy to find points on L'', for all we need is a P-vector field with vanishing divergence; no boundary condition need be satisfied. For example, we might choose any simple regular harmonic function u'' and set $p''_i = u''_{,i}$. Or we might set p''_i equal to the curl of an arbitrary vector field. As for L', we can take any (sufficiently smooth) function u' satisfying the boundary condition, and set p'_i equal to its gradient. One has an almost embarrassing freedom in choosing points in L' and L''.

But having got one point S' in L' and one point S'' in L'', what do we do next? At this point let us recall the two blind men in Section 1.

We have three points, S' in L', S'' in L'', and the solution S: we do not know what S is, but we do know that it is in L' and in L''. Moreover, the vectors $(S-S')$ and $(S-S'')$ lie in L' and L'' respectively, and so are orthogonal:

$$(S-S').(S-S'') = 0. \tag{3.3}$$

This is information about the unknown S, and indeed it is all the information we can extract from two chosen points, S' in L' and S'' in L''.

The simplest interpretation of (3.3) is to draw a diagram (Fig. 3),

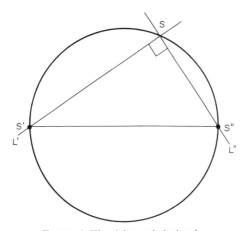

FIGURE 3. The right-angled triangle.

showing the three points **S**, **S′**, **S″**. Since these points are contained in a linear subspace of only two dimensions, it is quite permissible to draw such a diagram, although the points in it correspond, not to *P*-points but to *F*-points, i.e. to functions. This triangle has a right angle at **S**, and so we have the inequalities†

$$(\mathbf{S}-\mathbf{S}')^2 < (\mathbf{S}'-\mathbf{S}'')^2,$$
$$(\mathbf{S}-\mathbf{S}'')^2 < (\mathbf{S}'-\mathbf{S}'')^2. \tag{3.4}$$

The right hand side of (3.4) is a calculable number:

$$(\mathbf{S}'-\mathbf{S}'')^2 = \int (p_i'-p_i'')(p_i'-p_i'')\,\mathrm{d}V. \tag{3.5}$$

Thus (3.4) gives us information about **S** in integral form: if we put

$$\mathbf{S} \leftrightarrow p_i,\, p_i = u_{,i},\, (u)_B = f, \tag{3.6}$$

so that *u* is the solution of the Dirichlet problem, we have

$$(\mathbf{S}-\mathbf{S}')^2 = \int (u_{,i}-u_{,i}')(u_{,i}-u_{,i}')\,\mathrm{d}V,$$
$$(\mathbf{S}-\mathbf{S}'')^2 = \int (u_{,i}-p_i'')(u_{,i}-p_i'')\,\mathrm{d}V. \tag{3.7}$$

Substitution in (3.4) yields some information about *u*. But this information does not appear of much interest. It is better perhaps to recognize, in general terms, that (3.4) represent bounds on errors in a mean square sense.

But where is the hypercircle, which even the blind men in Section I were able to deduce? Here is the little trick. Using the distributive and commutative properties of the scalar product, we may write (3.3) in the form

$$\mathbf{S}^2 - \mathbf{S}.(\mathbf{S}'+\mathbf{S}'') + \mathbf{S}'.\mathbf{S}'' = 0. \tag{3.8}$$

Define the *F*-point

$$\mathbf{C} = \tfrac{1}{2}(\mathbf{S}'+\mathbf{S}''); \tag{3.9}$$

we recognize **C** as the mid-point of the line joining **S′** and **S″**. Then we have

$$\mathbf{S}^2 - 2\mathbf{S}.\mathbf{C} = -\mathbf{S}'.\mathbf{S}'', \tag{3.10}$$

or

$$(\mathbf{S}-\mathbf{C})^2 = R^2,\quad R^2 = \mathbf{C}^2 - \mathbf{S}'.\mathbf{S}'' = \tfrac{1}{4}(\mathbf{S}'-\mathbf{S}'')^2. \tag{3.11}$$

Now if we use **X** to denote a variable point in *F*-space, we recognize in

$$(\mathbf{X}-\mathbf{C})^2 = R^2 \tag{3.12}$$

the equation of a *hypersphere* with centre **C** and radius *R*. Thus once we have found a point on *L′* and a point on *L″*, we can locate the (unknown) solution **S** on a certain hypersphere (Fig. 4).

---

† We may disregard possible signs of equality, because they could occur only if we were lucky enough to hit **S** right on the nose in choosing **S′** or **S″**!

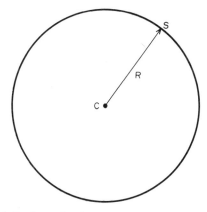

FIGURE 4. Solution located on hypersphere with centre C and radius R.

What is the scientific status of diagrams such as Figures 3 and 4? They are, if you like, merely symbolic sketches to help us follow the formulae. But actually they are more than that, for they may be regarded as accurate drawings of *projections* of elements in (infinite-dimensional) F-space on a linear 2-space contained in it, at least as far as points and lines are concerned. The circles representing hyperspheres are more symbolic, but if the interiors of the circles were shaded in, the shaded areas would be projections of the interiors of the hyperspheres.

IV. MORE GENERAL VIEW. THE HYPERCIRCLE

In offering a brief account of the hypercircle method, I deliberately avoid a closely reasoned logical approach. There is not time for that, and anyway the human mind does not perceive things in an ordered methodical way. If you ask me what I regard as the essential thing in this method, I would say without hesitation that it offers a geometrical interpretation of formulae which otherwise may appear as a muddle. The geometry is that of real Hilbert space, or, equivalently for our purposes, function-space or F-space. This geometry is the geometry of a Euclidean space with an infinite number of dimensions. The art of using this geometry consists in avoiding an essential feature — the infinity of dimensions — and to use, as far as possible, simple geometrical intuitions in this wider setting. Those who are not geometrically susceptible may find the whole thing a bore (preferring to deal with formulae as they come, without geometrical overtones), but I think it likely that there are those who appreciate the geometrical approach, even if only from an aesthetic standpoint.

In the last two Sections I have used a specific and famous problem as an illustration—the Dirichlet problem in a plane. But if you look at the final formulae, (3.8) to (3.12), you will observe that the Dirichlet problem has sunk out of sight. We are concerned with the unique intersection of two linear subspaces, completely orthogonal to each other. I introduced the Dirichlet problem in order to show that, when it comes to exploring the solutions of boundary-value problems, I was not talking through my hat.

Having thus established my credentials as a practically minded person, I can afford to look at the geometry directly, keeping the Dirichlet problem in the background as an illustration to be used if needed. And now I shall not put you off with a hypersphere—I shall get a *hypercircle*.

We have then an infinite-dimensional F-space, equipped with a scalar product so that it is a Hilbert space. And since we are dealing only with real functions, it is a *real* Hilbert space, a much easier thing to think about geometrically than the complex Hilbert space used in quantum theory.

In this F-space there exist two completely orthogonal linear subspaces, L' and L'', which have a unique point of intersection S. This F-point S is the solution of the problem, whatever that problem may be. We do not know what S is, nor do we hope to find it (although we might set up a systematic infinite process to this end). Our aim is rather to get information about S, to establish certain bounds.

We can, as in the Dirichlet problem, find points in L' and points in L''. Let us then find m points in L' and n points in L'': we may denote them as follows:

$$\text{In } L': \quad \mathbf{S}'_r, \quad r = 1, 2, \cdots m.$$
$$\text{In } L'': \quad \mathbf{S}''_s, \quad s = 1, 2, \cdots n. \tag{4.1}$$

It is a good idea to make a crude diagram to follow the argument (Fig. 5),

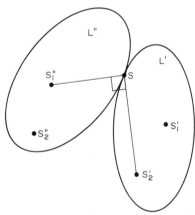

FIGURE 5. The completely orthogonal linear subspaces L', L'' and points on them.

and so we indicate in some symbolic way the linear subspaces L', L'', their unique intersection S, and the points as set out in (4.1).

Since the vector $S - S_r'$ lies in L' and $S - S_s''$ in L'', and the subspaces are completely orthogonal, we have at once a set of mn formulae

$$(S - S_r') \cdot (S - S_s'') = 0, \quad r = 1, 2, \cdots m; s = 1, 2, \cdots n. \tag{4.2}$$

Any one of these formulae puts S on a hypersphere, with centre and radius given by

$$C_{rs} = \tfrac{1}{2}(S_r' + S_s''), \quad R^2 = \tfrac{1}{4}(S_r' - S_s'')^2. \tag{4.3}$$

Now if, in ordinary geometry, we had a point lying on two spheres, we would at once simplify the description by saying that it lies on a circle, the intersection of the two spheres. We can do this in F-space also.

To carry this out, let us put $r = s = 1$ in (4.2). This gives us the hypersphere

$$(S - S_1') \cdot (S - S_1'') = 0. \tag{4.4}$$

(Although S is a certain point — the unknown solution — we regard it here as a variable F-point, so that (4.4) is not to be regarded as a relation between three points, but rather as the equation of a hypersphere. If this causes confusion, then change S to X in the argument.) Let us rewrite (4.4):

$$S^2 - S \cdot (S_1' + S_1'') + S_1' \cdot S_1'' = 0. \tag{4.5}$$

Now take $r = 1$, $s = 2$, and get

$$S^2 - S \cdot (S_1' + S_2'') + S_1' \cdot S_2'' = 0 \tag{4.6}$$

Subtract this from (4.5):

$$-S \cdot (S_1'' - S_2'') + S_1' \cdot (S_1'' - S_2'') = 0, \tag{4.7}$$

or

$$(S - S_1') \cdot (S_1'' - S_2'') = 0. \tag{4.8}$$

We have been playing with the distributive property of the scalar product, but rather stupidly, since we could have written down (4.8) at once: it merely says that the vector $S - S_1'$ (which lies in L') is orthogonal to $S_1'' - S_2''$ (which lies in L'')!

There is an essential difference between (4.4) and (4.8): (4.4) is *quadratic* in S (and represents a sphere, S being regarded as a variable point), whereas (4.8) is *linear* in S and represents — what? We need a name for such a locus in F-space (S being of course a variable point).

Consider the equation

$$(X - A) \cdot B = 0, \tag{4.9}$$

where X is a variable F-point and A, B given F-vectors. If you met this formula in 3-dimensional geometry, the dot indicating the usual scalar

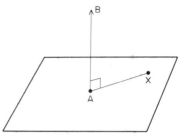

FIGURE 6. A hyperplane: the vector $(\mathbf{X} - \mathbf{A})$ is orthogonal to the vector $\mathbf{B}$.

product, you would say at once that it represents a certain plane, viz. the plane drawn through the point $\mathbf{A}$ at right angles to the vector $\mathbf{B}$ (Fig. 6). In F-space we might call (4.9) a plane, but this word has a two-dimensional implication, and (since (4.9) represents an infinite dimensional subspace) it is better to call it a *hyperplane*. It is actually a linear subspace, since all points of a straight line joining two points of it are in it.

A *hypercircle* is naturally defined as the intersection of a hypersphere with one or more hyperplanes. Thus, with $\mathbf{X}$ as variable F-point, the equations of a hypercircle are of the form

$$(\mathbf{X} - \mathbf{C})^2 = R^2,$$
$$(\mathbf{X} - \mathbf{A}_1).\mathbf{B}_1 = 0, \qquad (\mathbf{X} - \mathbf{A}_2).\mathbf{B}_2 = 0, \ldots \tag{4.10}$$

where R is some given number and $\mathbf{C}$, $\mathbf{A}$, $\mathbf{B}$ are given F-points (or vectors, if you like). But a word of warning. In Euclidean 3-space, a given sphere and a given plane may not intersect; likewise, if the data in (4.10) are improperly chosen, the hypercircle may not exist.

Returning now to (4.4), (4.8) and similar equations, we recognise that if we choose a number of points on the two completely orthogonal linear subspaces L', L'', *we can locate their (unknown) intersection* $\mathbf{S}$ *on a hypercircle*, the equations of the hypercircle being (4.4) (or something equivalent) and a number of (hyperplane) equations of the general form (4.8).

V. HISTORICAL NOTE

Mathematicians solve problems. They prove theorems. But, almost more important, they reduce complicated situations to simplicity by looking at those situations from a suitable angle, perhaps an unconventional angle. The hypercircle method originated in 1946 when W. Prager and I were discussing some rather complicated inequalities in the theory of elasticity.

We were delighted to find that these inequalities became much easier to understand in terms of the geometry of function-space.

It took me some time to see that this hypercircle method was a general method, not at all tied to the theory of elasticity, and indeed the scope of the method only became clear when A. J. McConnell showed, in 1951, its connection with the calculus of variations. In due course, with assistance from V. G. Hart, I wrote a book on the subject† and relevant references will be found there‡. But I should make it clear that during the past fifteen years I have made no systematic effort to follow whatever work may have been done along these or parallel lines. However, from time to time some questions have come up and I have been forced to re-read my own book in order to deal with them. This has made me realise how easily simple basic ideas get lost in the maze of details which the author of a book feels he should put in, in case they might be useful sometime. Therefore I welcome this opportunity to give a brief and deliberately sketchy account of the hypercircle method in the hope that some readers may feel the charm of this geometrical approach.

VI. THE SIGNIFICANCE OF A SMALL HYPERSPHERE

It may be said that the hypercircle method is concerned with boundary value problems for linear partial differential equations of certain types. It provides a useful intuition to establish certain inequalities, crisply and cleanly. It also provides the intuition for setting up a systematic approximation to the solution, establishing upper and lower bounds in a mean-square sense.

Suppose we have located the solution **S** on a hypersphere with centre **C** and radius R:

$$(\mathbf{S} - \mathbf{C})^2 = R^2. \tag{6.1}$$

Can we set bounds on $\mathbf{S}^2$ or $|\mathbf{S}|$, defined as the square root of the number $\mathbf{S}^2$?

Forget about F-space. Think of an ordinary sphere. Suppose R is less than $|\mathbf{C}|$. Then the origin lies outside the sphere, and it is clear intuitively

---

†J. L. Synge, *The hypercircle in mathematical physics*. Cambridge University Press 1957.

‡Let me add three: J. L. Synge, La géométrie élémentaire de l'espace fonctionnel avec des applications à la physique classique. *Colloque Henri Poincaré, Textes des Conférences,* Ed. provisoire (mimeographed) Paris 1954; V. G. Hart, *The method of the hypercircle in the solution of the biharmonic equation,* Thesis, National University of Ireland 1957; S. Haber, Numerical evaluation of multiple integrals, *SIAM Review* **12** (1970) 481–526, in particular pp. 504–505.

that

$$|\mathbf{C}| - R < |\mathbf{S}| < |\mathbf{C}| + R. \qquad (6.2)$$

On the other hand, if R is greater than $|\mathbf{C}|$, then the origin lies inside the sphere, and we have

$$R - |\mathbf{C}| < |\mathbf{S}| < R + |\mathbf{C}|. \qquad (6.3)$$

No one would dream of *proving* these inequalities in ordinary space, for they are intuitively obvious. But are they true in F-space? The power of the method lies in the fact that certain ordinary intuitions are valid in F-space. But which? That calls for a little experience, and, if one is in doubt, there is nothing for it but *proof*. So let us see.

The trick is simple. Define an F-vector $\mathbf{A}$ by

$$\mathbf{S} - \mathbf{C} = R\mathbf{A}, \qquad (6.6)$$

so that, by (6.1), $\mathbf{A}$ is a unit vector, $\mathbf{A}^2 = 1$. Now carry $\mathbf{C}$ to the right hand side in (6.6) and square:

$$\mathbf{S}^2 = \mathbf{C}^2 + 2R\mathbf{C}.\mathbf{A} + R^2. \qquad (6.7)$$

Next we appeal to the Schwarz inequality, based on the fact that in Hilbert space the norm is positive-definite, so that, for any F-vectors $\mathbf{P}, \mathbf{Q}$ and any real number x, we have

$$(\mathbf{P} - x\mathbf{Q})^2 \geqslant 0. \qquad (6.8)$$

But this is a quadratic in x:

$$x^2\mathbf{Q}^2 - 2x\mathbf{P}.\mathbf{Q} + \mathbf{P}^2 \geqslant 0, \qquad (6.9)$$

and so we have the Schwarz inequality:

$$(\mathbf{P}.\mathbf{Q})^2 \leqslant \mathbf{P}^2\mathbf{Q}^2 \quad \text{or} \quad |\mathbf{P}.\mathbf{Q}| \leqslant |\mathbf{P}|\,|\mathbf{Q}|, \qquad (6.10)$$

where the first modulus sign is interpreted in the ordinary way, the others as indicated after (6.1). Thus (6.7) gives

$$\mathbf{C}^2 - 2R\,|\mathbf{C}| + R^2 \leqslant \mathbf{S}^2 \leqslant \mathbf{C}^2 + 2R\,|\mathbf{C}| + R^2, \qquad (6.11)$$

and from this (6.2) and (6.3) follow on taking square roots and sharpening the inequalities. But what a bother simply because we could not trust our intuition!

Suppose now that we have located $\mathbf{S}$ on a *small* hypersphere, meaning by this that R is small. Then (6.2) holds, and $|\mathbf{S}|$ is clipped between two adjacent bounds; or, if you prefer it, by (6.11), $\mathbf{S}^2$ is clipped between two adjacent bounds. But what does this mean in terms of a boundary value problem?

In the Dirichlet problem,

$$\mathbf{S}^2 = \int u_{,i}u_{,i}\,dV, \qquad (6.12)$$

with integration over the domain V in which u is the unknown harmonic function satisfying the boundary condition. Once we have located **S** on a hypersphere, this integral is clipped between bounds, and, if the radius R of that hypersphere is small, then those bounds lie close to one another. That means that we have in C^2 a good approximation to the integral S^2. We do not get *pointwise* bounds (that is a much harder proposition), only bounds in this square-integral sense.

If we have located **S** on a hypercircle (and not merely on a hypersphere), there is more information about the solution **S**, but there is no point in going into that here. The general idea in approximating is to locate **S** on a *small* hypersphere.

VII. VERTICES

Let us now pretend that we are engaged on a serious effort to get a good approximation (always in the mean square sense). That means that we are trying to locate the solution **S** on a hypercircle of small radius. (Here the Dirichlet problem may be thought of as an illustration, but the argument is general.)

We assume the existence of two completely orthogonal linear subspaces L', L'', with **S** their unique point of intersection. Let us choose a point $\mathbf{S}_0'$ on L' and a point $\mathbf{S}_0''$ on L''. Next, choose r more points on L' and s more points on L''. It is convenient to use the following notation:

$$\begin{aligned} \text{points on } L': \quad & \mathbf{S}_0', \mathbf{S}_0' + \mathbf{T}_\rho' \ (\rho = 1, 2, \cdots r), \\ \text{points on } L'': \quad & \mathbf{S}_0'', \mathbf{S}_0'' + \mathbf{T}_\sigma'' \ (\sigma = 1, 2, \cdots s). \end{aligned} \tag{7.1}$$

Thus the F-vectors $\mathbf{T}_\rho'$ lie in L' and $\mathbf{T}_\sigma''$ lie in L'', and so, since L' and L'' are completely orthogonal, we have

$$\mathbf{T}_\rho' \cdot \mathbf{T}_\sigma'' = 0, \tag{7.2}$$

for the ranges of ρ and σ shown above.

Let us tidy up the vectors $\mathbf{T}_\rho'$, and also $\mathbf{T}_\sigma''$, by orthonormalisation, a routine operation. Then we have

$$\begin{aligned} \mathbf{T}_\rho' \cdot \mathbf{T}_\mu' &= \delta_{\rho\mu} \quad (\rho, \mu = 1, 2, \cdots r) \\ \mathbf{T}_\sigma'' \cdot \mathbf{T}_\nu'' &= \delta_{\sigma\nu} \quad (\sigma, \nu = 1, 2, \cdots s), \end{aligned} \tag{7.3}$$

while (7.2) of course still holds.

In ordinary 3-space, two points determine a line (linear 1-space) and three points determine a plane (linear 2-space), these being subspaces of the 3-space. Likewise the $(r+1)$ points in L', as in (7.1) or after orthonormalisation, determine a linear subspace of r dimensions, say L_r'; likewise the $(s+1)$ points in L'' determine a linear subspace of s dimensions, say L_s''. It is most important to distinguish between L' (of infinite dimensions) and L_r' (of only r dimensions); likewise between L'' and L_s''. L_r' is contained in L' and L_s'' in L''.

How closely do L' and L'' approach one another, in the sense of the metric of F-space, defined by the scalar product? A silly question! They meet at $\mathbf{S}$ and nowhere else. Try again: How closely do L_r' and L_s'' approach one another? This is a subtler question. If L_r' and L_s'' had a common point, that point would be common to L' and L'' — it would in fact be the solution $\mathbf{S}$ and nothing else. It would be an extraordinary stroke of luck if we had chosen our points so skilfully.

So let us lay aside the possibility that the finite-dimensional linear subspaces L_r' and L_s'' intersect. Let us rather seek their points of closest approach, say $\mathbf{V}'$ on L_r' and $\mathbf{V}''$ on L_s''; we call $\mathbf{V}'$ and $\mathbf{V}''$ *vertices*.

The general points on L_r' and L_s'' may be written

$$\mathbf{S}' = \mathbf{S}_0' + \sum_{\rho=1}^{r} a_\rho' \mathbf{T}_\rho', \quad \mathbf{S}'' = \mathbf{S}_0'' + \sum_{\sigma=1}^{s} a_\sigma'' \mathbf{T}_\sigma'', \tag{7.4}$$

where the a's are real numbers. The square of the F-distance between these two points is

$$(\mathbf{S}' - \mathbf{S}'')^2 = (\mathbf{S}_0' - \mathbf{S}_0'' + \sum_{\rho=1}^{r} a_\rho' \mathbf{T}_\rho' - \sum_{\sigma=1}^{s} a_\sigma'' \mathbf{T}_\sigma'')^2. \tag{7.5}$$

Working this out with use of (7.2) and (7.3), we get

$$(\mathbf{S}' - \mathbf{S}'')^2 = \mathbf{A}^2 + 2 \sum_{\rho=1}^{r} a_\rho' \mathbf{T}_\rho' . \mathbf{A} - 2 \sum_{\sigma=1}^{s} a_\sigma'' \mathbf{T}_\sigma'' . \mathbf{A} + \sum_{\rho=1}^{r} a_\rho'^2 + \sum_{\sigma=1}^{s} a_\sigma''^2, \tag{7.6}$$

where

$$\mathbf{A} = \mathbf{S}_0' - \mathbf{S}_0''. \tag{7.7}$$

To find the vertices $\mathbf{V}'$, $\mathbf{V}''$, we are to choose the constants a', a'' so as to minimise (7.6): thus, differentiating, we get

$$a_\rho' = -\mathbf{T}_\rho' . \mathbf{A}, \qquad a_\sigma'' = \mathbf{T}_\sigma'' . \mathbf{A}. \tag{7.8}$$

We are to substitute these values in (7.4) and write $\mathbf{V}'$ for $\mathbf{S}'$ and $\mathbf{V}''$ for $\mathbf{S}''$: the required vertices are

$$\mathbf{V}' = \mathbf{S}_0' - \sum_{\rho=1}^{r} \mathbf{T}_\rho' (\mathbf{T}_\rho' . \mathbf{A}), \quad \mathbf{V}'' = \mathbf{S}_0'' + \sum_{\sigma=1}^{s} \mathbf{T}_\sigma'' (\mathbf{T}_\sigma'' . \mathbf{A}). \tag{7.9}$$

In general terms we may say that this is best we can do with the chosen points (7.1). Although by differentiating we sought only a stationary value, it is easy to see that it is in fact a minimum.

Consider now what we have. First, there is the (unknown) solution $\mathbf{S}$.

Then there are two vertices $\mathbf{V}'$, $\mathbf{V}''$. $\mathbf{V}'$ is a point in L_r' and hence in L', since L_r' is contained in L'. Likewise $\mathbf{V}''$ is a point in L''. Since $\mathbf{S}$ is a point of both L' and L'', and these are completely orthogonal, we have

$$(\mathbf{S} - \mathbf{V}') . (\mathbf{S} - \mathbf{V}'') = 0. \tag{7.10}$$

This may be written in hypersphere form:

$$(\mathbf{S} - \mathbf{C})^2 = R^2, \tag{7.11}$$

where the centre is

$$\mathbf{C} = \tfrac{1}{2}(\mathbf{V}' + \mathbf{V}''), \tag{7.12}$$

the midpoint of the line joining the vertices, and the radius R is given by

$$R^2 = \tfrac{1}{4}(\mathbf{V}' - \mathbf{V}'')^2, \tag{7.13}$$

so that the diameter $2R$ is the distance between the vertices.

No matter how we choose the points (7.1) in L' and L'', we get the hypersphere (7.11), and hence the inequalities as in (6.2), (6.3), (6.11). If we are looking for a good approximation to $\mathbf{S}$, we must make R small, and this will require the use of a great number of points as in (7.1), unless we are extraordinarily lucky in the choice.

As for $\mathbf{C}$ and R, we turn back to (7.9) and get

$$\mathbf{C} = \tfrac{1}{2}\left[\mathbf{S}_0' + \mathbf{S}_0'' - \sum_{\rho=1}^{r} \mathbf{T}_\rho'(\mathbf{T}_\rho'.\mathbf{A}) + \sum_{\sigma=1}^{s} \mathbf{T}_\sigma''(\mathbf{T}_\sigma''.\mathbf{A})\right],$$

$$\mathbf{A} = \mathbf{S}_0' - \mathbf{S}_0''. \tag{7.14}$$

It would be most natural to accept $\mathbf{C}$ as the best available approximation to the solution $\mathbf{S}$. But R is the critical thing — we need it small for a good approximation. So we note that

$$\mathbf{V}' - \mathbf{V}'' = \mathbf{A} - \sum_{\rho=1}^{r} \mathbf{T}_\rho'(\mathbf{T}_\rho'.\mathbf{A}) - \sum_{\sigma=1}^{s} \mathbf{T}_\sigma''(\mathbf{T}_\sigma''.\mathbf{A}), \tag{7.15}$$

and hence

$$4R^2 = \mathbf{A}^2 - \sum_{\rho=1}^{r} (\mathbf{T}_\sigma'.\mathbf{A})^2 - \sum_{\sigma=1}^{s} (\mathbf{T}_\sigma''.\mathbf{A})^2. \tag{7.16}$$

We observe that, as we might expect, the inclusion of additional points in (7.1) steadily reduces the value of R. But we can be assured that R goes to zero when the number of points is infinite only if the vectors $\mathbf{T}'$ and $\mathbf{T}''$ form complete sets. In the book cited earlier, I have shown how so-called *pyramid functions* may be used to get close approximations to the solutions of Dirichlet and Neumann problems in a plane.

Lower Bounds for the Dirichlet Integral

J. J. McMahon

St. Patrick's College,
Maynooth, Ireland

The methods of numerical analysis have been employed to great advantage in approximating solutions to partial differential equations, and in particular to Laplace's equation $\Delta u = 0$. A paper entitled "Lower Bounds for the Electrostatic Capacity of a Cube" by the author in 1953[1], contained a procedure for obtaining lower bounds for the Dirichlet integral $\int_V (\nabla u)^2 dV$ for functions u satisfying Laplace's equation. The methods used in that paper were inspired by the work of J. L. Synge[2,3], and carried out with his encouragement. The purpose of this essay is to represent with some modification the methods of that paper. Again for the sake of comparison with previous results, attention will be confined to the problem of obtaining lower bounds for the electrostatic capacity of a cube.

I. THE PROBLEM

Let a cube be given in R^3, let B denote its surface, and V the part of R^3 exterior to the cube. Let the origin for coordinates (x_1,x_2,x_3) be the centre of the cube, and its edge length be 2, so that

$$B = \{(x_1,x_2,x_3); \max. \{|x_1|, |x_2|, |x_3|\}\} = 1, \tag{1.1}$$

$$V = \{(x_1,x_2,x_3); \max. \{|x_1|, |x_2|, |x_3|\}\} > 1. \tag{1.2}$$

We then seek a function u continuous in $V \cup B$, differentiable in V, and satisfying

$$\Delta u \equiv \frac{\partial^2 u}{\partial x_1{}^2} + \frac{\partial^2 u}{\partial x_2{}^2} + \frac{\partial^2 u}{\partial x_3{}^2} = 0 \quad \text{in } V \tag{1.3}$$

$$(u)_B = 1, \qquad (u)_\infty = O\left(\frac{1}{\sqrt{x_1{}^2 + x_2{}^2 + x_3{}^2}}\right). \tag{1.4}$$

The capacity C is then defined by

$$4\pi C = \int_V (\nabla u)^2 \, dV, \tag{1.5}$$

and we seek lower bounds for C.

II. ESSENTIAL INEQUALITY

Choose any piece-wise continuous and continuously differentiable vector-field $\mathbf{P}''$ in V, which has continuous normal component across the surfaces of discontinuity, which is divergent-free, and is of order $1/(x_1^2 + x_2^2 + x_3^2) = 1/r^2$ at infinity. Thus

$$\operatorname{div} \mathbf{P}'' = \nabla . \mathbf{P}'' = 0 \qquad \text{in } V \tag{2.1}$$

except for a finite number of piece-wise smooth surfaces in V,

$$\mathbf{P}''.\mathbf{n} \text{ is continuous} \tag{2.2}$$

across the surfaces of discontinuity, where $\mathbf{n}$ is normal to surface, and

$$|\mathbf{P}''| = O(1/r^2) \text{ for large } r. \tag{2.3}$$

From the inequality

$$\int_V (\nabla u - \mathbf{P}'')^2 \, dV \geqslant 0$$

it follows that

$$4\pi C \geqslant 2 \int_V \nabla u . \mathbf{P}'' \, dV - \int_V \mathbf{P}'' . \mathbf{P}'' \, dV. \tag{2.4}$$

By Green's theorem, the first term on the right hand side of (2.4) transforms as follows, when we take account of (2.1), (2.2) and (2.3);

$$\int_V \nabla u . \mathbf{P}'' \, dV = \int_B u\mathbf{P}'' . \mathbf{n} \, dB = \int_B \mathbf{P}'' . \mathbf{n} \, dB, \tag{2.5}$$

where $\mathbf{n}$ is the unit normal to B directed into the interior of the cube. Thus we have the inequality

$$4\pi C \geqslant 2 \int_B \mathbf{P}'' . \mathbf{n} \, dB - \int_V \mathbf{P}'' . \mathbf{P}'' \, dV. \tag{2.6}$$

For the vector-field $\mathbf{P}''$, a linear combination of vector-fields $\mathbf{P}''_r$, each satisfying the conditions (2.1), (2.2) and (2.3) will be chosen, so that

$$\mathbf{P}'' = \sum_{r=0}^n c_r \mathbf{P}''_r, \tag{2.7}$$

where the c_r are constants. The aim will be to maximize the expression

$$2 \int_B \mathbf{P}'' . \mathbf{n} \, dB - \int_V \mathbf{P}'' . \mathbf{P}'' \, dV \tag{2.8}$$

by choice of the constants.

III. METHOD OF CONSTRUCTING NORMALLY-CONTINUOUS DIVERGENT-FREE VECTOR-FIELDS

Let $f(x_1, x_2, x_3)$ be a continuous and piece-wise differentiable function defined on R^3. Thus, on a finite number of piece-wise smooth surfaces in R^3 f will not be differentiable, and these surfaces will be called exceptional surfaces. If (a, b, c) is a point on an exceptional surface at which there is a tangent plane to the surface, then f will define locally two functions, f_1 and f_2, where f_1 is the restriction of f to one side of the surface and f_2 its restriction to the other. By continuity we can define partial derivatives for f_1 and f_2 at (a, b, c), but these will not agree, as f is not differentiable at that point. However, if $\mathbf{1}$ is a tangent vector to the exceptional surface at (a,b,c), then

$$\mathbf{1}.\nabla f_1(a,b,c) = \mathbf{1}.\nabla f_2(a,b,c). \tag{3.1}$$

To see this, note that since the normal $\mathbf{n}$ is defined at (a,b,c), it is possible to represent locally the surface with one variable as a differentiable function of the other two. For instance, suppose that near the point (a,b,c) the exceptional surface is given by $x_3 = h(x_1, x_2)$. Then, since

$$f_1(x_1, x_2, h(x_1, x_2)) = f_2(x_1, x_2, h(x_1, x_2)),$$

we have

$$\lim_{x_1 = a} \frac{f_1(x_1, b, h(x_1, b)) - f_1(a, b, c)}{x_1 - a} = \lim_{x_1 = a} \frac{f_2(x_1, b, h(x_1, b)) - f_2(a, b, c)}{x_1 - a}$$

that is

$$\mathbf{1}.\nabla f_1(a, b, c) = \mathbf{1}.\nabla f_2(a, b, c),$$

where in this particular case $\mathbf{1} = (1, 0, (\partial h/\partial x_1)(a, b, c))$. A similar argument holds for any tangent vector to the surface at the point (a, b, c). Hence, excepting a finite number of curves on the exceptional surfaces, $\nabla f.\mathbf{1}$ will be defined and continuous across these surfaces.

Let g be a second function defined on R^3 with the same properties as f, so that the vector-field $f\nabla g$ is defined and continuously differentiable except on exceptional surfaces in R^3. Then, as may be verified by differentiation,

$$\text{curl } (f\nabla g) = \nabla f \times \nabla g, \tag{3.2}$$

where $\times$ denotes vector product. The vector-field $\nabla f \times \nabla g$ possesses some desirable properties. From its definition, it is defined and continuously differentiable on R^3, except on a finite number of exceptional surfaces. Since div curl of any vector-field is zero, then

$$\text{div } (\nabla f \times \nabla g) = \text{div curl } (f\nabla g) = 0. \tag{3.3}$$

Finally if $\mathbf{n}$ is unit normal to an exceptional surface at the point (a, b, c), then $\mathbf{n}.(\nabla f \times \nabla g)$ is defined at that point and is continuous across the surface. To prove this, take two unit, mutually orthogonal vectors $\mathbf{1}$ and $\mathbf{m}$ tangent to the surface at (a, b, c) and such that $\mathbf{n} = \mathbf{1} \times \mathbf{m}$. Then

$$\mathbf{n}.(\nabla f \times \nabla g) = (\mathbf{1} \times \mathbf{m}).(\nabla f \times \nabla g) = (\mathbf{1}.\nabla f)(\mathbf{m}.\nabla g) - (\mathbf{1}.\nabla g)(\mathbf{m}.\nabla f).$$
(3.4)

Now the right-hand side of (3.4) is defined and continuous across the exceptional surface by (3.1), can be used as the definition of $\mathbf{n}.(\nabla f \times \nabla g)$, and establishes its continuity across the surface.

The method outlined in this paragraph will be used to construct the vector-fields $\mathbf{P}_r''$.

IV. A CLASS OF FUNCTIONS AND VECTOR-FIELDS

Consider the cubic lattice in R^3

$$\{(la, ma, na); a > 0, l, m, n \text{ integers.}\}$$

At a typical lattice point ρ, eight lattice cubes meet. (See Figure 1)

Let (x, y, z) be a local coordinate system for which ρ is the origin and with the x, y and z-axes parallel to the edges of one of the lattice cubes which meet at ρ. Define a continuous and piece-wise continuously differentiable function P_ρ associated with the lattice-point ρ by

$$P_\rho(x, y, z) = \left(1 - \frac{|x|}{a}\right)\left(1 - \frac{|y|}{a}\right)\left(1 - \frac{|z|}{a}\right) \text{ for } 0 \le |x|, |y|, |z| \le a$$

$$P_\rho(x, y, z) = 0 \quad \text{for max. } \{|x|, |y|, |z|\} > a.$$
(4.2)

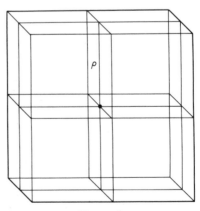

FIGURE 1

Then P_ρ is different from zero only within the eight lattice cubes meeting at ρ, and is continuously differentiable everywhere except on the faces of the eight lattice cubes which meet at ρ.

The symmetry group of the lattice cube $[(0, 0, 0), (a, a, a)]$ is of order 48, and is generated by the following reflections;

$$R_1: x' = a - x, y' = y, z' = z, \tag{4.3}$$

$$R_2: x' = y, y' = x, z' = z, \tag{4.3}$$

$$R_3: x' = x, y' = z, z' = y. \tag{4.4}$$

Any substitution of the symmetry group, when composed with P_ρ, will either leave it invariant or else transform it into $P_{\rho'}$, where ρ' is the vertex into which ρ is transformed by the substitution. For instance, when P_ρ is composed with the reflection R_1, we obtain

$$P_\rho o R_1 (x, y, z) = \left(1 - \frac{|a - x|}{a}\right)\left(1 - \frac{|y|}{a}\right)\left(1 - \frac{|z|}{a}\right),$$

which is the lattice function associated with the point $(a, 0, 0)$.

Although P_ρ is not differentiable on the faces of the eight lattice cubes meeting at ρ, yet it follows from (3.1) that if $\mathbf{1}$ is any vector lying in one of the faces, then $\mathbf{1}.\nabla P_\rho$ is defined and continuous across the face.

If σ is another lattice point and P_σ the associated lattice function, we can obtain, as in (3.2), the vector-field

$$\mathbf{P}''_{\rho\sigma} = \nabla P_\rho \times \nabla P_\sigma = \text{curl} (P_\rho \nabla P_\sigma) = -\mathbf{P}''_{\sigma\rho} \tag{4.5}$$

As $\mathbf{P}''_{\rho\rho}$ is identically zero, we may suppose that $\rho \neq \sigma$. Since ∇P_ρ is zero outside the 8 lattice cubes meeting at ρ, and ∇P_σ is zero outside the 8 lattice cubes meeting at σ, it follows that $\mathbf{P}''_{\rho\sigma}$ differs from zero if and only if the 8 cubes meeting at ρ and the 8 cubes meeting at σ have at least one lattice cube in common. Hence forth, therefore, it will be understood for the vector-field $\mathbf{P}''_{\rho\sigma}$ that ρ and σ are distinct vertices of the same lattice cube.

Three types of vector-fields arise:

1. If ρ and σ lie on the same edge of a lattice cube, then $\mathbf{P}''_{\rho\sigma}$ is non-zero within the four lattice cubes which have the edge $\rho\sigma$ in common.

2. If ρ and σ lie on the same face of a lattice cube, but do not lie on the same edge, then $\mathbf{P}''_{\rho\sigma}$ is non-zero only within the two cubes which have that face in common.

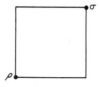

3. If ρ and σ are vertices of the same lattice cube, but do not lie on the same face, then $\mathbf{P}''_{\rho\sigma}$ is non-zero only within that lattice cube to which both ρ and σ belong.

V. PROPERTIES OF $\mathbf{P}''_{\rho\sigma}$

It will be convenient to tabulate the properties of $\mathbf{P}''_{\rho\sigma}$ as follows:

A. $\mathbf{P}''_{\rho\sigma}$ is a vector-field in R^3 which differs from zero only within the lattice cubes which have the vertices ρ and σ in common, and is continuously differentiable except on the faces of these lattice cubes.

B. The normal component $\mathbf{P}''_{\rho\sigma} . \mathbf{n}$ is defined and continuous across the faces of these lattice cubes.

C. div $\mathbf{P}''_{\rho\sigma} = 0$, wherever $\mathbf{P}''_{\rho\sigma}$ is differentiable.

D. If by a reflection in any plane of symmetry of the lattice cube to which both ρ and σ belong, ρ reflects to the vertex ρ' and σ to σ', then $\mathbf{P}''_{\rho\sigma}$ reflects to $-\mathbf{P}''_{\rho'\sigma'} = \mathbf{P}''_{\sigma'\rho'}$.

Property A follows from the definition and discussion in Section 4. of $\mathbf{P}''_{\rho\sigma}$. Properties B and C are an immediate consequence of (3.4) and (3.3) respectively. To establish property D, let R denote the reflection of a point $\mathbf{x} = (x,y,z)$ in the plane $a_1 x + a_2 y + a_3 z + d \equiv \mathbf{a} . \mathbf{x} + d = 0$. Thus

$$R(\mathbf{x}) = \mathbf{x} - 2\frac{(\mathbf{a} . \mathbf{x} + d)}{\mathbf{a}^2}\mathbf{a}. \qquad (5.1)$$

Let R' be the jacobian matrix of R, and so a vector $\mathbf{v}$ reflects in the same plane to the vector $R'(\mathbf{v})$, where

$$R'(\mathbf{v}) = \mathbf{v} - 2\frac{(\mathbf{a} . \mathbf{v})}{\mathbf{a}^2}\mathbf{a}. \qquad (5.2)$$

If **w** is also a vector, then

$$R'(\mathbf{v}) \times R'(\mathbf{w}) = (\mathbf{v} - 2(\mathbf{a}.\mathbf{v}/a^2)\mathbf{a}) \times (\mathbf{w} - 2(\mathbf{a}.\mathbf{w}/a^2)\mathbf{a})$$
$$= \mathbf{v} \times \mathbf{w} - 2/a^2(\mathbf{a} \times [(\mathbf{a}.\mathbf{v})\mathbf{w} - (\mathbf{a}.\mathbf{w})\mathbf{v}])$$
$$= \mathbf{v} \times \mathbf{w} - 2/a^2(\mathbf{a} \times [\mathbf{a} \times (\mathbf{w} \times \mathbf{v})])$$
$$= \mathbf{v} \times \mathbf{w} - 2/a^2(\mathbf{a}.(\mathbf{w} \times \mathbf{v})\mathbf{a} - a^2(\mathbf{w} \times \mathbf{v}))$$
$$= -\mathbf{v} \times \mathbf{w} + 2/a^2(\mathbf{a}.(\mathbf{v} \times \mathbf{w}))\mathbf{a} = -R'(\mathbf{v} \times \mathbf{w}). \quad (5.3)$$

The vector-field $\nabla P_\rho(x)$ reflects in the plane $\mathbf{a}.\mathbf{x} + \mathbf{d} = 0$ to the vector-field $R'(\nabla P_\rho)_oR(x)$. However, it is easily verified by differentiation that $R'(\nabla P_\rho)_oR(x) = \nabla(P_{\rho o}R)(x)$.

Now let R be a symmetry reflection of a lattice cube, and let $R(\rho) = \rho'$. Then, by section 4., $P_{\rho o}R = P_{\rho'}$, and hence $R'(\nabla P_\rho)_oR(x) = \nabla P_{\rho'}(x)$. In this case also, if $R(\rho) = \rho'$ and $R(\sigma) = \sigma'$, then

$$R'(\mathbf{P}''_{\rho\sigma})_oR = R'(\nabla P_\rho \times \nabla P_\sigma)_oR$$
$$= -(R'(\nabla P_\rho) \times R'(\nabla P_\sigma))_oR,$$

by (5.3)

$$= -(R'(\nabla P_\rho)_oR \times R'(\nabla P_\sigma)_oR)$$
$$= -(\nabla P_{\rho'} \times \nabla P_{\sigma'}) = -\mathbf{P}''_{\rho'\sigma'} = \mathbf{P}''_{\sigma'\rho'}. \quad (5.4)$$

This establishes property D.

VI. FORMULAE FOR THE THREE FUNDAMENTAL VECTOR-FIELDS

Consider a typical lattice cube with its local coordinate system, so that the point 1 has coordinates $(0,0,0)$, $2 = (a,0,0)$, $5 = (0,a,0)$, $8 = (a,a,a)$, etc., as indicated in the diagram. From property D) of $\mathbf{P}''_{\rho\sigma}$ it is possible to

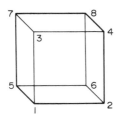

obtain, by an appropriate reflection in a plane of the lattice cube, the vector-fields $\mathbf{P}''_{15}$, $\mathbf{P}''_{13}$, etc., from the vector-field $\mathbf{P}''_{12}$, i.e. to obtain all vector-fields associated with edges of the lattice cube from the vector-field $\mathbf{P}''_{12}$. Similarly from $\mathbf{P}''_{16}$ we can obtain all vector-fields associated with diagonals of faces of the lattice cube, and finally, from $\mathbf{P}''_{18}$ we can obtain the vector-fields associated with transversals of the lattice cube. Hence it suffices to compute the three fundamental vector-fields $\mathbf{P}''_{12}$, $\mathbf{P}''_{16}$, and $\mathbf{P}''_{18}$, and each of the other 28 vector-fields associated with the lattice cube is obtainable from these.

Now

$$a^3 P_1(x,y,z) = (a - |x|)(a - |y|)(a - |z|) \quad \text{for } 0 \leqslant |x|, |y|, |z| \leqslant a,$$

$$a^3 P_2(x,y,z) = (a - |a - x|)(a - |y|)(a - |z|) \quad \text{for } 0 \leqslant |a - x|, |y|, |z| \leqslant a,$$

and $\mathbf{P}''_{12} = \nabla P_1 \times \nabla P_2$ differs from zero only when $0 < x, |y|, |z| < a$. Therefore, in calculating $\mathbf{P}''_{12}$ we may assume that $|x| = x$ and $|x - a| = a - x$. For these values of x,

$$a^3 \nabla P_1(x, y, z) = \left(-(a - |y|)(a - |z|), \frac{-y}{|y|}(a - x)(a - |z|), \right.$$
$$\left. \frac{-z}{|z|}(a - x)(a - |y|) \right)$$

$$a^3 \nabla P_2(x, y, z) = \left((a - |y|)(a - |z|), \frac{-y}{|y|}x(a - |z|), \frac{-z}{|z|}x(a - |y|) \right),$$

with $(\partial/\partial y)(|y|) = y/|y|$ for $y \neq 0$, and similarly for $(\partial/\partial z)(|z|)$. Hence

$$a^5 \mathbf{P}''_{12} = \left(0, \frac{-z}{|z|}(a - |z|)(a - |y|)^2, \frac{y}{|y|}(a - |y|)(a - |z|)^2 \right)$$

$$\text{for } 0 < x, |y|, |z| < a. \tag{6.1}$$

Similarly

$$a^5 \mathbf{P}''_{16} = \left(\frac{z}{|z|}x(a - x)(a - |z|), \frac{-z}{|z|}y(a - y)(a - |z|), (y - x)(a - |z|)^2 \right)$$

$$\text{for } 0 < x, y, |z| < a, \tag{6.2}$$

$$a^5 \mathbf{P}''_{18} = (x(a - x)(z - y), y(a - y)(x - z), z(a - z)(y - x))$$

$$\text{for } 0 < x, y, z < a. \tag{6.3}$$

As already noted, all vector-fields associated with the lattice cube may be obtained from the above three fundamental vector-fields by appropriate reflections. To obtain $\mathbf{P}''_{15}$, for instance, proceed as follows: the reflection R_1 of (4.3) transforms 1 to 1 and 2 to 5, which implies by (5.4) that

$$\mathbf{P}''_{51} = R'_1 (\mathbf{P}''_{12})_o R_1.$$

Thus to obtain $\mathbf{P}''_{51}$, interchange the first two components of $\mathbf{P}''_{12}$ and replace x by y. So as to facilitate the calculation of integrals such as $\int \mathbf{P}''_{\rho_1\sigma_1}.\mathbf{P}''_{\rho_2\sigma_2}\, dV$, it will be helpful to tabulate the following vector fields:

$$a^5\mathbf{P}''_{15} = \left(\frac{z}{|z|}(a-|z|)(a-|x|)^2, 0, \frac{-x}{|x|}(a-|x|)(a-|z|)^2 \right)$$

for $0 < |x|, y, |z| < a$,

$$a^5\mathbf{P}''_{56} = \left(0, \frac{-z}{|z|}(a-|z|)(a-|a-y|)^2, \frac{y-a}{|a-y|}(a-|a-y|)(a-|z|)^2 \right)$$

for $0 < x, |a-y|, |z| < a$,

$$a^5\mathbf{P}''_{57} = \left(\frac{a-y}{|a-y|}(a-|a-y|)(a-|x|)^2, \frac{x}{|x|}(a-|x|)(a-|a-y|)^2, 0 \right)$$

for $0 < |x|, |a-y|, z < a$,

$$a^5\mathbf{P}''_{78} = \left(0, \frac{a-z}{|a-z|}(a-|a-z|)(a-|a-y|)^2, \right.$$

$$\left. \frac{y-a}{|y-a|}(a-|a-y|)(a-|a-z|)^2 \right) \text{for } 0 < x, |a-y|, |a-z| < a,$$

$$a^5\mathbf{P}''_{37} = \left(\frac{z-a}{|a-z|}(a-|x|)^2(a-|a-z|), 0, \frac{-x}{|x|}(a-|x|)(a-|a-z|)^2 \right)$$

for $0 < |x|, y, |a-z| < a$,

$$a^5\mathbf{P}''_{25} = \left(\frac{z}{|z|}x(a-x)(a-|z|), \frac{z}{|z|}y(a-y)(a-|z|), (a-y-x)(a-|z|^2) \right)$$

for $0 < x, y, |z| < a$,

$$a^5\mathbf{P}''_{17} = \left((z-y)(a-|x|)^2, \frac{x}{|x|}y(a-y)(a-|x|), \frac{-x}{|x|}z(a-z)(a-|x|) \right)$$

for $0 < |x|, y, z < a$,

$$a^5\mathbf{P}''_{53} = \left((a-y-z)(a-|x|)^2, \frac{x}{|x|}y(a-y)(a-|x|), \frac{x}{|x|}z(a-z)(a-|x|) \right)$$

for $0 < |x|, y, z < a$,

$$a^5\mathbf{P}_{58}'' = \left(\frac{a-y}{|a-y|} x(a-x)(a-|a-y|), \ (x-z)(a-|a-y|)^2, \right.$$

$$\left. \frac{y-a}{|a-y|} z(a-z)(a-|a-y|) \right) \text{ for } 0 < x, |a-y|, z < a,$$

$$a^5\mathbf{P}_{38}'' = \left(\frac{z-a}{|a-z|} x(a-x)(a-|a-z|), \ \frac{a-z}{|a-z|} y(a-y)(a-|a-z|), \right.$$

$$\left. (y-x)(a-|a-z|)^2 \right) \text{ for } 0 < x, y, |a-z| < a,$$

$$a^5\mathbf{P}_{47}'' = \frac{z-a}{|z-a|} x(a-x)(a-|a-z|), \ \frac{z-a}{|a-z|} y(a-y)(a-|a-z|),$$

$$(a-x-y)(a-|a-z|)^2 \text{ for } 0 < x, y, |a-z| < a$$

$$a^5\mathbf{P}_{54}'' = (x(a-x)(a-y-z), \ y(a-y)(x-z), \ -z(a-z)(a-x-y))$$
$$\text{for } 0 < x, y, z < a,$$

$$a^5\mathbf{P}_{63}'' = (x(a-x)(a-y-z), \ -y(a-y)(a-x-z), \ z(a-z)(x-y))$$
$$\text{for } 0 < x, y, z < a.$$

VII. INTEGRALS OF VECTOR-FIELDS

It is necessary to construct $\mathbf{P}''$ for use in formula (2.6) as a linear combination of vector-fields with properties (2.1), (2.2) and (2.3). If, however, $\mathbf{P}''$ contained only vector-fields of the type $\mathbf{P}_{\rho\sigma}''$, then the maximum value of (2.8) would be zero, for, by Stokes' theorem, if $\mathbf{f}$ is any tangentially continuous vector-field in $V \cup B$, then $\int_B \operatorname{curl} \mathbf{f}.\mathbf{n}\, dB = 0$. However, $\mathbf{P}_{\rho\sigma}''$ is the curl of a tangentially continuous vector-field, namely $P_\rho \nabla P_\sigma$, and so for every $\mathbf{P}_{\rho\sigma}''$

$$\int_B \mathbf{P}_{\rho\sigma}''.\mathbf{n}\, dB = 0. \tag{7.1}$$

Thus $\int \mathbf{P}''.\mathbf{n}\, dB = 0$, and the maximum value of (2.8) is, indeed, zero. It will be necessary, therefore, to choose at least one vector-field $\mathbf{P}_r''$ among the component vector-fields of $\mathbf{P}''$, which is not the curl of a vector-field, to ensure that

$$\int_B \mathbf{P}''.\mathbf{n}\, dB \neq 0.$$

Therefore, with $\mathbf{P}'' = \sum\limits_{r=0}^{n} c_r \mathbf{P}''_r$, we shall choose $\mathbf{P}''_0$ to be a vector-field satisfying conditions (2.1), (2.2), and (2.3), and such that $\int \mathbf{P}''_0.\mathbf{n} dB \neq 0$, while $\mathbf{P}''_r$ will be of the type $\mathbf{P}''_{\rho\sigma}$ for $r > 0$. Then,

$$\int_B \mathbf{P}''.\mathbf{n}\,dB = c_0 \int_B \mathbf{P}''_0.\mathbf{n}\,dB. \tag{7.2}$$

For (2.8) we also need $\int_V (\mathbf{P}'')^2\,dV = \sum\limits_{r=0}^{n} \sum\limits_{s=0}^{n} c_r c_s \int_V \mathbf{P}''_r.\mathbf{P}''_s\,dV$, so that the following integrals must be evaluated

$$\int_V (\mathbf{P}''_0)^2\,dV, \quad \int_V \mathbf{P}''_0.\mathbf{P}''_{\rho\sigma}\,dV, \quad \int_V \mathbf{P}''_{\rho_1\sigma_1}.\mathbf{P}''_{\rho_2\sigma_2}\,dV, \tag{7.3}$$

for each vector field $\mathbf{P}''_{\rho\sigma}$ and each pair of vector-fields $\mathbf{P}''_{\rho_1\sigma_1}$ and $\mathbf{P}''_{\rho_2\sigma_2}$. As no choice has yet been made for $\mathbf{P}''_0$, we shall confine ourselves in this section to evaluating the last integrals in (7.3).

To evaluate these integrals, first note that $\mathbf{P}''_{\rho_1\sigma_1}.\mathbf{P}''_{\rho_2\sigma_2} = 0$ unless the lattice points $\rho_1, \rho_2, \sigma_1, \sigma_2$ all belong to the same lattice cube. For any lattice cube there are $\frac{1}{2}28^2$ such integrals, but it suffices to evaluate 23 of them, since any integral is unaltered in value when both vector-fields are reflected in a plane of symmetry of the lattice cube to which $\rho_1, \rho_2, \sigma_1, \sigma_2$ belong. Moreover, to obtain an integral such as $\int (\mathbf{P}''_{12})^2 dV$, we need only integrate $(\mathbf{P}''_{12})^2$ over any one of the four lattice cubes that meet at the edge 12, and then multiply the result by four. Thus

$$a^{10} \int_V (\mathbf{P}''_{12})^2\,dV = 4 \int_0^a \int_0^a \int_0^a \{(a-z)^2(a-y)^4$$

$$+ (a-y)^2(a-z)^4\}\,dx\,dy\,dz = \frac{8a^9}{15}.$$

It suffices to tabulate the following 23 integrals:

$$a \int_V (\mathbf{P}''_{12})^2\,dV = \tfrac{8}{15},$$

$$a \int_V \mathbf{P}''_{12}.\mathbf{P}''_{15}\,dV = -\tfrac{1}{10}, \quad a \int_V \mathbf{P}''_{12}.\mathbf{P}''_{56}\,dV = -\tfrac{4}{90},$$

$$a \int_V \mathbf{P}''_{12}.\mathbf{P}''_{57}\,dV = -\tfrac{1}{120}, \quad a \int_V \mathbf{P}''_{12}.\mathbf{P}''_{78}\,dV = -\tfrac{1}{90},$$

$$a \int_V \mathbf{P}''_{12}.\mathbf{P}''_{37}\,dV = -\tfrac{1}{120}, \quad a \int_V \mathbf{P}''_{12}.\mathbf{P}''_{16}\,dV = 0,$$

$$a \int_V \mathbf{P}''_{12}.\mathbf{P}''_{17}\,dV = -\tfrac{1}{40}, \quad a \int_V \mathbf{P}''_{12}.\mathbf{P}''_{53}\,dV = 0,$$

$$a \int_V \mathbf{P}''_{12}.\mathbf{P}''_{58}\,dV = -\tfrac{1}{90}, \quad a \int_V \mathbf{P}''_{12}.\mathbf{P}''_{18}\,dV = -\tfrac{1}{120}, \quad a \int_V \mathbf{P}''_{12}.\mathbf{P}''_{54}\,dV = -\tfrac{1}{120}$$

$$a \int_V (\mathbf{P}''_{16})^2\,dV = \tfrac{1}{9},$$

$$a \int_V \mathbf{P}''_{16}.\mathbf{P}''_{25}\,dV = 0, \qquad a \int_V \mathbf{P}''_{16}.\mathbf{P}''_{17}\,dV = -\tfrac{1}{60},$$

$$a \int_V \mathbf{P}''_{16}.\mathbf{P}''_{53}\,dV = 0, \qquad a \int_V \mathbf{P}''_{16}.\mathbf{P}''_{38}\,dV = -\tfrac{1}{180},$$

$$a \int_V \mathbf{P}''_{16}.\mathbf{P}''_{47}\,dV = 0, \qquad a \int_V \mathbf{P}''_{16}.\mathbf{P}''_{18}\,dV = \tfrac{1}{360}, \quad a \int_V \mathbf{P}''_{16}.\mathbf{P}''_{54}\,dV = 0.$$

$$a \int_V (\mathbf{P}''_{18})^2\,dV = \tfrac{1}{60},$$

$$a \int \mathbf{P}''_{18}.\mathbf{P}''_{54}\,dV = \tfrac{1}{180}, \qquad a \int_V \mathbf{P}''_{18}.\mathbf{P}''_{36}\,dV = \tfrac{1}{180}. \tag{7.4}$$

VIII. USE OF SYMMETRY

The symmetry group G of the basic cube $= \{(x_1, x_2, x_3); \max. |x_1|, |x_2|, |x_3| \leqslant 1\}$ is of order 48, and is generated by reflections in the three planes

$$x_1 = 0, \quad x_1 = x_2, \quad x_2 = x_3.$$

A fundamental domain for this group is the domain bounded by these three planes. Let V_{48} be the closure of the part of this fundamental domain which lies in $V \cup B$, i.e.

$$V_{48} = \{(x_1, x_2, x_3); x_3 \geqslant 1 \quad \text{and} \quad 0 \leqslant x_1 \leqslant x_2 \leqslant x_3\}. \tag{8.1}$$

The potential function u, which is a solution to the Dirichlet problem for the basic cube, is transformed by the elements of the symmetry group G into a potential function which also satisfies the boundary conditions. As the solution is unique, it follows that u is invariant under G. It will be convenient, therefore, to define a vector-field $\mathbf{Q}$, which is a linear combination of vector-fields $\mathbf{P}''_{\rho\sigma}$ and whose support is contained in $V_{48} \cup V^*_{48} \cup V^{**}_{48}$, where V^*_{48} is the reflection of V_{48} in the plane $x_1 = x_2$, and V^{**}_{48} its reflection in $x_2 = x_3$. We shall then set

$$\mathbf{P}'' = c_0 \mathbf{P}''_0 + \sum_G R'_g(\mathbf{Q}) o R_g \tag{8.2}$$

where $\mathbf{P}''_0$ is a vector-field satisfying the conditions of (7.2), and the summation is over the symmetry group G. If $\mathbf{Q}^*$ is the reflection of $\mathbf{Q}$ in $x_1 = x_2$, and $\mathbf{Q}^{**}$ its reflection in $x_2 = x_3$, then of the 48 vector-fields obtained from $\mathbf{Q}$ by reflections, $\mathbf{Q}$ will have non-zero scalar-product only with $\mathbf{Q}^*$ and $\mathbf{Q}^{**}$. As scalar-products are invariant under reflections, this implies that

$$\int_V (\mathbf{P}'')^2 \, dV = c_0^2 \int_V (\mathbf{P}_0'')^2 \, dV + 96 c_0 \int_V \mathbf{P}_0'' . \mathbf{Q} \, dV + 48 \int_V (\mathbf{Q}^2 + \mathbf{Q} . \mathbf{Q}^* +$$
$$+ \mathbf{Q} . \mathbf{Q}^{**}) \, dV. \qquad (8.3)$$

IX. A CALCULATION WITH LATTICE EDGE a = 1/2

Take a cubic lattice in R^3 with $a = \frac{1}{2}$ as edge length of lattice cube. Select the following lattice points, all of which lie in V_{48}:

$1_0 = (0,0,1), \quad 2_0 = (0,\frac{1}{2},1), \quad 3_0 = (0,1,1), \quad 4_0 = (\frac{1}{2},\frac{1}{2},1), \quad 5_0 = (\frac{1}{2},1,1),$

$6_0 = (1,1,1),$

$1_1 = (0,0,\frac{3}{2}), \quad 2_1 = (0,\frac{1}{2},\frac{3}{2}), \quad 3_1 = (0,1,\frac{3}{2}), \quad 4_1 = (\frac{1}{2},\frac{1}{2},\frac{3}{2}),$

$5_1 = (\frac{1}{2},1,\frac{3}{2}), \quad 6_1 = (1,1,\frac{3}{2}),$

and let $1_0^* = 1_0$, 2_0^*, 3_0^*, $4_0^* = 4_0$, 5_0^*, $6_0^* = 6_0$, $1_1^* = 1_1$, 2_1^*, 3_1^*, $4_1^* = 4_1$, $5_1^*, 6_1^* = 6_1$ be their reflections in the plane $x_1 = x_2$. (See diagram.)

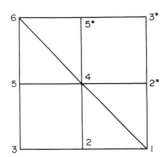

It will be necessary to choose a vector-field $\mathbf{P}_0''$ of the type specified in section 7., and we now define $\mathbf{P}_0''$ as follows: — For points (x_1,x_2,x_3) such that $x_3 \geqslant 1$, $|x_1| < x_3$, $|x_2| < x_3$

$$\mathbf{P}_0''(x_1,x_2,x_3) = r^3/x_3^3 \, (\text{grad } 1/r) = (-x_1/x_3^3, -x_2/x_3^3, -x_3/x_3^3), \quad (9.1)$$

and the value of $\mathbf{P}_0''$ at other points of $V \cup B$, excepting the exceptional surfaces, is obtained by symmetry. $\mathbf{P}_0''$ is normally continuous across these exceptional surfaces. Then

$$\int_B \mathbf{P}_0'' . \mathbf{n} \, dB = \int 1 \, dB = 24,$$

and by (7.2)

$$\int_B \mathbf{P}''.\mathbf{n}\,dB = c_0 \int_B \mathbf{P}_0''.\mathbf{n}\,dB = 24c_0. \tag{9.2}$$

Also

$$\int_V (\mathbf{P}_0'')^2\,dV = 6 \int_{x_3=1}^{\infty}\int_{x_1=-x_3}^{x_3}\int_{x_2=-x_3}^{x_3} \frac{x_1^2+x_2^2+x_3^2}{x_3^6}\,dx_1 dx_2 dx_3 = 40. \tag{9.3}$$

Choose $\mathbf{Q}$ as follows:-

$$\mathbf{Q} = c_1\mathbf{P}''_{4_0 2_0} + c_2\mathbf{P}''_{4_0 5_0} + c_3\mathbf{P}''_{4_0 3_0} + c_4\mathbf{P}''_{2_0 5_0} + c_5\mathbf{P}''_{2_1 5_0} + c_6\mathbf{P}''_{2_0 5_1} + c_7\mathbf{P}''_{4_1 3_0} + c_8\mathbf{P}''_{4_0 3_1},$$

so that

$$\mathbf{Q}^* = c_1\mathbf{P}''_{2_0^* 4_0^*} + c_2\mathbf{P}''_{5_0^* 4_0^*} + c_3\mathbf{P}''_{3_0^* 4_0^*} + c_4\mathbf{P}''_{5_0^* 2_0^*} + c_5\mathbf{P}''_{5_0^* 2_1^*}$$

$$+ c_6\mathbf{P}''_{5_1^* 2_0^*} + c_7\mathbf{P}''_{3_0^* 4_1^*} + c_8\mathbf{P}''_{3_1^* 4_0^*}.$$

As the support of $\mathbf{Q}$ is contained in $V_{48} \cup V_{48}^*$, then by (8.3)

$$\int_V (\mathbf{P}'')^2\,dV = c_0^2 \int_V (\mathbf{P}_0'')^2\,dV + 96c_0 \int_V \mathbf{P}_0''.\mathbf{Q}\,dV + 48 \int_V (\mathbf{Q}^2+\mathbf{Q}.\mathbf{Q}^*)\,dV. \tag{9.4}$$

To evaluate $\int_V \mathbf{P}_0''.\mathbf{Q}\,dV$, we need to obtain integrals such as $\int_V \mathbf{P}_0''.\mathbf{P}''_{4_0 2_0}\,dV$. For this calculation it is convenient to adopt a local coordinate frame in which $4_0 2_0$ is the x-direction and $4_0 2_0^*$ is the y-direction, so that (x_1,x_2,x_3) are related to (x,y,z) by

$$x_1 = \tfrac{1}{2}-x, \quad x_2 = \tfrac{1}{2}-y, \quad x_3 = 1+z.$$

If the last transformation is denoted by T and T' is its jacobian, then $\mathbf{P}_0''$ in this local frame becomes $T'(\mathbf{P}_0'')_oT$, i.e.

$$\mathbf{P}_0''(x,y,z) = \left(\frac{\tfrac{1}{2}-x}{(1+z)^3}, \frac{\tfrac{1}{2}-y}{(1+z)^3}, \frac{-(1+z)}{(1+z)^3}\right).$$

Therefore, by (6.1)

$$a^5 \int_V \mathbf{P}_0''.\mathbf{P}''_{4_0 2_0}\,dV = \int_0^{1/2} dz \int_0^{1/2} dx \int_{-1/2}^{1/2}$$

$$-\frac{(\tfrac{1}{2}-y)(\tfrac{1}{2}-y)^2(\tfrac{1}{2}-z) - y/|y|(\tfrac{1}{2}-y)(1+z)(\tfrac{1}{2}-z)^2}{(1+z)^3}\,dy$$

$$= \int_0^{1/2} dz \int_0^{1/2} dx \int_{-1/2}^0 \frac{-(\tfrac{1}{2}-y)(\tfrac{1}{2}+y)^2(\tfrac{1}{2}-z) + (\tfrac{1}{2}+y)(1+z)(\tfrac{1}{2}-z)^2}{(1+z)^3}\,dy$$

$$+ \int_0^{1/2} dz \int_0^{1/2} dx \int_0^{1/2} \frac{-(\tfrac{1}{2}-y)^3(\tfrac{1}{2}-z) - (\tfrac{1}{2}-y)(1+z)(\tfrac{1}{2}-z)^2}{(1+z)^3}\,dy.$$

Replace y by $-y$ in the first integral above, and this integral becomes

$$\int_0^{1/2} dz \int_0^{1/2} dx \int_0^{1/2} \frac{-(\frac{1}{2}+y)(\frac{1}{2}-y)^2(\frac{1}{2}-z) + (\frac{1}{2}-y)(1+z)(\frac{1}{2}-z)^2}{(1+z)^3} dy.$$

Thus both integrals of (9.4) combine to give

$$a^5 \int_V \mathbf{P}_0''.\mathbf{P}_{4020}'' \, dV = \int_0^{1/2} \int_0^{1/2} \int_0^{1/2} - \{(\tfrac{1}{2}+y)(\tfrac{1}{2}-y)^2 + (\tfrac{1}{2}-y)^3\}$$

$$\times (\tfrac{1}{2}-z)/(1+z)^3 \, dx \, dy \, dz$$

$$= -\tfrac{1}{48} \int_0^{1/2} (\tfrac{1}{2}-z)/(1+z)^3 \, dz = -\tfrac{1}{576}.$$

Hence, and by similar calculations, we obtain

$$\int_V \mathbf{P}_0''.\mathbf{P}_{4020}'' \, dV = -\tfrac{1}{18}, \quad \int_V \mathbf{P}_0''.\mathbf{P}_{4050}'' \, dV = -\tfrac{1}{18}, \quad \int_V \mathbf{P}_0''.\mathbf{P}_{4030}'' \, dV = -\tfrac{1}{36},$$

$$\int_V \mathbf{P}_0''.\mathbf{P}_{2050}'' \, dV = \tfrac{1}{72}, \quad \int_V \mathbf{P}_0''.\mathbf{P}_{2150}'' \, dV = -\tfrac{1}{432}, \quad \int_V \mathbf{P}_0''.\mathbf{P}_{2051}'' \, dV = -\tfrac{1}{432},$$

$$\int_V \mathbf{P}_0''.\mathbf{P}_{4130}'' \, dV = -\tfrac{1}{216}, \quad \int_V \mathbf{P}_0''.\mathbf{P}_{4031}'' \, dV = -\tfrac{1}{216}.$$

Finally consider the integral $\int_V (\mathbf{Q}^2 + \mathbf{Q}.\mathbf{Q}^*) \, dV$ in (9.4). Since $\mathbf{P}_{\rho_1\sigma_1}''.\mathbf{P}_{\rho_2\sigma_2}'' = 0$ unless $\rho_1, \sigma_1, \rho_2, \sigma_2$ all belong to the same lattice cube, then the only non-zero terms in $\mathbf{Q}.\mathbf{Q}^*$ are $c_1{}^2\mathbf{P}_{4_020}''.\mathbf{P}_{2_0^*4_0^*}''$ and $c_2{}^2\mathbf{P}_{4_05_0}''.\mathbf{P}_{5_0^*4_0^*}''$. To evaluate their integrals, and the integrals of other products which occur in $\mathbf{Q}^2$, use the results of Section 7, remembering, however, that if some of the lattice cubes upon which $\mathbf{P}_{\rho\sigma}''$ is defined, lie outside V, then the tabulated result will have to be correspondingly modified. Hence

$$\int_V (\mathbf{Q}^2 + \mathbf{Q}.\mathbf{Q}^*) \, dV = 2 \{ (\tfrac{4}{15}+\tfrac{1}{20})(c_1{}^2+c_2{}^2) + \tfrac{1}{18}(c_3{}^2+c_4{}^2)$$

$$+ \tfrac{1}{60}(c_5{}^2+c_6{}^2+c_7{}^2+c_8{}^2) - \tfrac{2}{20}(c_1c_2)$$

$$+ \tfrac{2}{120}(c_1c_5+c_1c_6-c_1c_7-c_1c_8)$$

$$- \tfrac{2}{120}(c_2c_5+c_2c_6+c_2c_7+c_2c_8)$$

$$+ \tfrac{2}{360}(c_3c_7+c_3c_8+c_4c_5+c_4c_6)$$

$$+ \tfrac{2}{180}(c_5c_6+c_5c_7-c_5c_8-c_6c_7+c_6c_8+c_7c_8) \}.$$

$$(9.6)$$

The inequality (2.6) now reads

$$4\pi C \geqslant 48c_0 - 80c_0{}^2 - 96c_0 \left(\frac{-c_1}{18} - \frac{c_2}{18} - \frac{c_3}{36} + \frac{c_4}{72} - \frac{c_5}{432} - \frac{c_6}{432} - \frac{c_7}{216} - \frac{c_8}{216} \right)$$

$$- 48 \int_V (\mathbf{Q}^2 + \mathbf{Q}.\mathbf{Q}^*) \, dV,$$

$$(9.7)$$

where the integral is the quadratic form in the c's of (9.6). Maximizing (9.7) with respect to the c's, leads to the following equations:

$$360c_0 - 24c_1 - 24c_2 - 12c_3 + 6c_4 - c_5 - c_6 - 2c_7 - 2c_8 = 216$$

$$114c_1 - 18c_2 \qquad\qquad + 3c_5 + 3c_6 - 3c_7 - 3c_8 = 10c_0$$

$$- 18c_1 + 114c_2 \qquad\qquad - 3c_5 - 3c_6 - 3c_7 - 3c_8 = 10c_0$$

$$20c_3 \qquad\qquad + c_7 + c_8 = 5c_0$$

$$20c_4 + c_5 + c_6 \qquad\qquad = -\tfrac{5}{2}c_0$$

$$3c_1 - 3c_2 \qquad + c_4 + 2c_5 + 6c_6 + 2c_7 - 2c_8 = \tfrac{5}{12}c_0$$

$$3c_1 - 3c_2 \qquad + c_4 + 6c_5 + 2c_6 - 2c_7 + 2c_8 = \tfrac{5}{12}c_0.$$

$$- 3c_1 - 3c_2 + c_3 \qquad + 2c_5 - 2c_6 + 6c_7 + 2c_8 = \tfrac{5}{6}c_0$$

$$- 3c_1 - 3c_2 + c_3 \qquad - 2c_5 + 2c_6 + 2c_7 + 6c_8 = \tfrac{5}{6}c_0.$$

Solving these linear equations, yields by (9.7) the estimate

$$C \geqslant 1.176$$

Certain observations concerning this result may be helpful. The exact value for C is estimated to be 1.31. Letting $c_5 = c_6 = c_7 = c_8 = 0$, there remain only four linear equations, and the estimate $C \geqslant 1.142$ is obtained. Replacing $\mathbf{P}_0''$ in the above calculation by grad $1/r$, leads to the much sharper estimate $C \geqslant 1.258$. In this case, however, the integrals $\int_V$ grad $1/r.\mathbf{P}_{\rho\sigma}'' \, dV$ can be rather nasty.

REFERENCES

1. James J. McMahon, 1953 Lower Bounds for the Electrostatic Capacity of a Cube, *Proc. R. Ir. Acad.*, **A55**, 133–167.
2. J. L. Synge. 1952 Triangulation in the hypercircle method for plane problems, *Proc. R. Ir. Acad.*, **A54**, 341–367.
3. J. L. Synge. 1957 *The hypercircle in mathematical physics*, Cambridge Uni. Press.

The Koenig-Hadamard Theorem again

ALSTON S. HOUSEHOLDER

University of Tennessee
and
Oak Ridge National Laboratory, Tennessee, U.S.A.

The Koenig-Hadamard theorem can be formulated in either of several equivalent ways. For present purposes the following seems most convenient:

Let $h(z)$, $h(0) \neq 0$, be meromorphic in the disk

$$|z| \leq R,$$

within which it possesses p poles, counting multiplicities. Let these be $r_1, r_2, \cdots, r_p,$ *with*

$$|r_j| < R, \quad j = 1, 2, \cdots, p.$$

Let

$$\psi(z) = (1 - z/r_1)(1 - z/r_2) \cdots (1 - z/r_p).$$

If in the column of the Padé table for which the denominators are $\psi_\nu(z)$, each of degree p at most with zeros $r_1^{(\nu)}, r_2^{(\nu)}, \cdots r_p^{(\nu)}$, then, with a suitable ordering,

$$\lim_{\nu \to \infty} r_j^{(\nu)} = r_j.$$

More precisely, it is possible to take $\psi_\nu(0) = 1$ for every ν sufficiently large. Then

$$\lim_{\nu \to \infty} \psi_\nu(z) = \psi(z).$$

In fact, if

$$|r_i| < \sigma R < R,$$

then

$$|\psi_\nu(z) - \psi(z)| = o(\sigma^\nu).$$

This is the fundamental convergence theorem that justifies the use of the qd algorithm for computing the zeros of a polynomial or of a transcendental function in some domain of analyticity, or, as it is here phrased,

235

the poles of a function in a region throughout which it is meromorphic. It also justifies the use of Aitken's extension of the Bernoulli method, which, however, is in fact a forerunner of the qd algorithm. And it justifies the use of the algorithm of Sebastião e Silva[9] since it has been shown[7] that after a transformation this becomes mathematically equivalent to the qd algorithm.

The case $p = 1$ is due to Koenig[8], and the proof is quite elementary. For arbitrary p the theorem is a consequence of a somewhat more general one due to Hadamard[3], for which the proof requires rather sophisticated analysis. Direct proofs of the theorem itself have appeared, by Golomb[1], by Henrici[4], and by Gragg and Householder[2,5], the last making use of matrix norms, the first two requiring quite intricate determinantal manipulations. The theorem can also be obtained as a consequence of a very general convergence theorem recently published by Stewart[10]. The proof to be given here is thought to be new, and quite elementary. It depends mainly upon an explicit representation of the entries in the Padé table for an arbitrary power series, and this is easily stated and readily understood. Moreover, the qd algorithm can hardly be understood except in terms of the Padé table. For more details, however, concerning the Padé table, and its connection with the qd algorithm, reference may be made to[5,6].

The explicit representation alluded to makes use of determinants of a class known as bigradients. Consider the three formal power series

$$f(z) = \sum_0^\infty a_\nu z^\nu,$$

$$g(z) = \sum_0^\infty b_\nu z^\nu, \quad a_0 b_0 \neq 0,$$

$$h(z) = g(z)/f(z) = \sum_0^\infty c_\nu z^\nu.$$

From the coefficients of $f(z)$ and $g(z)$ can be formed the following determinants:

$$\delta(a, b)_{\nu,p} = \delta \begin{pmatrix} a_0 & a_1 & a_2 \cdots a_{\nu+p-2} & a_{\nu+p-1} \\ 0 & a_0 & a_1 \cdots a_{\nu+p-3} & a_{\nu+p-2} \\ \hline 0 & b_0 & b_1 \cdots b_{\nu+p-3} & b_{\nu+p-2} \\ b_0 & b_1 & b_2 \cdots b_{\nu+p-2} & b_{\nu+p-1} \end{pmatrix}$$

$$\delta(f, g)_{\nu,p} = \delta \begin{pmatrix} a_0 & a_1 & a_2 & \cdots & a_{\nu+p-2} & f(z) \\ 0 & a_0 & a_1 & \cdots & a_{\nu+p-3} & zf(z) \\ \hline 0 & b_0 & b_1 & \cdots & b_{\nu+p-3} & zg(z) \\ b_0 & b_1 & b_2 & \cdots & b_{\nu+p-2} & g(z) \end{pmatrix}$$

Here the "δ" signifies the determinant. Each of the determinants exhibited is of order $\nu + p$, and they are identical except in the last column. There are ν rows formed from $f(z)$ and its coefficients, p from $g(z)$ and its coefficients, and in the last column of the second determinant, in going down the column the exponents of the powers of z multiplying $f(z)$ increase those multiplying $g(z)$ decrease. Expansion of the second determinant by elements of its last column shows that

$$\delta(f, g)_{\nu,p} = \Psi_{\nu-1,p-1}(z)g(z) - \Phi_{\nu-1,p-1}(z)f(z), \tag{1}$$

where $\Psi_{\nu-1,p-1}(z)$ is a polynomial of degree $p - 1$ at most, and $\Phi_{\nu-1,p-1}(z)$ is a polynomial of degree $\nu - 1$ at most. On the other hand, if, before expansion, the last column is replaced by the result of subtracting the first column, z times the second, z^2 times the third, $\cdots$, it is clear that

$$\delta(f,g)_{\nu,p} = z^{\nu+p-1}\delta(a,b)_{\nu,p} + \cdots,$$

where omitted terms are of degrees higher than $\nu + p - 1$. It is also important to note, and is easily verified, that

$$\Phi_{\nu-1,p-1}(0) = \Psi_{\nu-1,p-1}(0) = \delta(a, b)_{\nu-1,p-1}. \tag{2}$$

There is no restriction in supposing that

$$f(0) = g(0) = h(0) = 1.$$

It will then be convenient to write (raising the indices)

$$\phi_{\nu,p}(z) = \Phi_{\nu,p}(z)/\delta(a, b)_{\nu,p}, \quad \psi_{\nu,p}(z) = \Psi_{\nu,p}(z)/\delta(a, b)_{\nu,p}, \tag{3}$$

provided $\delta(a, b)_{\nu,p} \neq 0$. Then

$$\phi_{\nu,p}(0) = \psi_{\nu,p}(0) = 1,$$

and

$$\psi_{\nu,p}(z)g(z) - \phi_{\nu,p}(z)f(z) = z^{\nu+p+1}\omega_{\nu,p}(z) \tag{4}$$

where, as has been seen, $\omega_{\nu,p}(z)$ is a power series in z. Consequently, whenever $\delta(a,b)_{\nu,p} \neq 0$, the entry in the (ν,p) cell of the Padé table for $h(z)$ is $\phi_{\nu,p}(z)/\psi_{\nu,p}(z)$, the explicit determinantal representation for these polynomials is easily seen, and this is valid for any two expansions $f(z)$ and $g(z)$ whose quotient, $g(z)/f(z)$, has the same expansion as $h(z)$.

It will appear in the course of the argument to be given, that for the case of interest for given p, at most a finite number of these determinants $\delta(a, b)_{\nu,p}$ can vanish. This and the fact that $f(z)$ and $g(z)$ can be taken arbitrarily subject only to the requirement that $g(z)/f(z) = h(z)$, and that they satisfy suitable analyticity conditions, will be sufficient for the following development.

For $f(z)$, take $\psi(z)$ as defined in the theorem. Let

$$g(z) = h(z)\psi(z). \tag{5}$$

Then $g(z)$ is analytic for $|z| \leq R$. Since p is fixed, it is not needed as an index, and (4) can be written

$$\psi_\nu(z)g(z) - \phi_\nu(z)\psi(z) = z^{\nu+p+1}\omega_\nu(z), \tag{6}$$

where each $\psi_\nu(z)$ is of degree p, at most, and $\phi_\nu(z)$ is of degree ν at most. Let

$$\psi(z) = 1 + a_1 z + \cdots + a_p z^p,$$

$$g(z) = 1 + b_1 z + b_2 z^2 + \cdots \tag{7}$$

$$= g_{\nu-1}(z) + z^\nu \gamma_\nu(z),$$

$$\gamma_\nu(z) = b_\nu + b_{\nu+1} z + b_{\nu+2} z^2 + \cdots$$

Since $g(z)$ is analytic in the closed disk $|z| \leq R$, the series converges absolutely for $z = R$, hence has a term of maximal modulus. Let its value be γ. Then

$$|b_\nu| \leq \gamma R^{-\nu},$$

and

$$|\gamma_\nu(z)| \leq \gamma R^{-\nu}(1 + |z|R^{-1} + |z|^2 R^{-2} + \cdots) = \gamma R^{-\nu}/(1 - |z|/R). \tag{8}$$

For illustration, consider the case $p = 2$, $\nu = 3$, and $r_1 \neq r_2$. Then

$$\psi_3(z) = \delta \begin{pmatrix} 1 & a_1 & a_2 & 0 & 0 & 0 & 0 \\ 0 & 1 & a_1 & a_2 & 0 & 0 & 0 \\ 0 & 0 & 1 & a_1 & a_2 & 0 & 0 \\ 0 & 0 & 0 & 1 & a_1 & a_2 & 0 \\ 0 & 0 & 1 & b_1 & b_2 & b_3 & z^2 \\ 0 & 1 & b_1 & b_2 & b_3 & b_4 & z \\ 1 & b_1 & b_2 & b_3 & b_4 & b_5 & 1 \end{pmatrix} / \delta(a,b)_{3,2}.$$

We wish first to evaluate $\delta(a,b)_{3,2}$. Form the product

$$\delta(a,b)_{3,2}\delta\begin{pmatrix} 1 & 1 \\ r_1 & r_2 \end{pmatrix} = \delta \begin{pmatrix} 1 & a_1 & a_2 & 0 & 0 \\ 0 & 1 & a_1 & a_2 & 0 \\ 0 & 0 & 1 & a_1 & a_2 \\ 0 & 1 & b_1 & b_2 & b_3 \\ 1 & b_1 & b_2 & b_3 & b_4 \end{pmatrix} \delta \begin{pmatrix} 1 & 1 & 0 & 0 & 0 \\ r_1 & r_2 & 0 & 0 & 0 \\ r_1{}^2 & r_2{}^2 & 1 & 0 & 0 \\ r_1{}^3 & r_2{}^3 & 0 & 1 & 0 \\ r_1{}^4 & r_2{}^4 & 0 & 0 & 1 \end{pmatrix}.$$

This gives

$$\zeta^{1/2}(r_1,r_2)\delta(a,b)_{3,2} = a_2{}^3\delta \begin{pmatrix} r_1 g_3(r_1) & r_2 g_3(r_2) \\ g_4(r_1) & g_4(r_2) \end{pmatrix},$$

where $\zeta^{1/2}(r_1,r_2)$ *is the Vandermonde.*

Next, apply the same device to the evaluation of $\Psi_3(z)$ except that the last row of the multiplying determinant has zeros everywhere except in the last place. The result is:

$$\zeta^{1/2}(r_1,r_2)\Psi_3(z) = a_2{}^4\delta\begin{pmatrix} r_1{}^2g_3(r_1) & r_2{}^2g_3(r_2) & z^2 \\ r_1g_4(r_1) & r_2g_4(r_2) & z \\ g_5(r_1) & g_5(r_2) & 1 \end{pmatrix}.$$

This is perhaps sufficient to make clear that in general

$$\zeta^{1/2}(r_1,r_2)\delta(a,b)_{\nu,2} = a_2{}^\nu\delta\begin{pmatrix} r_1g_\nu(r_1) & r_2g_\nu(r_2) \\ g_{\nu+1}(r_1) & g_{\nu+1}(r_2) \end{pmatrix},$$

This proves, for the case $p = 2$, the assertion made above that at most a finite number of the $\delta(a,b)_{\nu,p}$ can vanish. Likewise

$$\zeta^{1/2}(r_1, r_2)\Psi_\nu(z) = a_2{}^{\nu+1}\delta\begin{pmatrix} r_1{}^2g_\nu(r_1) & r_2{}^2g_\nu(r_2) & z^2 \\ r_1g_{\nu+1}(r_1) & r_2g_{\nu+1}(r_2) & z \\ g_{\nu+2}(r_1) & g_{\nu+2}(r_2) & 1 \end{pmatrix}.$$

Note that $a_2 = r_1{}^{-1}r_2{}^{-1}$.

The proof is now essentially complete for the case $p = 2$, $r_1 \neq r_2$. If

$$1 > \sigma > \max\left(|r_1|/R, |r_2|/R\right),$$

then for r either r_1 or r_2,

$$g_\nu(r) = g(r) + o(\sigma^\nu),$$

hence,

$$\delta(a, b)_{\nu,2} = r_1{}^{-\nu} r_2{}^{-\nu} g(r_1) g(r_2) \left[1 + o(\sigma^\nu)\right],$$

hence,

$$a_{\nu,1} = -\left(r_1{}^{-1} + r_2{}^{-1}\right)\left[1 + o(\sigma^\nu)\right],$$
$$a_{\nu,2} = r_1{}^{-1} r_2{}^{-1}\left[1 + o(\sigma^\nu)\right]. \tag{9}$$

The confluent case, $r_1 = r_2$, is usually the one that is hardest to handle, but in view of (9) the difficulty does not arise, and the relation continues to hold even when $r_1 = r_2$.

The proof for arbitrary p is completely parallel, and it should hardly be necessary to spell out the details. The determinant used for reducing $\delta(a, b)_{\nu,p}$ has for its first p columns the powers of the r_i, and the remaining columns are those of the identity. The value of this determinant is that of the Vandermonde of the r_i. To reduce $\Psi_\nu(z)$, a similar determinant, having the same value, is used, but with zeros in the last row everywhere but in the final position. In the quotient, the Vandermonde cancels out, as do the powers of a_p, leaving only the first power in the numerator.

It is also true, of course, that if $h(z)$ is analytic for $|z| \leq R$, and has p

zeros within the disk but none on the boundary, then the polynomials $\phi_{p,\nu}(z)$ with fixed p have a limit, and this is the polynomial $\phi_p(z)$ for which $\phi_p(0) = 1$ and for which the zeros are those of $h(z)$ within the disk. In its original, and usual, formulation, the qd algorithm for finding the zeros of $f(z)$ is stated in terms of the poles of $1/f(z)$, or of $g(z)/f(z)$, where $g(z)$ is analytic throughout the domain of interest and has no zero in common with $f(z)$. Perhaps this was because the original Koenig-Hadamard theorem had been stated in terms of poles rather than zeros.

ACKNOWLEDGEMENT

This work was done in part while the author was a guest of the Technische Hochschule, München. The author is indebted to Dr. C. Reinsch for valuable suggestions.

REFERENCES

1. Golomb, M. (1943) Zeros and poles of functions defined by Taylor series. *Bull. Amer. Math. Soc.* **49**, 581–92.
2. Gragg, W. B. and Householder, A. S. (1966) On a theorem of Koenig. *Numer. Math.* **8**, 465–8.
3. Hadamard, J. (1892) Essai sur l'étude des fonctions données par leur dé-veloppement de Taylor. *J. de Math.* (4) **8**, 101–86.
4. Henrici, Peter (1958) The quotient-difference algorithm. *Nat. Bur. Standards Appl. Math. Ser.* **49**, 23–46.
5. Householder, A. S. (1970) *The numerical treatment of a single nonlinear equation.* McGraw-Hill Book Co., New York.
6. Householder, A. S. (1971) The Padé table, the Frobenius identities, and the qd algorithm. *Linear Algebra and Appl.* **4**, 161–74.
7. Householder, A. S. and Stewart, G. W. (1971) The numerical factorization of a polynomial. *SIAM Rev.* **13**, 38–46.
8. König, J. (1884) Ueber eine Eigenschaft der Potenzreihen. *Math. Ann.* **23**, 447–9.
9. Sebastião e Silva, José (1941) Sur une méthode d'approximation semblable a celle de Graeffe. *Portugal Math.* **2**, 271–9.
10. Stewart, G. W. (1971) On a companion operator for analytic functions. *Numer. Math.* **18**, 26–43.

Significance Arithmetic* —
On the Algebra of Binary Strings

N. METROPOLIS AND GIAN-CARLO ROTA

Los Alamos Scientific Laboratory
University of California
Los Alamos, New Mexico, U.S.A.

Since the days of antiquity it has been the privilege of the mathematician to engrave his conclusions, expressed in a rarefied and esoteric language, upon the rocks of eternity. While this method is excellent for the codification of mathematical results, it is not so acceptable to the many addicts of mathematics, for whom the science of mathematics is not a logical game, but the language in which the physical universe speaks to us, and whose mastery is inevitable for the comprehension of natural phenomena.

Cornelius Lanczos, 1961

I. INTRODUCTION

It has often been observed that most mathematicians are Platonists. Mathematical objects are for them endowed with an unshakable, if only ideal, existence, and their properties are to be investigated in the same spirit of objectivity that guides the naturalist in his study of the surrounding world. The challenges which periodically come forth to shake this faith are quickly neutralized by acclamation into the mainstream of mathematics, much like the latter-day Roman Senate would tame a Barbarian by acclaiming him as one of their own. This sort befell symbolic logic at its beginning, to name but one example out of the remote past.

We are now on the threshold of another drastic challenge, the so-called "computer revolution," whose manifold effects upon all sciences have been variously decried. In this note, and in those that will hopefully

* Work performed under the auspices of the U.S. Atomic Energy Commission.

241

follow it, we propose to focus upon a small aspect of this "revolution," namely, upon some revisions in classical, or Platonic, mathematical thinking that are forced by a candid appraisal of what the computer can and cannot do in operating with the real number field.

Our main objective can best be presented by an example. To a classical mathematician, even to a strict constructivist, the product xy of two real numbers x and y is a well-defined and self-evident operation. A computer, however, when instructed to perform such a multiplication, must perforce recognize only the presentation of the numbers x and y as strings of binary digits, and carry out instructions that are given in the form of an algorithm based upon Boolean operations upon strings of binary digits. Such an algorithm is presented explicitly in Section III; as remarked there, it is far from unique, and the form we have chosen contains obvious redundancies. Why not take this explicit algorithm as the definition of multiplication of real numbers? In a similar vein, we present an explicit Boolean algorithm for the addition of any number of summands (Section II). Together, these two algorithms provide a novel description of the real number field, and one that is far more "constructive" than the usual ones, as will be shown in detail elsewhere, in collaboration with Stephen Tanny.

The notable feature of the two Boolean algorithms is the explicit role played by the *carries*, an old concept familiar to every schoolboy (at least before the new math). The evidence from these and other algorithms on infinite binary strings suggests that the structure of the carries is crucial to any notion of complexity.

Other concepts of real variable theory can be analyzed in a similar vein, as we hope to do; for example, the notion of a continuous function. A computer must of necessity compute the values of a function by a Boolean algorithm, and such an algorithm alone may yield a measure of complexity of the function that realistically mirrors the computer's difficulties.

It would be remiss to omit a mention of a second motivation for the present approach. No string of binary digits resulting from physical measurement is ever given in its entirety; instead, all digits are unknown from a certain point on. One way to take this fact into account is to study the arithmetic of binary strings containing random digits. The study of efficient algorithms for computation under uncertainty conditions leads to an analysis of carries, much as sketched above. Again, we leave the detailed development of this point of view to a later note.

The presentation we have chosen is heuristic, in order to better display the simplicity of the results. Because of the large and growing literature on the subjects, we are forced to limit the bibliography to a few basic references.

II. THE ARITHMETIC OF BINARY STRINGS

In today's computations, the real numbers are represented by binary strings. Why not invert the process of definition, and reconstruct the real number field starting with the primary notion of a binary string? Our objective is to define sum, product, and division by constructive "Boolean" algorithms. We shall sketch how such a program can be carried out, and how it leads to unsuspected problems in Boolean algebra and even classical arithmetic.

By a binary string we mean a double infinite sequence

$$a = \cdots a_{-n}a_{-n+1}a_{-n+2} \cdots a_0 . a_1 a_2 a_3 \cdots$$

with a dot arbitrarily placed; the entries or *digits* a_i are members of the field GF(2) with two elements, and written as usual 0 or 1, often called *Boolean variables*. It is further assumed that for sufficiently small i, all digits a_i are equal. If they equal zero, the string is called *positive*; if they equal one, the string is called *negative*.

Two strings, a and b, are *equivalent* when there is an index n such that:

(i) $a_i = a_j$ for all $i, j > n$;
(ii) $b_i = b_j$ for all $i, j > n$;
(iii) $a_i b_i = 0$ and $a_i + b_i = 1$ for all $i > n$;
(iv) $a_n a_i = 0$ and $a_n + a_i = 1$ for all $i > n$;
(v) $b_n b_i = 0$ and $b_n + b_i = 1$ for all $i > n$.

It is easily verified that this definition gives an equivalence relation. An equivalence class of binary strings will be called a *real number*.

We shall now outline the main steps of a proof that the real numbers, in this somewhat exotic definition, form a field. We shall often improperly use the term "binary string" in place of "real number," leaving it to the reader to make the required logical adjustments.

Let us begin with the definition of addition of two binary strings a and b, taking as motivation the classical operation. If $c = a + b$, then the n-th digit c_n of the binary string c is a function of the n-th digits a_n and b_n of a and b, and of all the succeeding digits on the right. It is convenient to write

$$c_n = a_n + b_n + \phi_n(a,b),$$

when ϕ_n is a binary function of a doubly infinite sequence of binary variables. Furthermore, ϕ_n is to be well-defined for all values of the arguments. Thus, ϕ_n is a *Boolean function*, that is, a function taking values in GF(2), depending on infinitely many variables. Its explicit expression is easily found. Write ϕ for ϕ_0, and assume (as is suggested by ordinary arithmetic) that ϕ_n differs from ϕ only by a shift of the indices in the arguments:

$$c_0 = a_0 + b_0 + \phi(a,b).$$

The value of ϕ, or the *carry function* as it is called, will be 1 if and only if at least two of the three Boolean variables a_1, b_1, and $\phi_1(a,b)$ equal 1. Elementary Boolean interpolation gives the symmetric Boolean function

$$\phi(a,b) = a_1 b_1 + (a_1 + b_1)\phi_1(a,b) \qquad (1)$$

and in view of the fact that ϕ_1 has the same functional form that ϕ has, this determines the value of ϕ by iteration:

$$\phi(a,b) = a_1 b_1 + (a_1 + b_1)(a_2 b_2 + (a_2 + b_2)(\cdots)). \qquad (2)$$

All additions and multiplications indicated on the right side of this formula are to be carried out in the field with two elements; in other words, they are Boolean operations.

The remarkable fact about this formula is that the right side is *well-defined* despite the fact that it depends upon an infinite number of variables. This can be seen as follows.

Case 1. For some index i, we have $a_i + b_i = 0$. Then since $a_i + b_i = 0$ only those terms appear on the right side whose indices are lower than i. Thus, the sum on the right contains only a finite number of terms.

Case 2. $a_i + b_i = 1$ for all i. Then the right side reduces to the infinite product $\prod\limits_{i \geq 1} (a_i + b_i) = 1$, and again the expression is well-defined.

Thus, formula (2) can be used as a definition of addition without any previous notions about addition of real numbers. The proof that (2) defines an Abelian group law on the set of all binary strings is not trivial. In fact, it is best carried out after deriving an analogous expression for the sum of any number — finite or infinite! — of binary strings. This expression, which we believe to be new, requires the use of the *binary elementary symmetric functions* of an infinite sum of binary variables:

$$S^n(x, y, z, \cdots) = \text{sum of all products of } n \text{ distinct}$$
$$\text{variables from among } x, y, z, \cdots;$$

for example

$$S^1(x, y, z, \cdots) = x + y + z + \cdots$$

$$S^2(x, y, z, \cdots) = xy + xz + yz + \cdots,$$

etc.

We next define a doubly infinite sequence of Boolean functions $A_j^{(k)}$, $j, k \geq 0$, called the *carries*. Intuitively, the digit $A_j^{(k)}$ stands for the carry — that is, 0 or 1 — *into* the j-th entry *coming from* the sum of the k-th digits, as well as from all carries into the k-th entry. More precisely, set

$$A_0^{(0)} = S^1(x_0, y_0, z_0, \cdots; A_0^{(1)}, A_0^{(2)}, \cdots)$$

$$A_0^{(1)} = S^2(x_1, y_1, z_1, \cdots; A_1^{(2)}, A_1^{(3)}, \cdots)$$

$$A_0^{(2)} = S^4(x_2, y_2, z_2, \cdots; A_2^{(3)}, A_2^{(4)}, \cdots) \tag{3}$$

$$\cdots$$

$$A_0^{(k)} = S^{2^k}(x_k, y_k, z_k, \cdots; A_k^{(k+1)}, A_k^{(k+2)}, \cdots)$$

and similarly

$$A_1^{(1)} = S^1(x_1, y_1, z_1, \cdots, A_1^{(2)}, A_1^{(3)}, \cdots).$$

The expression for $A_j^{(k)}$ is obtained from the expression for $A_0^{(k-j)}$ by increasing all indices within the parentheses by the amount j. Clearly $A_j^{(k)} = 0$ unless $k \geqslant j$.

The zeroth digit of the *sum* of any number of binary strings $x, y, z, \cdots$ is $A_0^{(0)}$.

Before examining this definition in detail, let us consider the case of two summands x and y. Then $A_j^{(k)} = 0$ if $k - j > 1$ (since the corresponding symmetric functions identically vanish) and all other preceding expressions for carries reduce to the two single ones,

$$A_0^{(0)} = S^1(x_0, y_0, A_0^{(1)})$$

$$A_0^{(1)} = S^2(x_1, y_1, A_1^{(2)}).$$

All other carries are obtained from these by shifting indices. Expanding, we again find, as in (2)

$$A_0^{(0)} = x_0 + y_0 + A_0^{(1)}$$

$$A_0^{(1)} = x_1 y_1 + (x_1 + y_1) A_1^{(2)}$$

$$= x_1 y_1 + (x_1 + y_1)(x_2 y_2 + (x_2 + y_2) A_2^{(3)})$$

$$= x_1 y_1 + (x_1 + y_1)(x_2 y_2 + (x_2 + y_2)(\cdots)).$$

For clarity, we next derive from (3) the expression for the sum of four binary strings x, y, z, w. One first computes the carries into the zeroth digit, and one remarks that $A_0^{(k)} = 0$ if $k > 2$, and more generally $A_j^{(k)} = 0$ if $k - j > 2$. We can therefore write down a simplified form of (3) as follows:

$$A_0^{(0)} = x_0 + y_0 + z_0 + w_0 + A_0^{(1)} + A_0^{(2)}$$

$$A_0^{(1)} = x_1 y_1 + x_1 z_1 + x_1 w_1 + y_1 z_1 + y_1 w_1 + z_1 w_1$$

$$+ (x_1 + y_1 + z_1 + w_1) A_1^{(2)} + (x_1 + y_1 + z_1 + w_1) A_1^{(3)}$$

$$+ A_1^{(2)} A_1^{(3)}. \tag{4}$$

$$A_0^{(2)} = x_2y_2z_2w_2 + (y_2z_2w_2 + x_2z_2w_2 + x_2y_2z_2)(A_2^{(3)} + A_2^{(4)})$$
$$+ (x_2y_2 + x_2z_2 + x_2w_2 + y_2z_2 + y_2w_2)A_2^{(3)}A_2^{(4)}.$$

The reader may verify by examples that (4) does indeed give the classical algorithm for the sum of four real numbers.

From the preceding examples we can glean the rule for taking the sum of n summands. One sets all carries $A_j^{(k)} = 0$ whenever $k - j$ exceeds n, and one then proceeds to obtain the corresponding simplified form of (3) as we have done in the case of four summands. The complexity of the formula thus obtained depends on the biggest power of 2 closest to the number n of summands. For example, the formula for 8 summands is appreciably more complex than that for 7 summands.

The proof that the formula that we have described does indeed give the structure of an Abelian group to the set of binary strings, and the one that coincides with ordinary addition, is rather complex. This difficulty may be expected, in view of the fact that the definition we have given does away almost completely with questions of convergence. While we shall leave to a later work the detailed development of the algebraic properties of binary symmetric functions that are required for carrying out such a proof, we can give here a heuristic argument based upon previous knowledge of the properties of addition of real numbers.

The carry $A_0^{(k)}$ can be interpreted as the digit (0 or 1) that is carried to the zeroth entry after all digits at the k-th entry are summed together with the carries into the k-th entry. Thus for example in the addition of

$$0{\cdot}01 + 0{\cdot}01 + 0{\cdot}01 + 0{\cdot}01$$

The addition of the digits at the second entries, namely, $1 + 1 + 1 + 1$, give a carry into the zeroth-entry, that is, $A_0^{(2)} = 1$. Similarly, in the addition of

$$0{\cdot}01 + 0{\cdot}01 + 0{\cdot}001 + 0{\cdot}001 + 0{\cdot}01$$

there is a carry from the third entry to the second entry due to the addition of the third and fourth summands, that is, $A_2^{(3)} = 1$. The carry into the first digit is $A_0^{(2)} = 1$, because the carry $A_2^{(3)}$ contributes a unit as in the preceding example.

Once the idea of a carry is unravelled, it is easy to explain the role of symmetric functions in (3) by giving them a Boolean interpretation. The binary symmetric function of an infinite number of variables

$$S^{2^k}(x, y, z, \cdots)$$

equals one if and only if the number n of non-zero entries among the variables $x, y, z, \cdots$ can be written in dyadic form as

$$n = \sum_{i=2}^{\infty} a_i 2^i$$

with $a_k = 1$. This fact can be easily verified directly or else gleaned from a theorem on binomial coefficients due to N. J. Fine. This gives the "Boolean" interpretation of the carry $A_0{}^{(k)}$. It will equal one if and only if, in adding all digits at the k-th entry together with all carries into the kth entry, exactly one unit is carried to the zeroth entry.

III. MULTIPLICATION

Using the preceding Boolean formula (3) for addition of binary strings, one can derive in a similar vein formulas for multiplication, say of two binary strings x and y. We shall assume for simplicity that all digits to the left of x_0 are 0, and similarly for y. Then, using digitwise multiplication, the problem is that of adding the infinite sequence of binary strings

$$
\begin{aligned}
u &= (x_0 y_0) \cdot (x_0 y_1)(x_0 y_2)(x_0 y_3)(x_0 y_4) \cdots \\
v &= \quad 0 \quad \cdot (x_1 y_0)(x_1 y_1)(x_1 y_2)(x_1 y_3) \cdots \\
w &= \cdots \cdot \quad 0 \quad (x_2 y_0)(x_2 y_1)(x_2 y_2) \cdots \\
t &= \cdots \cdot \quad 0 \quad\quad 0 \quad\quad \cdots
\end{aligned}
\tag{5}
$$

and this can be done using the formula for summation of an infinite number of summands which has been discussed in the preceding Section. Because of the triangular pattern of zeros, enough of the carry terms vanish to make the sum well-defined.

The proof that this operation is associative and distributive with addition can be carried out on a purely "Boolean" level; division by a nonzero string can be treated similarly, and in this way one obtains a construction of the real number field.

By way of example, we derive a few terms of xy, using the notation $A_j{}^{(k)}$ for the carries as in (3) under a different interpretation. We have

$$
\begin{aligned}
A_0{}^{(0)} &= S^1(x_0 y_0, A_0{}^{(1)}, A_0{}^{(2)}, \cdots) \\
&= x_0 y_0 + A_0{}^{(1)} + A_0{}^{(2)} + \cdots.
\end{aligned}
$$

$$
\begin{aligned}
A_0{}^{(1)} &= S^2(x_0 y_1, x_1 y_0, A_1{}^{(2)}, A_1{}^{(3)}, \cdots) \\
&= x_0 x_1 y_0 y_1 + (x_0 y_1 + x_1 y_0)(A_1{}^{(2)} + A_1{}^{(3)} + \cdots) \\
&\quad + A_1{}^{(2)} A_1{}^{(3)} + \cdots.
\end{aligned}
$$

$$
\begin{aligned}
A_0{}^{(2)} &= S^4(x_0 y_2, x_1 y_1, x_2 y_0, A_2{}^{(3)}, A_2{}^{(4)}, \cdots) \\
&= x_0 x_1 x_2 y_0 y_1 y_2 (A_2{}^{(3)} + A_2{}^{(4)} + \cdots) \\
&\quad + (x_0 x_1 y_1 y_2 + x_0 x_2 y_0 y_2 + x_1 x_2 y_0 y_1)(A_2{}^{(3)} A_2{}^{(4)} + \cdots) \\
&\quad + (x_0 y_2 + x_1 y_1 + x_2 y_0)(A_2{}^{(3)} A_2{}^{(4)} A_2{}^{(5)} + \cdots).
\end{aligned}
$$

These displays amply show that multiplication is a far more complex operation than addition, as we shall have occasion to remark later.

Note that, in contrast to addition, the carries for multiplication contain all the preceding digits; for example, the carry $A_0^{(k)}$ is a function depending crucially upon the Boolean variable x_1, for every k. It is probably impossible to obtain a Boolean formula for multiplication where the dependence upon each variable is "isolated," as it is in the addition formulas, but this impossibility remains to be demonstrated. At any rate, the expression for multiplication as a Boolean function is redundant, like the expression for addition. The reason for this redundancy lies in the fact that the terms occurring in different columns of (5) are "coupled." This is intuitively obvious by inspection of (5): for example, the first and second columns both contain the variable x_0, and are therefore not "independent." This notion can be made precise in many ways, of which perhaps the simplest is by use of probability. Suppose all digits appearing in (5) are independent random variables with values in GF(2) and taking the values 0 and 1 with probability $\frac{1}{2}$ each. Let

$$C_0 = x_0 y_0, \quad C_1 = x_0 y_1 + x_1 y_0, \cdots$$

Then

$$\text{Prob }(C_0 = 1) = \tfrac{1}{4}$$
$$\text{Prob }(C_1 = 1) = \tfrac{3}{8}$$
$$\text{Prob }(C_0 = 1 \text{ and } C_1 = 1) = \text{Prob }(C_0 C_1 = 1) = \tfrac{1}{8},$$

so that the random variables C_0 and C_1 are not independent.

Thus, it is to be conjectured that more efficient algorithms for multiplication can be derived by efficient use of canonical forms for Boolean functions (for example, prime implicants). The techniques considered so far in the literature are applicable only to multiplication of finite strings, representing integers, and disregard the difficulties of infinite carrying, which motivate us at present. On the other hand, from the present point of view it seems that the efficiency of an algorithm for multiplication should be measured by broader criteria than simply the reduction in number of "operations," as has been customary so far. A Boolean expression which explicitly displays the dependence upon a given digit – as is the case in addition – is, from the present point of view, more "efficient" (or perhaps the word might be "more desirable"), as it better lends itself to error analysis.

Similar remarks apply to other "operations" on binary strings, for example, $f(x) = x^2$, which easily gives the carries

$$A_0^{(0)} = x_0 + A_0^{(1)} + A_0^{(2)} + \cdots$$
$$A_0^{(1)} = x_0 x_1 + A_1^{(2)} A_1^{(3)} + \cdots$$

$$A_0^{(2)} = x_0 x_1 x_2 (A_2^{(3)} + A_2^{(4)} + \cdots)$$
$$+ x_0 x_2 (A_2^{(3)} A_2^{(4)} + \cdots)$$
$$+ x_1 (A_2^{(3)} A_2^{(4)} A_2^{(5)} + \cdots).$$

$\cdots$

These carry expressions can be iterated and a particularly pleasing formula for x^2 can be obtained.

Continuing in this vein, we are led to formulate tentatively the problem in greater generality. Let $f(x, y, \cdots, z)$ be a function of binary strings, with values in GF(2), and *continuous* in the usual sense. Notice that the following definition of continuity: "f is near to g whenever the first n digits of the entries in f coincide with the first n digits of the entries of g, for n large" is insufficient, in view of the double representation of certain real numbers. This imperfection of the real number system is the source of a great many computational difficulties.

To such a function we may associate *binary complexity expansions*, in imitation of the above examples. For simplicity we consider only the case of functions $f(x)$ of a single string x—whereas addition and multiplication, as considered above, depend on two or more strings. A binary complexity expansion of $f(x)$ will then be a sequence of *carry functions* C_{jk}, each carry function being a Boolean function of the variable $x_0, x_1, x_2, \cdots$, with the following properties:

(a) $f(x) = C_{00}(x_0, C_{01}, C_{02}, \cdots)$

   $\quad C_{01} = C_{01}(x_0, x_1, C_{12}, C_{13}, \cdots)$

   $\cdots$

   $C_{jk} = C_{jk}(x_j, x_{j+1}, \cdots, x_k, C_{k,k+1}, C_{k,k+2}, \cdots)$;
   with $j \leqslant k$.

(b) C_{jk} vanishes identically if x_k is set to zero.

Leaving a detailed study of binary complexity expansion to a later occasion, we now remark the following:

(1) A binary complexity expansion of a continuous function is in general not unique; it is an open question whether further conditions can be imposed upon the structure of the expansion so that it can be made unique.

(2) It is at times possible to obtain a binary complexity expansion when C_{jk} does not depend on $x_j, x_{j+1}, \cdots, x_{k-1}$, but only upon another proper subset (as in the above example of addition). It would be desirable to know which continuous functions possess such a simpler expansion.

(3) The continuity of $f(x)$ insures that the representation by "carries" breaks; that is, that the operations are well-defined.

(4) The structure of the carries C_{jk} should lead to a more satisfactory notion of complexity of a function $f(x)$ than any of the ones considered so far.

IV. BINARY FUNCTIONS

By the preceding analysis of addition and multiplication we hope we have brought into light some of the objectives of significance arithmetic. An ordinary continuous function $F(x)$ of a real variable, with real values, when considered from the point of view of digital computers, is a sequence of binary functions

$$F(x) = \cdots F_0(x) \cdot F_1(x) F_2(x) \cdots$$

where $F_n(x)$ represents the n-th digit of $F(x)$ in binary notation. The problem of formulating an algorithm for the computation of $F(x)$ is therefore reduced to the formulation of an algorithm for the computation of a *Boolean* function, say $F_0(x)$, of a *binary string x*. Once the real axis is identified with the set of binary strings in the well-known manner, all properties of continuous functions that are relevant for computation must be expressible by properties of (binary) Boolean functions. Such properties of classical interest as differentiability and periodicity become irrelevant in this context, while on the other hand, the complexity of representation of a Boolean function by carry functions becomes of prime relevance, as well as other features of such representation.

It is unfortunate that the theory of Boolean functions of infinitely many variables is still in its infancy. We can only surmise that some of the classical notions developed for Boolean functions of a finite number of variables can be carried over, at least under the assumption of continuity. It would be particularly interesting to carry over Quine's theory of *prime implicants*. An easy argument shows that prime implicants exist for the sum function of any number of summands, but are of course infinite disjunctions. It would be of the utmost interest to determine the prime implicants for multiplication and the other elementary functions.

We further surmise that the representation of addition and multiplication as an (infinite) disjunction of their prime implicants will lead to simple "Boolean" proofs of the laws of arithmetic and thereby of the "existence" of the real number field.

It would be remiss to conclude without mentioning another approach to the analysis of Boolean functions, namely, the Walsh functions. Recall that for real x the *Rademacher functions* $\phi_n(x)$ are defined by (v. Fine)

$$\phi_0(x) = 1, \quad 0 \leqslant x < 1/2; \quad \phi_0(x) = -1, \quad 1/2 \leqslant x < 1,$$

$$\phi_0(x+1) = \phi_0(x), \quad \phi_n(x) = \phi_0(2^n x), \quad n \geqslant 1,$$

and then x is restricted to the interval $[0, 1]$. Thus $\phi_n(x)$ takes the value $+1$ if the n-th digit of the binary expansion of the real number x is 0, and -1 if such a digit is 1. The products

$$\psi_n(x) = \phi_{n_1}(x) \, \phi_{n_2}(x) \cdots \phi_{n_k}(x),$$

when $n = 2^{n_1} + 2^{n_2} + \cdots + 2^{n_k}$, are the well-known *Walsh functions*, and are a complete orthonormal set for $L_2(0, 1)$.

Now suppose we wish to analyze a function such as $x+y$ in terms of its binary components. Using a two-variable expansion into Walsh functions, we have

$$w(m, n) = \int_0^1 \int_0^1 (x+y)\psi_m(x)\psi_n(y) \, \mathrm{d}x \, \mathrm{d}y,$$

and

$$\sum_{m,n \geqslant 0} w(m, n)\psi_m(x)\psi_n(y) = x+y. \tag{6}$$

The coefficient $\psi_m(x)\psi_n(y)$ depends only upon a finite number of digits of x and of digits of y. Since $w(m, n)$ is fixed once and for all, formula (6) can be interpreted as giving an algorithm for addition "without carrying." A similar analysis can be carried out for the other arithmetic operations. The computational weaknesses of this method are almost self-evident, although it is surprising that it should not have been used for the study of prime implicants of Boolean functions, as we hope will be done.

Finally, we remark that the carry formulas we stated in the preceding Section are examples of functions that might be called "antirecursive." It would be interesting to make the notion precise and compare it with the ordinary notion of recursive functions.

BIBLIOGRAPHY

Ádám, A., *Truth Functions*, Akademiai Kiadó, Budapest, 1968.

Bishop, E., *Constructive Analysis*, McGraw-Hill, New York, 1968.

Hammer, P., *Boolean Methods in Operations Research*, Springer, Heidelberg, 1968.

Fine, N. J., Binomial Coefficients modulo a prime, *American Mathematical Monthly*, 54 (1947), pp. 589–592.

Fine, N. J., On the Walsh functions, *Trans. Am. Math. Soc.*, 65 (1949), pp. 372–414.

Knuth, D., *The Art of Computer Programming*, Vols. 1 & 2, Addison-Wesley, Reading, Mass.

Metropolis, N. and R. L. Ashenhurst, Significant digit computer arithmetic, *IRE Trans. Electronic Computers*, EC-7, (1958), pp. 265–267.

Practical Algorithms for Finding the Type of a Polynomial

JOHN J. H. MILLER

*Trinity College,
Dublin, Ireland*

I. INTRODUCTION

This paper presents a survey of the algorithms we have already published for determining the type of a polynomial relative either to the unit circle in the complex plane or to the imaginary axis. It also contains a number of new results for which the proofs will appear elsewhere.

A polynomial is said to be of type (p_1, p_2, p_3) relative to the unit circle if, counting multiplicities, the number of its zeros inside, on, and outside the unit circle is respectively p_1, p_2 and p_3. Similarly a polynomial is said to be of type (p_1, p_2, p_3) relative to the imaginary axis if, counting multiplicities, the number of its zeros to the left of, on, and to the right of the imaginary axis is respectively p_1, p_2, and p_3. The unit circle and the imaginary axis are the cases which arise most often in practice. There is no difficulty of course in generalizing our definitions and theorems to arbitrary circles and lines in the complex plane.

The practical importance of finding the type of a polynomial needs little comment here, since most stability problems of numerical analysis and applied mathematics reduce ultimately to the problem of showing that some polynomial is either of type $(p_1, p_2, 0)$ or of type $(p_1, 0, 0)$. Furthermore, recent work on mixed initial boundary value problems for hyperbolic systems of differential and difference equations has shown the importance of unstable polynomials of type $(0, p_2, p_3)$.

At this stage it is appropriate to point out how the present results differ from those stemming from the classical work of Schur and Cohn for the unit circle and of Routh and Hurwitz for the imaginary axis. These classical algorithms generally deal only with polynomials of type $(p_1, 0, 0)$ and frequently some or all of the coefficients are restricted to being real. By a careful reappraisal of the classical results and by finding new methods of proof, we have been able to extend the classical algorithms to polynomials of arbitrary type and to remove all restrictions on the coefficients. Furthermore, the introduction of the simple concept of the "type of a polynomial" and the related notation prove extremely useful in discussing problems of stability and in stating the theorems, old and new, in both a concise and unambiguous manner.

The algorithms presented here are simple to use in practice. They have been applied with success in a variety of situations by both the author and other workers. They are easily memorized and are surprisingly useful even for polynomials of the second degree. They are particularly important in cases where the coefficients of the polynomial depend on one or more parameters and it is required to find the range of values of the parameters for which the polynomial is of a given type.

In the next Section we summarize our results for the unit circle. In Section III we present the algorithms for the imaginary axis and, finally, in Section IV we give a number of examples using the algorithms of Section III.

II. THE TYPE OF A POLYNOMIAL RELATIVE TO THE UNIT CIRCLE

In this section we state the theorems which enable us to find the type of any polynomial relative to the unit circle. Since an ample number of applications of these to stability problems for finite difference schemes for parabolic and hyperbolic partial differential equations is given in [1], we do not illustrate the use of these algorithms here.

Let $f(z)$ be a polynomial of degree n with real or complex coefficients. Without loss of generality we may assume throughout the rest of the paper that $f(0) \neq 0$, since zeros at the origin can be recognized by inspection, recorded and then discarded without difficulty. The inverse of f in the unit circle is defined as $f^*(z) = z^n \overline{f(z^*)}$ where $z^* = 1/\bar{z}$ is the inverse of z in the unit circle. If $f(z) = a_0 + a_1 z + \cdots + a_n z^n$ then clearly $f^*(z) = \bar{a}_n + \bar{a}_{n-1} z + \cdots + \bar{a}_0 z^n$. Since f is of degree n and $f(0) \neq 0$ the same is true of f^*. We call f self-inversive if the zeros of f and their multiplicities are symmetric with respect to inversion in the unit circle. Equivalently f is self-inversive if f and f^* have the same set of zeros and each particular zero has the same multiplicity in f^* as in f. It is easy to show that this is also equivalent to the condition $|f^*(z)| = |f(z)|$ for all z. We define the reduced polynomial corresponding to f as $\check{f}(z) =$

$(f^*(0)f(z) - f(0)f^*(z))/z$. Self-inversiveness is clearly equivalent to $\check{f}(z) \equiv 0$.

We can now state the fundamental result for the unit circle.

Theorem 2.1

Assume f is a polynomial such that $f(0) \neq 0$ and $|f^*(0)| \neq |f(0)|$. Then f is of type (p_1, p_2, p_3) iff $\check{f}$ is of type $(p_1 - 1, p_2, p_3)$ if $|f^*(0)| > |f(0)|$ and of type $(p_3 - 1, p_2, p_1)$ if $|f^*(0)| < |f(0)|$.

This theorem relates the type of f to the type of $\check{f}$. The condition $|f^*(0)| \neq |f(0)|$ guarantees that the degree of $\check{f}$ is one less than that of f, and thus repeated application of the result reduces the original problem of degree n, stepwise, to a trivial problem for a polynomial of the first degree. Notice however that this procedure breaks down if, at any stage of the algorithm, the condition $|f^*(0)| \neq |f(0)|$ is violated by one of the polynomials generated by the reduction process. For example the condition is certainly violated by a self-inversive polynomial. Hence, in such a case, the reduction procedure must be modified. In the self-inversive case, instead of constructing $\check{f}$ (which is in fact identically zero) we construct the derivative f' of f. This is certainly of degree one less than that of f and the relation between the types of f and f' is given by

Theorem 2.2

Assume f is a self-inversive polynomial. Then f is of type (p_1, p_2, p_3) iff f' is of type $(p_1 + k - 1, p_2 - k, p_3)$ where k is the number of distinct zeros of f on the unit circle.

This theorem enables us to find not only the type of a self-inversive polynomial, but also the number of distinct zeros it has on the unit circle. Note that the introduction of the new unknown k does not increase the number of unknowns, since for self-inversive polynomials $p_1 = p_3$.

Intermediate cases can also occur where the condition $|f^*(0)| \neq |f(0)|$ is violated but the polynomial is not self-inversive. In such cases neither of the above theorems is applicable, and a new approach must be sought. If this occurs it is easy to show that $\check{f}$ is of the form $\check{f}(z) = z^q g(z)$ where g is a self-inversive polynomial such that $g(0) \neq 0$ and q is some integer satisfying $0 \leqslant q \leqslant [(n-2)/2]$.

We then construct $F(z) = (z^{q+1} + \theta)f(z)$ where $\theta = \frac{1}{2}(f^*(0)f(0)/|f^*(0)f(0)|)(|g(0)|/g(0))$. Clearly $|\theta| = \frac{1}{2}$ so that F is of type $(p_1 + q + 1, p_2, p_3)$ if f is of type (p_1, p_2, p_3). It is easy to check that $|F^*(0)| > |F(0)|$ so that, by Theorem 2.1, $\check{F}$ is of type $(p_1 + q, p_2, p_3)$. Indeed $\check{F}$ is of the form $\check{F}(z) = z^q h(z)$, where h is a polynomial of type (p_1, p_2, p_3), the same type as f. Furthermore $|h^*(0)| < |h(0)|$ and thus Theorem 2.1 can be applied to h whereas it cannot be applied to f. Obviously the degree of h

is one less than that of h, and we have therefore the modified reduction procedure given by

Theorem 2.3
Assume f is not a self-inversive polynomial but $|f^*(0)| = |f(0)|$. Then f is of type (p_1, p_2, p_3) iff $\check{h}$ is of type $(p_3 - 1, p_2, p_1)$.

Theorems 2.1, 2.2 and 2.3 cover all possibilities, and by combining them an algorithm is obtained whereby the type relative to the unit circle of an arbitrary polynomial may be determined. Further details, proofs and examples illustrating the use of Theorems 2.1 and 2.2 may be found in [1] and [2]. The proof of Theorem 2.3 has been outlined above.

III. THE TYPE OF A POLYNOMIAL RELATIVE TO THE IMAGINARY AXIS

In this Section we state the theorems which are used to determine the type of a polynomial relative to the imaginary axis. We warn the reader that the notation used here is the same as in Section II but its meaning is different.

Let $f(z)$ be a polynomial of degree n with real or complex coefficients. As before we may assume without loss of generality that $f(0) \neq 0$. The inverse of f in the imaginary axis is defined as $f^*(z) = \overline{f(z^*)}$ where $z^* = -\bar{z}$ is the inverse (reflection) of z in the imaginary axis. If $f(z) = a_0 + a_1 z + \cdots + a_n z^n$ then clearly $f^*(z) = \bar{a}_0 - \bar{a}_1 z + \cdots + (-1)^n \bar{a}_n z^n$. Obviously f^* is also of degree n and $f^*(0) \neq 0$. We call f self-inversive if the zeros of f and their multiplicities are symmetric with respect to inversion in the imaginary axis. Equivalently f is self-inversive if f and f^* have the same set of zeros and each particular zero has the same multiplicity in f^* as in f. It is easy to show that this is also equivalent to the condition $|f^*(z)| = |f(z)|$ for all z.

Unlike the case for the unit circle the reduced polynomial f may be defined in several ways. We construct three different reduced polynomials and discuss the relations between them. We begin with the definition $\check{f}_M(z) = (f^*(w)f(z) - f(w)f^*(z))/(z - w)$, where w is regarded as a parameter and it is assumed that w does not lie on the imaginary axis, that is Re $w \neq 0$. $\check{f}_M$ is the natural analogue of the reduced polynomial of Section II, and it can be arrived at by a suitable conformal transformation of the complex plane mapping the unit circle into the imaginary axis. However, in practice, it has the serious disadvantage that division by $z - w$ is required, which makes the calculations difficult. The relation between the types of f and $\check{f}_M$ is given in

Theorem 3.1

Assume that for the polynomial f a complex number w can be chosen such that $|f^*(w)| \neq |f(w)|$ and Re $w \neq 0$. *Then f is of type (p_1, p_2, p_3) iff $\check{f}_M$ is of type*

$$
\begin{aligned}
&(p_1 - 1, p_2, p_3) && \textit{if} && \left.|f^*(w)| > |f(w)|\right\} \\
&(p_3 - 1, p_2, p_1) && \textit{if} && \left.|f^*(w)| < |f(w)|\right\} \textit{ and } \text{Re } w < 0
\end{aligned}
$$

$$
\begin{aligned}
&(p_1, p_2, p_3 - 1) && \textit{if} && \left.|f^*(w)| > |f(w)|\right\} \\
&(p_3, p_2, p_1 - 1) && \textit{if} && \left.|f^*(w)| < |f(w)|\right\} \textit{ and } \text{Re } w > 0.
\end{aligned}
$$

Note that a w satisfying the hypotheses of the above theorem can be found for all non-self-inversive polynomials f. Furthermore the condition $|f^*(w)| \neq |f(w)|$ guarantees that the degree of $\check{f}_M$ is one less than that of f. Thus repeated application of the result reduces the original problem for a polynomial of the nth degree, stepwise, to a trivial problem for a polynomial of the first degree, provided of course that the conditions of the theorem are met at each stage of the reduction procedure. Because of the computational difficulties however this algorithm is not recommended for practical purposes.

To obtain a simpler algorithm we proceed as follows. From the definition it is clear that $\check{f}_M$ may be regarded as a polynomial in the parameter w with coefficients which are polynomials in z. Thus we may write $\check{f}_M(z) = F_0(z) + F_1(z)w + \cdots + F_{n-1}(z)w^{n-1}$. The coefficients $F_j(z)$ are then determined by multiplying across by $z - w$ and comparing coefficients of w. It is not hard to show then that

$$F_0(z) = (f^*(0)f(z) - f(0)f^*(z))/z$$

$$F_1(z) = ((f^*)'(0)f(z) - f'(0)f^*(z))/z + (f^*(0)f(z) - f(0)f^*(z))/z^2.$$

It now transpires that a satisfactory algorithm can be obtained by truncating the above expression for $\check{f}_M$ at the second term. This leads us to the definition of our second reduced polynomial as $\check{f}_S(z) = F_0(z) + F_1(z)w$. Using the above explicit formulas for F_0 and F_1 this may also be written in the form

$$
\begin{aligned}
\check{f}_S(z) = {}& w((f^*)'(0)f(z) - f'(0)f^*(z))/z \\
& + (z + w)(f^*(0)f(z) - f(0)f^*(z))/z^2.
\end{aligned}
$$

This has the important property that division by z only is required, so that $\check{f}_S$ is much easier to construct than is $\check{f}_M$. The relation between the types of f and $\check{f}_S$ is given in

Theorem 3.2

Assume f is a polynomial such that Re $f^*(0) f'(0) \neq 0$ *and choose a number w such that* Re w *and* Re $f^*(0) f'(0)$ *are of opposite sign. Then f*

is of type (p_1, p_2, p_3) *iff* $\check{f}_S$ *is of type* $(p_1 - 1, p_2, p_3)$ *if* $\operatorname{Re} f^*(0) f'(0) > 0$ *and of type* $(p_1, p_2, p_3 - 1)$ *if* $\operatorname{Re} f^*(0) f'(0) < 0$.

The condition $\operatorname{Re} f^*(0) f'(0) \neq 0$ guarantees that the degree of $\check{f}$ is one less than that of f and it also ensures that a suitable w can be found. Thus the original problem may be reduced, stepwise, in the usual way, provided the above condition is fulfilled at each stage.

In the above theorem it is obvious that there is considerable latitude in the choice of the parameter w. One special choice of w leads to our third, and last, definition of the reduced polynomial.

We begin by writing $\check{f}_S$ in the form

$$\check{f}_S(z) = [(f^*(0) + w(f^*)'(0)) f(z) - (f(0) + wf'(0)) f^*(z)]/z$$
$$+ w(f^*(0) f(z) - f(0) f^*(z))/z^2.$$

We note now that this expression is considerably simplified if w is chosen to satisfy $f(0) + wf'(0) = 0$ or $w = -f(0)/f'(0)$. Clearly then

$$\operatorname{Re} w = |w|^2 \operatorname{Re}(1/\overline{w}) = |w|^2 \operatorname{Re}(1/w)$$
$$= -|w|^2 \operatorname{Re}(f'(0)/f(0)) = -(|w|/|f(0)|)^2 \operatorname{Re} f^*(0) f'(0),$$

so that $\operatorname{Re} w$ and $\operatorname{Re} f^*(0) f'(0)$ are of opposite sign, which is consistent with the requirement of the above theorem. Substituting this value of w into $\check{f}_S$ and multiplying by $f'(0)$ leads to the reduced polynomial

$$\check{f}_D(z) = \frac{f^*(0) f'(0) - f(0) (f^*)'(0)}{z} f(z) - f(0) \frac{f^*(0) f(z) - f(0) f^*(z)}{z^2}.$$

$\check{f}_D$ is of course simply a special case of $\check{f}_S$. The relation between the types of f and $\check{f}_D$ is then given by

Theorem 3.3
Assume f is a polynomial such that $\operatorname{Re} f^*(0) f'(0) \neq 0$. *Then f is of type* (p_1, p_2, p_3) *iff* $\check{f}_D$ *is of type* $(p_1 - 1, p_2, p_3)$ *if* $\operatorname{Re} f^*(0) f'(0) > 0$ *and of type* $(p_1, p_2, p_3 - 1)$ *if* $\operatorname{Re} f^*(0) f'(0) < 0$.

As before the condition $\operatorname{Re} f^*(0) f'(0) \neq 0$ guarantees that the degree of $\check{f}_D$ is one less than that of f. Once again therefore we have a stepwise reduction process, provided that at each stage the above condition is fulfilled.

We turn now to the cases where the conditions $|f^*(w)| \neq |f(w)|$ and $\operatorname{Re} f^*(0) f'(0) \neq 0$ are violated. This certainly happens for example if f is self-inversive. Indeed, as is to be expected from the results of Section II, self-inversiveness with respect to the imaginary axis is equivalent to the reduced polynomial vanishing identically, where the reduced polynomial may be taken as am one of $\check{f}_M, \check{f}_S$ and $\check{f}_D$. In the self-inversive case therefore the reduction procedure must again be modified. Instead of the reduced polynomial we construct the polar derivative p of f, where if f is of the nth degree $p(z) = nf(z) - (z-1)f'(z)$. The relation between the types of f and p is then given by

Theorem 3.4
Assume f is a self-inversive polynomial. Then f is of type (p_1, p_2, p_3) iff p is of type $(p_1 + k - 1, p_2 - k, p_3)$, where k is the number of distinct zeros of f on the imaginary axis.

The degree of p is easily seen to be one less than that of f. Furthermore, as before, the number of unknowns is not increased by the presence of the new unknown k, since $p_1 = p_3$ for all self-inversive polynomials. In the self-inversive case therefore we can find not only the type of f but also the number of distinct zeros it has on the imaginary axis.

Again there are intermediate cases where $|f^*(w)| = |f(w)|$ or Re $f^*(0)f'(0) = 0$ but f is not self-inversive. Possibly a result analogous to Theorem 2.3 could be found by a conformal transformation. However, such a result would be based on $\check{f}_M$ rather than $\check{f}_S$ or $\check{f}_D$, and it would not necessarily be the most satisfactory method if one or other of the latter reduced polynomials was being used. A useful result based on $\check{f}_D$ is given at the end of this Section for a special case. It is hoped to return to the general case in a subsequent paper.

In many practical applications the coefficients of f are all real. It is thus of interest to see if the computations can be simplified in this case. This is indeed so for $\check{f}_S$ and $\check{f}_D$. If we write $f = \text{odd } f + \text{even } f$, where odd f and even f denote the terms in f involving the odd and even powers of z respectively, then $\check{f}_S$ and $\check{f}_D$ take the form

$$\check{f}_S(z) = -wf'(0) \text{ even } f(z)/z + (z+w)f(0) \text{ odd } f(z)/z^2$$

$$\check{f}_D(z) = (f'(0)f(z) - f(0) \text{ odd } f(z)/z)/z$$

where we have cancelled some non-zero constant factors. Furthermore Re $f^*(0)f'(0) = f(0)f'(0)$ so that, in the real case, we never have to construct $f^*(z)$ and the calculations are greatly simplified.

Also in the case of real coefficients we can obtain a simple partial analogue of Theorem 2.3. Consider the intermediate case where $f(0)f'(0) = 0$ but f is not self-inversive. Since $f(0) \neq 0$ it follows that $f'(0) = 0$ and it is easy to show that $\check{f}_D(z) = -f(0)zg(z)$, where g is self-inverse and $g(0) = f'''(0)/6 \neq 0$ provided we assume that $f'''(0) \neq 0$. We construct $F(z) = (z+1)f(z)$, which is of type $(p_1 + 1, p_2, p_3)$ if f is of type (p_1, p_2, p_3). Then $F(0)\check{F}'(0) > 0$ so that by Theorem 3.3 $\check{F}_D$ is of type (p_1, p_2, p_3). Indeed $\check{F}_D(z) = f(0)h(z)$ where $h(z) = f(z) + z(z-1)g(z)$. Furthermore $h(0)h'(0) = -f(0)g(0) \neq 0$ and thus Theorem 3.3 can be applied to h whereas it cannot be applied to f. We have therefore

Theorem 3.5

Assume that the coefficients of f are real and that $f(0) \neq 0$, $f'(0) = 0$, $f'''(0) \neq 0$. Then f is of type (p_1, p_2, p_3) iff $\check{h}$ is of type $(p_1 - 1, p_2, p_3)$ if $f(0)f'''(0) < 0$ and of type $(p_1, p_2, p_3 - 1)$ if $f(0)f'''(0) > 0$.

Further details and the proofs of Theorems 3.1 and 3.4 may be found in[3], where examples are also given of their use. The proof of Theorem 3.2 will appear in a subsequent paper, and Theorem 3.3 is of course a corollary of Theorem 3.2. The proof of Theorem 3.5. is outlined above.

For practical purposes it is advisable to use $\check{f}_D$ rather than $\check{f}_M$. For this reason the examples given in the next Section make use only of $\check{f}_D$.

IV. EXAMPLES

In this Section we give examples of some of the above algorithms for finding the type of a polynomial relative to the imaginary axis. We do not illustrate the unit circle case, since this has been dealt with adequately in [1].

The stepwise reduction of an arbitrary polynomial to a polynomial of the first degree is carried out by constructing $\check{f}_D$, p or $\check{h}$ at each step, depending on which of the Theorems 3.3, 3.4 or 3.5 is appropriate for that step.

In each example considered below the sequence of polynomials of descending degree, which is generated by the repeated application of these theorems, is given. The highest common factor of the coefficients of each polynomial has been cancelled, and in the case of real coefficients the simplified expression for $\check{f}_D$ has always been used.

The first polynomial in each sequence is assumed to be of type (p_1, p_2, p_3), and the type of each subsequent polynomial in the sequence is known in terms of p_1, p_2, p_3 and possibly k from the theorems used. However, the type of the last polynomial of the sequence may also be determined by inspection. A comparison of the two expressions for the type of this final polynomial is then made and the type of the original polynomial found.

Example 1

$f(z) = z^5 + 2z^4 + 4z^3 + 4z^2 + 2z + 1$. f has real coefficients and is a stable polynomial.

$f_5(z) = z^5 + 2z^4 + 4z^3 + 4z^2 + 2z + 1$ (p_1, p_2, p_3)	$f_5(0)f_5'(0) = 2$
$f_4(z) = 2z^4 + 3z^3 + 8z^2 + 4z + 4$ $(p_1 - 1, p_2, p_3)$	$f_4(0)f_4'(0) = 16$
$f_3(z) = 2z^3 + 3z^2 + 5z + 4$ $(p_1 - 2, p_2, p_3)$	$f_3(0)f_3'(0) = 20$

$f_2(z) = 10z^2 + 7z + 25$ $(p_1 - 3, p_2, p_3)$ $f_2(0)f_2'(0) = 175$

$f_1(z) = 10z + 7$ $(p_1 - 4, p_2, p_3) = (1, 0, 0)$

f is of type $(5, 0, 0)$.

Example 2

$f(z) = 9z^5 + 4z^4 + 5z^3 + 4z^2 + 2z + 1$. f has real coefficients and is an unstable polynomial.

$f_5(z) = 9z^5 + 4z^4 + 5z^3 + 4z^2 + 2z + 1$

         (p_1, p_2, p_3) $f_5(0)f_5'(0) = 2$

$f_4(z) = 18z^4 - z^3 + 10z^2 + 3z + 4$

         $(p_1 - 1, p_2, p_3)$ $f_4(0)f_4'(0) = 12$

$f_3(z) = 54z^3 - 3z^2 + 34z + 9$

         $(p_1 - 2, p_2, p_3)$ $f_3(0)f_3'(0) = 306$

$f_2(z) = 459z^2 - 147z + 289$

         $(p_1 - 3, p_2, p_3)$ $f_2(0)f_2'(0) = -42483$

$f_1(z) = 153z - 49$

         $(p_1 - 3, p_2, p_3 - 1) = (0, 0, 1)$

f is of type $(3, 0, 2)$.

Example 3

$f(z) = z^3 - 5z^2 + (9 + 4i)z + (-1 - 8i)$. f has complex coefficients and is unstable.

$f_3(z) = z^3 - 5z^2 + (9 + 4i)z + (-1 - 8i)$

         (p_1, p_2, p_3) $\operatorname{Re} f_3^*(0)f_3'(0) = -41$

$f_2(z) = 41z^2 + (-204 + 8i)z + (49 + 204i)$

         $(p_1\ p_2, p_3 - 1)$ $\operatorname{Re} f_2^*(0)f_2'(0) = -8364$

$f_1(z) = z - i$

         $(p_1, p_2, p_3 - 1) = (0, 1, 0)$

f is of type $(0, 1, 2)$.

Example 4

$f(z) = z^5 + (1 + 2i)z^4 + (-1 + 2i)z^3 + (1 - 2i)z^2 + (-2 + 2i)z - 4i$. f has complex coefficients. It has zeros in all three localities. f_3 is self-inversive.

$f_5(z) = z^5 + (1 + 2i)z^4 + (-1 + 2i)z^3 + (1 - 2i)z^2 + (-2 + 2i)z - 4i$

         (p_1, p_2, p_3) $\operatorname{Re} f_5^*(0)f_5'(0) = -8$

$f_4(z) = z^4 + (1 + 2i)z^3 + (1 + 2i)z^2 + (1 + 2i)z - 2i$

         $(p_1, p_2, p_3 - 1)$ $\operatorname{Re} f_4^*(0)f_4'(0) = 4$

$f_3(z) = z^3 + 2iz^2 + z + 2i$

         $(p_1 - 1, p_2, p_3 - 1)$ $p_1 = p_3$

$f_2(z) = (3+2i)z^2 + (2+4i)z + (1+6i)$

$$(p_1+k-2, p_2-k, p_1-1) \quad \text{Re} f_2^* (0) f_2' (0) = 26$$

$f_1(z) = (39+26i)z + (-22+60i)$

$$(p_1+k-3, p_2-k, p_1-1) = (1,0,0)$$

f is of type $(1,3,1)$ and all three zeros on the imaginary axis are distinct.

Example 5

$f(z) = z^3 + 30z^2 + (200+\alpha)z + 40\alpha$. f has real coefficients depending on a real parameter α. The critical values of the parameter (those values at which the hypothesis of Theorem 3.3 is violated) are found by first constructing the basic sequence. The various values of the parameter between and at the critical values are then considered separately. This example arises from a problem in chemical engineering.

$f_3(z) = z^3 + 30z^2 + (200+\alpha)z + 40\alpha$

$$f_3(0)f_3'(0) = 40\alpha(200+\alpha)$$

$f_2(z) = (200+\alpha)z^2 + 10(600-\alpha)z + (200+\alpha)^2$

$$f_2(0)f_2'(0) = 10(600-\alpha)(200+\alpha)^2$$

$f_1(z) = (600-\alpha)[(200+\alpha)z + 10(600-\alpha)]$.

The critical values of α are clearly $-200, 0$ and 600.

(i) $\alpha < -200$:

    f_3 (p_1, p_2, p_3) $f_3(0)f_3'(0) > 0$

    f_2 (p_1-1, p_2, p_3) $f_2(0)f_2'(0) > 0$

    f_1 $(p_1-2, p_2, p_3) = (0,0,1)$

    f is of type $(2, 0, 1)$.

(ii) $-200 < \alpha < 0$:

    f_3 (p_1, p_2, p_3) $f_3(0)f_3'(0) < 0$

    f_2 (p_1, p_2, p_3-1) $f_2(0)f_2'(0) > 0$

    f_1 $(p_1-1, p_2, p_3-1) = (1,0,0)$

    f is of type $(2, 0, 1)$.

(iii) $0 < \alpha < 600$:

    f_3 (p_1, p_2, p_3) $f_3(0)f_3'(0) > 0$

    f_2 (p_1-1, p_2, p_3) $f_2(0)f_2'(0) > 0$

    f_1 $(p_1-2, p_2, p_3) = (1,0,0)$

    f is of type $(3, 0, 0)$.

(iv) $\alpha > 600$:

    f_3 (p_1, p_2, p_3) $f_3(0)f_3'(0) > 0$

    f_2 (p_1-1, p_2, p_3) $f_2(0)f_2'(0) < 0$

$f_1 \qquad (p_1-1, p_2, p_3-1) = (0, 0, 1)$

f is of type $(1, 0, 2)$.

(v) $\alpha = -200$:

$f_3(z) = z^3 + 30z^2 - 8000$	(p_1, p_2, p_3)	$f_3(0) f_3'(0) = 0$
$f_2(z) = -z^2 + 7969z + 1$	(p_1-1, p_2, p_3)	$f_2(0) f_2'(0) = 7969$
$f_1(z) = z - 7969$	$(p_1-2, p_2, p_3) = (0, 0, 1)$	

f is of type $(2, 0, 1)$.

The step from f_3 to f_2 illustrates an intermediate case. The hypotheses of Theorem 3.5 are satisfied and f_2 is constructed by the process associated with that theorem.

(vi) $\alpha = 0$:

$f_3(z) = z(z^2 + 30z + 200)$	(p_1, p_2, p_3)	Zero at the origin
$f_2(z) = z^2 + 30z + 200$	(p_1, p_2-1, p_3)	$f_2(0) f_2'(0) = 6000$
$f_1(z) = z + 30$	$(p_1-1, p_2-1, p_3) = (1, 0, 0)$	

f is of type $(2, 1, 0)$.

(vii) $\alpha = 600$:

$f_3(z) = z^3 + 30z^2 + 800z + 24000$

$\qquad\qquad\qquad (p_1, p_2, p_3) \qquad\qquad\qquad\qquad f_3(0)f_3'(0) = 19 \cdot 2 \times 10^6$

$f_2(z) = z^2 + 800$

$\qquad\qquad\qquad (p_1-1, p_2, p_3) \qquad\qquad\qquad\qquad p_1 - 1 = p_3$

$f_1(z) = z + 800$

$\qquad\qquad\qquad (p_1+k-2, p_2-k, p_1-1) = (1, 0, 0)$

f is of type $(1,2,0)$ and both zeros on the imaginary axis are distinct.

The step from f_2 to f_1 illustrates a self-inversive case. f_2 is self-inversive and f_1 is constructed from the polar derivative in accordance with Theorem 3.4.

All possibilities are covered by (i) to (vii) and we have thus found the type of f for all real values of the parameter α.

REFERENCES

For further details on the above results and for references to earlier work by other authors the following papers should be consulted:

1. Miller, John J. H., "On the location of zeros of certain classes of polynomials with applications to numerical analysis". *J. Inst. Maths. Applics.*, **8**, (1971), 397–406.

2. Miller, John J. H., "On weak stability, stability and the type of a polynomial". *Conference on Applications of Numerical Analysis, Lecture Notes in Mathematics* Vol. 228, Springer-Verlag (1971), 316–320.
3. Miller, John J. H., "On the stability of differential equations" *SIAM J. Control*, **10**, (1972) 639–648.

The Edge-Function Method in Elastostatics

P. M. QUINLAN

University College,
Cork, Ireland

INTRODUCTION

The relative paucity of methods capable of giving reliable numerical results in elastostatic problems led the author into attempting to apply the edge-function method—Quinlan (1964)—in this field. The powerful theory of singular integral equations has been extended by Fredholm (1905) and Lauricella (1909) to elasticity problems. This was especially successful in the more recent complex variable approach of Sherman (1941), Mikhlin (1957) and Muskhelishvili (1953) to plane elasticity problems, whether in plane strain or generalised plane stress—now referred to as elastostatic problems.

Recently Jaswon (1963) and Rizzo (1967) have adopted a real variable approach, based on a knowledge of singular solution corresponding to a line-load. The method, however, gets down in numerical cases to a type of either point or segment matching on the boundary, and results so far are not very impressive.

In the present paper a direct approach is taken to solving the equations of equilibrium in terms of displacement components, without the introduction of a stress function. The boundary conditions are dealt with directly, without the intervention of an integration—which leaves as a legacy constants of integration that prove very troublesome in the case of multiply connected regions.

The functions introduced may appear very complicated, but the author has developed over the past five years a tensorized, or algorithmic, method called the computer form method—Quinlan (1965)—of controlling the resulting algebra. Accordingly, edge functions can now be as readily applied in a computer program as, say, the trigonometric or

265

hyperbolic functions, merely requiring their own subroutine. Of special significance is the ease with which these can be generalised to corresponding functions with single or double symmetry.

The touchstone of any new method for elastostatic problems must be the ease with which the relevant theory is developed in a form especially suited to programming, since most elastostatic problems eventually end up on a computer; the provision of a simple yet comprehensive reliability test for the engineer to decide whether to accept the results or not; and most important of all, the acid test of results for difficult cases other than the conventional squares and circles. The edge function method is presented for critical evaluation on all three points, and especially on its success in solving the problems in the appendices which, to the author's knowledge, are not solved elsewhere.

A companion paper will take the work on Polar Elasto-functions, via a new concept of conformal mapping—which only requires a very rough mapping of the region onto a circle—into the realm of regions bounded by polygonal or curved outer boundaries with any number of curved holes. In this it is following the general lines of development of the author's work on plates.

I. THE ELASTOSTATIC PROBLEM

The displacement components (u, v) in an isotropic elastostatic problem satisfy the equations

$$(\lambda + 2\mu)\frac{\partial^2 u}{\partial x^2} + \mu\frac{\partial^2 u}{\partial y^2} + (\lambda + \mu)\frac{\partial^2 v}{\partial x \partial y} + \rho X = 0$$

$$(\lambda + \mu)\frac{\partial^2 u}{\partial x \partial y} + \mu\frac{\partial^2 v}{\partial x^2} + (\lambda + 2\mu)\frac{\partial^2 v}{\partial y^2} + \rho Y = 0,$$

(1.1)

in the corresponding domain D, where the body-force components per unit mass are (X, Y), ρ is the mass-density, and the coordinate axes are $0x$ and $0y$ in Figure 1.

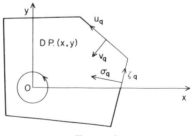

FIGURE 1

The usual boundary conditions involve specifying on the boundary of D either the displacement components

$$(u_q, v_q), \tag{1.2}$$

or the stress components

$$(\zeta_q, \sigma_q), \tag{1.3}$$

as shown in Figure 1. Different conditions may be specified on the different parts of the boundary.

If (u^p, v^p) is a particular solution of equation (1.1) a more general solution is given by

$$u = u^c + u^p$$
$$v = v^c + v^p \tag{1.4}$$

where the complementary part of the solution satisfies the equations

$$(\lambda + 2\mu) \frac{\partial^2 u^c}{\partial x^2} + \mu \frac{\partial^2 u^c}{\partial y^2} + (\lambda + \mu) \frac{\partial^2 v^c}{\partial x \partial y} = 0$$

$$(\lambda + \mu) \frac{\partial^2 u^c}{\partial x \partial y} + \mu \frac{\partial^2 v^c}{\partial x^2} + (\lambda + 2\mu) \frac{\partial^2 v^c}{\partial y^2} = 0 \tag{1.5}$$

Accordingly, the problem can be split, as in the present approach, into the twin problems of determining
  (i) the particular integral for the specified body-forces, and
  (ii) suitable complementary functions, which lead to a well-conditioned system of simultaneous equations when adjusted to satisfy the prescribed boundary conditions.

Singular functions are obviously required when the body-force (X, Y) contains a singular part, as arises when a concentrated line-load acts in the interior of the beam D, as at P in Figure 1, or when a concentrated force, or moment, acts at a point on the boundary.

II. EDGE-FUNCTIONS FOR A POLYGONAL REGION

The edge-function idea, as developed in Quinlan (1968), can be extended to systems of simultaneous partial differential equations with constant coefficients, as in (1.5), by transforming both displacement components and coordinates to the corresponding edge-displacement components* (u'_j, v'_j) and edge-coordinates* (x'_j, y'_j).

Let $P_j P_{j+1}$ be the jth side of any polygon and P any point (x, y), which may be either in the interior or on the boundary of the region. On denoting

---

* Quantities referring to edge-axes will be denoted by a prime, i.e., x'_j, y'_j.

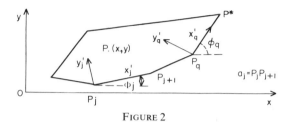

FIGURE 2

the coordinates of P_j by (x_j, y_j) in the x-y system it follows that the edge-coordinates (x'_j, y'_j) of P satisfy the relations:

$$x = x_j + x'_j \cos \phi_j - y'_j \sin \phi_j$$
$$y = y_j + x'_j \sin \phi_j + y'_j \cos \phi_j,$$

(2.1)

while the edge-displacement components (u'_j, v'_j) have resolutes (u^c_j, v^c_j) along the x-y axes given by:

$$u_j^{c} = u'_j \cos \phi_j - v'_j \sin \phi_j$$
$$v_j^{c} = u'_j \sin \phi_j + v'_j \cos \phi_j$$

(2.2)

Since equations (1.5) are linear, it follows that the superposition principle applies. Accordingly, if we obtain complementary functions (u_j^c, v_j^c) satisfying equations (1.5) that are especially related to the jth edge-axes, we can superpose such complementary functions from all sides of the polygon to give more general complementary functions:

$$u^c = \sum u_j^c; \quad v^c = \sum v_j^c.$$

(2.3)

We now proceed to determine u_j^c and v_j^c by transforming both the independent and the dependent variables in (1.5), which requires the directional derivatives

$$\frac{\partial V(x'_j, y'_j)}{\partial x} = \mathbf{x}.\nabla V = \cos \phi_j \frac{\partial V}{\partial x'_j} - \sin \phi_j \frac{\partial V}{\partial y'_j},$$

or the operators

$$\frac{\partial}{\partial x} = \cos \phi_j \frac{\partial}{\partial x'_j} - \sin \phi_j \frac{\partial}{\partial y'_j}$$

$$\frac{\partial}{\partial y} = \sin \phi_j \frac{\partial}{\partial x'_j} + \cos \phi_j \frac{\partial}{\partial y'_j}$$

(2.4)

It follows on transforming equations (1.5) that

$$(\lambda + 2\mu) \frac{\partial^2 u'_j}{\partial x'^2_j} + \mu \frac{\partial^2 u'_j}{\partial y'^2_j} + (\lambda + \mu) \frac{\partial^2 v'_j}{\partial x'_j \partial y'_j} = 0$$

$$(\lambda + \mu)\frac{\partial^2 u'_j}{\partial x'_j \partial y'_j} + \mu \frac{\partial^2 v'_j}{\partial x'^2_j} + (\lambda + 2\mu)\frac{\partial^2 v'_j}{\partial y'^2_j} = 0 \qquad (2.5)$$

The same result could, of course, be obtained from physical reasoning, since the medium being isotropic it follows that equations (1.5) must be invariant with respect to any rotation of axes. However, it is instructive to proceed as above with a view to extension in a later paper to media that are not isotropic.

On eliminating v'_j from equations (2.5), it follows that u'_j satisfies the biharmonic equation

$$\left(\frac{\partial^2}{\partial x'^2_j} + \frac{\partial^2}{\partial y'^2_j}\right)^2 u'_j = 0, \qquad (2.6)$$

and simple solutions of the edge-function type are given by

$$u'_j = (A_M{}^j + m_j y'_j B_M{}^j) e^{-m_j y'_j} \sin(m_j x'_j + \alpha), \qquad (2.7)$$

where α is arbitrary and set m_j are constants of separation ordered by the index M. The subscript j in m_j denotes that the set may vary with the different edge-axes.

On substituting for u'_j in the first of equations (2.5), integrating, and suppressing any arbitrary functions of integration, we obtain:

$$v'_j = A_M{}^j + B_M{}^j(T_3 + m_j y'_j) e^{-m_j y'_j} \cos(m_j x'_j + \alpha), \qquad (2.8)$$

where

$$T_3 = (\lambda + 3\mu)/(\lambda + \mu).$$

III. DERIVED EDGE-FUNCTIONS

Preparatory to formulating the boundary value problems in Section V, we require formulae, to be combined in a single computer formula, for the displacement components $(u_q{}^j, v_q{}^j)$ and stress components $(\zeta_q{}^j, \sigma_q{}^j)$ produced by displacement (u'_j, v'_j) at any point P on the qth side of slope ϕ_q. These components are along and perpendicular to the qth side.

On resolving components, it follows that

$$u_q{}^j = u'_j \cos \phi_{qj} + v'_j \sin \phi_{qj}$$

$$v_q{}^j = -u'_j \sin \phi_{qj} + v'_j \cos \phi_{qj}; \quad \phi_{qj} = \phi_q - \phi_j, \qquad (3.1)$$

and the stress-components at P on qth side* are related to the corres-

---

*Formulae (3.2) also give stress components at an internal point P associated with a plane of slope ϕ_q through P.

ponding displacement components by the formulae:

$$\zeta_q^{\ j} = \mu\left(\frac{\partial u_q^{\ j}}{\partial y_q'} + \frac{\partial v_q^{\ j}}{\partial x_q'}\right)$$

$$\sigma_q^{\ j} = (\lambda + 2\mu)\frac{\partial v_q^{\ j}}{\partial y_q'} + \lambda\frac{\partial u_q^{\ j}}{\partial x_q'} \tag{3.2}$$

As in equation (2.4), we obtain the operators

$$\frac{\partial}{\partial x_q'} = \cos\phi_{qj}\frac{\partial}{\partial x_j'} + \sin\phi_{qj}\frac{\partial}{\partial y_q'}$$

$$\frac{\partial}{\partial y_q'} = -\sin\phi_{qj}\frac{\partial}{\partial x_j'} + \cos\phi_{qj}\frac{\partial}{\partial y_j'}, \tag{3.3}$$

and on applying to equation (3.2), using (3.1), (2.7), (2.8) and simplifying, it follows that $\zeta_q^{\ j}$ reduces to the form under, obtained in Quinlan (1968), and given the name *Computer Formula for Edge-Functions*:

$$Q_t E_M^{\ j} = m_j^{\ s}\big[(h_1\sin\chi_M + h_3\cos\chi_M)A_M^{\ j} + \{(h_2 + h_1 m_j y_j')\sin\chi_M + (h_4 + h_3 m_j y_j')\cos\chi_M\}B_M^{\ j}\big]\,e^{-m_j u_j'} \tag{3.4}$$

where

$$\chi_M = m_j x_j' + \alpha \tag{3.5}$$

$$s = 1; \quad h_1 = -2\,\mu\cos 2\phi_{qj}; \quad h_2 = -(T_3 - 1)\mu\cos 2\phi_{qj}$$

$$h_3 = -2\,\mu\sin 2\phi_{qj}; \quad h_4 = -(T_3 - 1)\mu\sin 2\phi_{qj}, \tag{3.6}$$

and the index $t = 3$ is set to correspond to $\zeta_q^{\ j}$.

Similar forms are obtained for $\sigma_q^{\ j}$, $u_q^{\ j}$ and $v_q^{\ j}$, and all can be combined in the single Computer Formula for Edge-Functions given by (3.4), with parameters as given in Table I.

Table I
Edge-Function Parameters

Function	$u_q^{\ j}$	$v_q^{\ j}$	$\zeta_q^{\ j}$	$\sigma_q^{\ j}$
t	1	2	3	4
h_1	$\cos\phi_{qj}$	$-\sin\phi_{qj}$	$-2\mu\cos 2\phi_{qj}$	$2\mu\sin 2\phi_{qj}$
h_2	0	0	$-(T_3 - 1)\mu\cos 2\phi_{qj}$	$(T_3 - 1)\mu\sin 2\phi_{qj}$
h_3	$\sin\phi_{qj}$	$\cos\phi_{qj}$	$-2\mu\sin 2\phi_{qj}$	$-2\mu\cos 2\phi_{qj}$
h_4	$T_3\sin\phi_{qj}$	$T_3\cos\phi_{qj}$	$(1 - T_3)\mu\sin 2\phi_{qj}$	$(1 - T_3)(\lambda + 2\mu\cos^2\phi_{qj})$
s	0	0	1	1

IV. SOME PARTICULAR INTEGRALS

Particular integrals of system (1.1), denoted by u^p and v^p, are easily obtained when the body forces are due to gravity or to rotary inertia. If $X = X_0$ and $Y = Y_0$ both constants, inspection of equations (1.1) shows that these can be replaced by the simpler equations:

$$(\lambda + \mu) \frac{\partial^2 v^p}{\partial x \partial y} + \rho X_0 = 0$$

$$(\lambda + \mu) \frac{\partial^2 u^p}{\partial x \partial y} + \rho Y_0 = 0,$$

(4.1)

if we set

$$\frac{\partial^2 u^p}{\partial x^2} = \frac{\partial^2 u^p}{\partial y^2} = \frac{\partial^2 v^p}{\partial x^2} = \frac{\partial^2 v^p}{\partial y^2} = 0,$$

and accordingly on integrating, we obtain

$$v^p = -\frac{\rho X_0}{\lambda + \mu} xy; \quad u^p = -\frac{\rho Y_0}{\lambda + \mu} xy \tag{4.2}$$

Similarly, if the body is rotating around some fixed point 0 at ω rad/sec, the reversed acceleration force is

$$\bar{F} = (X,Y) = \omega^2 \bar{r}, \tag{4.3}$$

and a corresponding particular integral is found to be

$$u^p = -\frac{\omega^2 x r^2}{8(\lambda + 2\mu)}; \quad v^p = -\frac{\omega^2 y r^2}{8(\lambda + 2\mu)} \tag{4.4}$$

an integral that is symmetrical about both the x and y axes. On introducing

$$U_x = \frac{\rho X_0}{\lambda + \mu}; \quad U_y = \frac{\rho Y_0}{\lambda + \mu}; \quad U_r = \frac{\rho \omega^2}{8(\lambda + 2\mu)}, \tag{4.5}$$

we obtain the corresponding particular integral as:

$$u^p = -U_y xy - U_r r^2 x$$

$$v^p = -U_x xy - U_r r^2 y$$

(4.6)

As in Section III, we require the following boundary quantities at any point P on the qth side:

$$u_q^p = u^p \cos \phi_q + v^p \sin \phi_q$$

$$v_q^p = -u^p \sin \phi_q + v^p \cos \phi_q$$

$$\zeta_q^p = \mu\left(\frac{\partial u^p}{\partial y_q'} + \frac{\partial v^p}{\partial x_q'}\right)$$

$$\sigma_q^p = (\lambda + 2\mu)\frac{\partial v^p}{\partial y_q'} + \frac{\partial u^p}{\partial x_q'}. \tag{4.7}$$

and on substituting from equation (4.6) and using the operators

$$\frac{\partial}{\partial x_q'} = \cos\phi_q\frac{\partial}{\partial x} + \sin\phi_q\frac{\partial}{\partial y}$$

$$\frac{\partial}{\partial y_q'} = -\sin\phi_q\frac{\partial}{\partial x} + \cos\phi_q\frac{\partial}{\partial y} \tag{4.8}$$

we obtain

$$u_q^p = -(U_y\cos\phi_q + U_x\sin\phi_q)xy - U_r r^2(x\cos\phi_q + y\sin\phi_q) \equiv Q_1^p$$

$$v_q^p = (U_y\sin\phi_q - U_x\cos\phi_q)xy + U_r r^2(x\sin\phi_q - y\cos\phi_q) \equiv Q_2^p$$

$$\zeta_q^p = \mu U_y(y\sin 2\phi_q - x\cos 2\phi_q) - \mu U_x(y\cos 2\phi_q + x\sin 2\phi_q)$$
$$\quad + 2\mu U_r\{(x^2 - y^2)\sin 2\phi_q + 2xy\cos 2\phi_q\} \equiv Q_3^p \tag{4.9}$$

$$\sigma_q^p = U_y\{-y(\lambda + \mu - \mu\cos 2\phi_q) + \mu x\sin 2\phi_q\}$$
$$\quad + U_x\{\mu y\sin 2\phi_q - x(\lambda + \mu + \mu\cos 2\phi_q)\}$$
$$\quad + 2U_r\{-2(\lambda + \mu)r^2 + \mu(x^2 - y^2)\cos 2\phi_q + 2\mu xy\sin 2\phi_q)\} \equiv Q_4^p.$$

Again, as in Table I, the derived functions are ordered by the index t and denoted by Q_t^p for $t = 1, 2, 3, 4$ respectively. *Note* that if gravity acts along the negative y-axis, then

$$X_0 = 0, \quad Y_0 = -g; \quad U_x = 0; \quad U_y = -\frac{\rho g}{\lambda + \mu}, \tag{4.10}$$

and the deformation is symmetrical around $0Y$.

V. POLYGONAL BEAM: THE BOUNDARY IDENTITY PROBLEM

Consider an isotropic elastic beam of polygonal cross-section of j' sides. On using solutions (1.4) and substituting from equations (2.3), (3.4) and (4.9), we can write the derived function, with index t, on the qth boundary as

$$Q_t = Q_t^p + \sum_{j=1}^{j'}\sum_{(M)} Q_t E_M^j; \quad t = 1, 2, 3, 4. \tag{5.1}$$

The boundary conditions require that two of these functions $t = t_1$ and $t = t_2$ be specified on each side of the polygon. Physically meaningful problems limit the choice to either displacements $t = 1, 2$, or stresses $t = 3, 4$ on each side, but displacements can be given on some sides and stresses on others.

Let P_q^* be any point $(x_q', 0)$ on the qth side, and $Q_t(P_q^*)$ the function as specified on this side for $t = t_1$ and $t = t_2$. Accordingly, from equation (5.1), the boundary conditions on the qth side lead to the boundary identities for the polygon

$$Q_t(P_q^*) = Q_{t^p}(P_q^*) + \sum_{j=1}^{j'} \sum_{(M)} Q_t E_M{}^j (P_q^*) \quad (t = t_1: t = t_2; q = 1, j'). \quad (5.2)$$

For all points P_q^* on the qth side, or each is an identity in the variable x_q' in the range $(0, a_q)$.

To determine the set m_j required for any convex polygon, we note that for the qth member of the summation $j = 1, j'$ the quantity y_i' is zero, this then being the only member without a negative exponential to diminish boundary effects. Consequently, on bringing this dominant term to one side of equation (5.2) it reads*:

$$\sum_{(M)} m_q{}^s (h_1 A_m{}^q + h_2 B_M{}^Q) \sin (m_q x_q' + \alpha) + (h_3 A_M{}^q + h_4 B_M{}^q) \cos (m_q x_q' + \alpha)$$

$$= Q_t(P_q^*) - Q_{t^p}(P_q^*) - \sum_{j=1}^{j'}{}' Q_t E_M{}^j (P_q^*), \quad (5.3)$$

an identity that must be satisfied for x_q' in the range $(0, a_q)$.

The right-hand side is some function of x_q', say $G(x_q')$, and the identity can then be satisfied provided that $G(x_q')$ can be expanded in the trigonometric form on the left-hand side. On examining Table I we note that when $j = q$:

(i) $h_3 = h_4 = 0$ when $t = 1$ or $t = 3$ (t odd), in which case the left-hand side of identity (5.3) reduces to a Fourier sine series for the interval a_q, if we set

$$\alpha_M = 0; \quad m_q = \pi M/a_q, \quad (5.4)$$

and take the summation in M from one to infinity. The coefficient of sin $m_q x_q'$ converges only as $1/M$ unless we set

$$G(0) = G(a_q) = 0, \quad (5.5)$$

at the end points of the interval, in which case the convergence is then of order $1/M^3$.

(ii) $h_1 = h_2 = 0$ when $t = 2$ or $t = 4$ (t even), and the identity then reduces to a Fourier cosine series on using (5.4) and taking the summation

---

*The prime in the symbol $\sum_{j=1}^{j'}{}'$ denotes that the member $j = q$ is omitted.

in M from 0 to infinity. The coefficient of $\cos m_q x'_q$ converges only as $1/M^2$ unless we set

$$\frac{dG(0)}{dx'_q} = \frac{dG(a_q)}{dx'_q} = 0, \tag{5.6}$$

when the convergence becomes of order $1/M^4$. Difficulties are encountered in the degeneracy of the quantities $Q_t Q_M{}^j$ when $M = 0$ implying $m_q = 0$. This has been overcome by omitting the $M = 0$ term from the left-hand side, which requires the provision of an additional unknown per side, say A^q_{-1}, on the right-hand side to enable its zeroth harmonic to be set exactly to zero. A^q_{-1} is the coefficient of one of the set of edge-functions (2.8), not already used in the summation, and is designated by the index $M = -1$, where definition (5.4) for m_q and α is generalised to

$$m_q = \pi Z_M / a_q; \quad \alpha = \alpha_M,$$

where

$$Z_M = M, \alpha_M = 0; \quad M = 1, 2, \cdots \infty \tag{5.7}$$

$$= Z_M, \text{ a non-integer}; \quad \alpha_M \neq 0; \quad M = 0, -1, -2, \cdots$$

Best results have been obtained by taking Z_M in the region (0.2, 0.4), these being not very sensitive to the actual value chosen for Z_M. Frequently we distinguish edge-functions corresponding to fractional values for Z_M by referring to them as Fractional Edge-Functions.

Equations (5.5) and (5.6) will be referred to as Vertex Equations due to their role in sharpening the convergence of the edge-unknowns $A_M{}^j$ and $B_M{}^j$ by imposing vertex conditions.

Simultaneous equations for boundary conditions

We have seen that identity (5.3) can be interpreted either as a Fourier sine or cosine series, depending on the value of t. Accordingly, if $t = 1$ it follows from the formula for the M'^{th} harmonic in a sine series that

$$m^{'s}_q (h_1 A^q_{M'} + h_2 B^q_{M'}) = M'^{\text{th}} \text{ harmonic of } G(x'_q)$$

$$= \frac{2}{a_q} \int_0^{a_q} G(x'_q) \sin \frac{\pi M'}{a_q} x'_q \, dx'_q. \tag{5.8}$$

More conveniently, this could be got from boundary identity (5.2) on multiplying by $\sin m'_q x'_q$ and integrating over x'_q from 0 to a_q. Likewise, the equations involving a cosine series expansion follow by replacing $\sin m'_q x'_q$ by $\cos m'_q x'_q$. Accordingly, on defining

$$F(m'_q x'_q) = \sin(m'_q x'_q); \quad t = 1, 3$$

$$= \cos(m'_q x'_q); \quad t = 2, 4, \tag{5.9}$$

the simultaneous equations arising from the Fourier coefficients in identity (5.2) for boundary identity (5.3) follow in the form

$$\int_0^{a_q} Q_t(P_q^*)F(m_q'x_q')\,dx_q' = \int_0^{a_q}(Q_t^p(P_q^*) + \sum_{j=1}^{j'} Q_t E_M^j)F(m_q'x_q')\,dx_q' \qquad (5.10)$$

$$t = t_1; \quad t = t_2; \quad q = 1, j'; \quad M' = 0/1, 2, \cdots, \infty,$$

when M' set starts at 0 or 1 depending on whether t is even or odd.

The above leads to an infinite set of simultaneous equations and, in practice, these must be truncated at L_q harmonics. The subscript q is used in L_q to indicate that different truncation points may be used on different edges. Kelly (1967) has shown the advantage of taking L_q proportional to the corresponding side length a_q

On taking $M' = 0/1$, L_q in (5.10) a sufficient number of primary unknowns for the resulting set of simultaneous equations is provided by taking the edge-function summation in (5.2) as

$$\sum_{M=0/1}^{L_q}.$$

Since the boundary effects decay as we move inwards from the boundary, it is obviously necessary that the exponentials in solution (5.2) should be always negative or zero, or that the polygon in Figure 2 should be convex.

Additional Unknowns

Each vertex equation obviously requires the introduction of an additional two unknowns into the system for each side, and these are readily provided by A_M^j and B_M^j arising from one of the fractional edge-functions (5.7) on each side. For convenience in programming, the pairs corresponding to equations $t = t_1$ and $t = t_2$ in equations (5.5) and (5.6) respectively will be ordered to correspond to $M = 0$ and $M = -2$, since the zero harmonic arising in the cosine series case has already been set to correspond to $M = -1$. Accordingly, the summation in equation (5.2) becomes $\sum_{M=-2}^{L_q}$.

In all cases of fractional edge-functions better conditioning is obtained by taking α different from zero, say $\alpha = \pi/3$. Care is required in the computer program to eliminate identical vertex equations, which arise: (1) by continuity at any vertex on either side of which displacement conditions are specified, and (2) from the principle of reciprocal shears, for function $t = 3$, at a right angle corner, on either side of which stress conditions are specified.

Uniqueness

If stress conditions are specified on all sides, the resulting displacement is indeterminate to the extent of a rigid body rotation, and consequently displacements (1.4) become

$$u = A + Cy + u^c + u^p$$
$$v = B - Cx + v^c + v^p,$$

(5.11)

and three further equations must be added to the system to determine A, B and C. In practice we have set $u = v = 0$ at the centroid of the polygon and taken $u = 0$ at one other prescribed point, which is read in.

Care must also be taken, when stress conditions are prescribed on all edges, to ensure that the forces on the system satisfy the requirements of rigid body equilibrium.

VI. TRAPEZOIDAL BEAM — ILLUSTRATIVE EXAMPLES

A master computer program has been developed by the author for a very wide range of Laplace, plate and elastostatic problems. It is based on a main program which sets up, and solves, the matrix for system (5.10) with the aid of several specialist subroutines. The approximate root mean square of the boundary residuals — the difference between the computed boundary quantities corresponding to the solution obtained and the specified boundary values — are then obtained as a practical measure of how well the mathematical model* obtained corresponds to the actual physical problem.

The program is arranged so that solutions for any specified number of truncation levels, up to a maximum of eight, are obtained. This uses little additional computer time, beyond that required for a single solution at the maximum truncation level. Accordingly, a succession of mathematical models is obtained, and this enables the residuals to be studied together with their rate of decrease as the truncation level increases.

In the production program, as seen from the examples herein, the results

---

*The solution (1.4) or (5.11) obtained consisting, as it does, of a finite number of functions, each of which satisfies equations (1.1), is therefore an exact mathematical solution to the prescribed elastostatic problem, with slightly different boundary conditions. Hence, the r.m.s. of the boundary residuals provides sufficient information for the engineer to enable him to decide whether or not to accept the proffered mathematical model.

as given for different truncation levels show how the boundary residuals affect the results at the interior points.

As a check, we solved the problem of a beam of square cross-section under the action of unit normal pressures on all its faces. This checked very satisfactorily against the known solution in Timoshenko and Goodier, (1951) given by the stress-function

$$\chi = -\tfrac{1}{2}(x^2 + y^2),$$

the r.m.s. of the residual normal and shear stresses being less than $1/10^4$ for truncation at $L = 8$ harmonics. This involved 76 equations, but these reduce to 21 when full account is taken of the double symmetry of the problem, involving the use of the symmetrical edge-functions of Section XI.

Example

A steel beam $\lambda = 8100$, $\mu = 3400$ and $g = 0\cdot1333 \times 10^{-3}$ of cross-section ABCD in Figure 3, where B(1, 0), C(1·5, 1) and D(0, 2), is subject to normal surface displacements

$$v_q = z(1-z)^2/1000; \quad u_q = 0; \quad z = x'_q/a_q, \tag{6.1}$$

on each edge where x'_q is along each in turn.

The boundary residuals are given in Appendix A. Since the maximum specified normal edge-displacement is 0·2E-03 and the root mean square of the residual at 8 harmonics — involving 72 equations — is only 0·4E-06, or 1/500 of maximum, the residuals are negligible from a practical standpoint. Another 4 harmonics, involving an additional 32 equations, would reduce the residuals by a further factor of 10. The r.m.s. of the shear displacements is 0·3E-05, or about three times better than the normal displacement.

Typical results for displacements and stresses at any specified number of equidistant points on any specified line are given in Appendix A. The consistency of the results is evident, the difference between harmonics being compatible with the boundary residuals reported. It can be

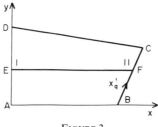

FIGURE 3

legitimately deduced that at 8 harmonics both displacements and stresses are correct to the order of the boundary residuals or 1/500. If anything, the stability and accuracy of the stress components is even better than that of the displacements.

VII. POLAR ELASTO-FUNCTIONS

On eliminating v^c from equations (1.5) we obtain

$$\left(\frac{\partial^2}{\partial x^2}+\frac{\partial^2}{\partial y^2}\right)^2 u^c = 0 \tag{7.1}$$

and hence both u^c and v^c are biharmonic functions.

As in Quinlan (1968) polar form biharmonic solutions are given in complex variable form by the expression

$$u^c = \mathrm{Re}[\,Ez^p + F\bar{z}z^{p+1}\,]; \quad z = x + iy = re^{i\theta} \tag{7.2}$$

However, a more balanced solution to equations (1.5) is provided by setting

$$u^c = \mathrm{Re}[\,Ez^p + F\bar{z}z^{p+1} + Gz^{p+2}\,]$$
$$v^c = \mathrm{Re}[\,E'z^p + F'\bar{z}z^{p+1} + G'z^{p+2}\,] \tag{7.3}$$

where

$$E = E_1 + iE_2; \quad F = F_1 + iF_2, \tag{7.4}$$

p is any real exponent, E and F are the four primary unknowns in the pth solution, and G is arbitrary. E and F can alter with p, and where desired, this can be emphasised by writing them as E^p and F^p.

The complex variable form for the first of equations (1.5) follows easily on using the operators

$$\frac{\partial}{\partial x}=\frac{\partial}{\partial z}+\frac{\partial}{\partial \bar{z}}; \quad \frac{\partial}{\partial y}=i\left(\frac{\partial}{\partial z}-\frac{\partial}{\partial \bar{z}}\right), \tag{7.5}$$

and T_3 as in definition (2.8). Thus

$$\frac{\partial^2 u^c}{\partial z^2}+\frac{\partial^2 u^c}{\partial \bar{z}^2}+2T_3\frac{\partial^2 u^c}{\partial z\partial \bar{z}}+i\left(\frac{\partial^2}{\partial z^2}-\frac{\partial^2}{\partial \bar{z}^2}\right)v^c = 0, \tag{7.6}$$

and on substituting expressions from (7.3), the resulting identify gives

$$E' = iE; \quad F' = iF, \tag{7.7}$$

and

$$G' = iG + 2iFT_3/(p+2). \tag{7.8}$$

Multi-valued Solutions

We note that there may be singular displacement components, associated with $F^{(-2)}$ when $p + 2 = 0$, and examine the corresponding solutions involving F by putting $p + 2 = \epsilon$ where ϵ approaches zero.

We now examine two possibilities of simplifying equation (7.8) consequent on setting either G or G' to zero:

(1) On putting $G = 0$, equation (7.3) shows that the corresponding u^c is finite, while v_c contains, on using equation (7.8), the limiting part

$$
\bar{v}_c = 2T_3 \operatorname*{Lt}_{\epsilon \to 0} \operatorname{Re} \left[\frac{iF}{\epsilon} z^\epsilon \right]
$$

$$
= 2T_3 \operatorname*{Lt}_{\epsilon \to 0} r^\epsilon \left[-F_1 \frac{\sin \epsilon\theta}{\epsilon} + F_2 \frac{\cos \epsilon\theta}{\epsilon} \right]
$$

$$
= -2T_3 F_1 \theta, \tag{7.9}
$$

provided we remove the infinite part by setting $F_2 = 0$.

Accordingly, for $p = -2$, the F part of equations (7.3) leads to the regular solution

$$
u^* = F_1 \cos 2\theta
$$
$$
v^* = F_1 (\sin 2\theta - 2T_3\theta). \tag{7.10}
$$

(2) Similarly, on setting $G' = 0$, the possible singular part of u^c is

$$
\bar{u}_c = -T_3 \operatorname*{Lt}_{\epsilon \to 0} \left[\frac{F z^\epsilon}{\epsilon} \right]
$$

$$
= 2T_3 F_2 \theta; \quad F_1 = 0.
$$

and analogous to solution (7.10), we obtain

$$
u^{**} = F_2 (\sin 2\theta + 2T_3\theta)
$$
$$
v^{**} = -F_2 \cos 2\theta. \tag{7.11}
$$

Accordingly, when $p = -2$, the pair of primary unknowns in (7.3) corresponding to $F^{(-2)}$ as obtained in expressions (7.10) and (7.11), are multivalued since θ is multivalued.

Another multivalued displacement is given by the usual logarithmic solution of equation (7.1):

$$
\bar{u} = \operatorname{Re}[D \log z]
$$
$$
\bar{v} = \operatorname{Re}[D' \log z], \tag{7.12}
$$

where, on substituting in equation (7.6), we obtain

$$
D' = iD,
$$

or

$$\bar{u} = D_1 \log r - D_2 \theta$$
$$\bar{v} = -D_2 \log r - D_1 \theta. \tag{7.13}$$

On superposing displacements (7.10), (7.11) and (7.13), the θ terms can be suppressed by taking $D_2 = 2T_3F_2$; $D_1 = -2T_3F_1$, giving

$$u_c^* = F_1(\cos 2\theta - 2T_3 \log r) + F_2 \sin 2\theta$$
$$v_c^* = F_1 \sin 2\theta - F_2(\cos 2\theta + 2T_3 \log r), \tag{7.14}$$

or in complex variable form:

$$u_c^* = \text{Re}[\, F(\bar{z}/z - T_3 \log z\bar{z}) \,]$$
$$v_c^* = \text{Re}[\, iF(\bar{z}/z + T_3 \log z\bar{z}) \,] \tag{7.15}$$

there being a singularity at $z = 0$.

In all other cases $p \neq -2$, G in equation (7.8) is arbitrary. The desirability of having the forms (7.3) balanced leads, in equation (7.8), to setting

$$G = -FT_3/(p+2); \quad G' = iFT_3/(p+2), \tag{7.16}$$

and the pth Polar Elasto-Functions (7.3) then become

$$u_c = \text{Re}\left[Ez^p + F\left(zz^{p+1} - \frac{T_3}{p+2} z^{p+2}\right) \right]$$
$$v_c = \text{Re}\left[iEz^p + F\left(zz^{p+1} + \frac{T_3}{p+2} z^{p+2}\right) \right], \quad p \neq -2. \tag{7.17}$$

When $p = -2$, the part involving F should be replaced by displacement (7.15).

Curvilinear Forms for the Polar Elasto-Functions

If C is part of the boundary it is necessary to obtain expressions for the displacements and stresses at any point P_q, corresponding to $z = x + iy$, on C. We will call these the curvilinear forms for the polar elasto-static functions, being similar to the derived edge-functions in Section III, and easily combined in the single computer form given in

$$Q_t = \text{Re}\,[AEz^{p*} + F\,(Bzz^{p*+1} + Cz^{p*+2})] \tag{7.18}$$

On resolving the displacement components u_c, v_c it follows that

$$u_q^* = u_c \cos \phi_q + v_c \sin \phi_q,$$

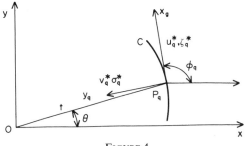

and, on substituting from (7.17), it reduces to form (7.18) wherein

$$A = B = e^{i\phi_q}; \quad C = -\frac{T_3}{p+2}e^{-i\phi_q}; \quad p^* = p. \tag{7.19}$$

Similarly,

$$v_q^* = -u_c \sin \phi_q + v_c \cos \phi_q,$$

giving the parameters

$$A = B = ie^{i\phi_q}; \quad C = \frac{iT_3}{p+2}e^{i\phi_q}; \quad p^* = p. \tag{7.20}$$

On taking physical axes coinciding with x_q' and y_q' at P_q the corresponding stress components are given by equations (3.2) on replacing the superscript j by $*$. Accordingly,

$$\zeta_q^* = \mu\left(\frac{\partial u_q'}{\partial y_q'} + \frac{\partial v_q'}{\partial x_q'}\right)$$

$$\sigma_q^* = (\lambda + 2\mu)\frac{\partial v_q^*}{\partial y_q'} + \lambda\frac{\partial u_q^*}{\partial x_q'}. \tag{7.21}$$

On substituting expressions for ζ_q^*, u_q^* and v_q^* from (7.19) and (7.20), and using the operators

$$\frac{\partial}{\partial x_q'} = e^{i\phi_q}\frac{\partial}{\partial z} + e^{-i\phi_q}\frac{\partial}{\partial \bar{z}}$$

$$\frac{\partial}{\partial y_q'} = ie^{i\phi_q}\frac{\partial}{\partial z} - ie^{-i\phi_q}\frac{\partial}{\partial \bar{z}}, \tag{7.22}$$

we obtain ζ_q^* in form (7.18), wherein

$$A = p\mu i e^{2i\phi_q}; \quad B = (p+1)\mu i e^{2i\phi_q}$$

$$C = 0; \quad p^* = p - 1. \tag{7.23}$$

The parameters for σ_q^* are similarly obtained, and all are presented in Table II below.

Table II
Polar Elasto-Parameters

Function	u_q^*	v_q^*	ζ_q^*	σ_q^*
t	1	2	3	4
p^*	p	p	$p-1$	$p-1$
A	$e^{i\phi_q}$	$i\,e^{i\phi_q}$	$p\mu i\,e^{2i\phi_q}$	$-2p\mu\,e^{2i\phi_q}$
B	$e^{i\phi_q}$	$i\,e^{i\phi_q}$	$(p+1)\mu i\,e^{2i\phi_q}$	$-2(p+1)\mu\,e^{2i\phi_q}$
C	$-\dfrac{T_3\,e^{-i\phi_q}}{p+2}$	$\dfrac{iT_3\,e^{-i\phi_q}}{p+2}$	0	-4μ

The above functions will be used, with the aid of conformal mapping, in a companion paper on "Elastostatic Problems in Multiply Connected Regions" in solving problems of beams of polygonal or curved cross-section with cavities or curved cross-section. The above part of the development is given to facilitate the introduction of the singular solutions (7.15), which are used to represent singular loadings in the next Section.

VIII. SINGULAR LOADINGS

We now proceed to develop from the singular solutions (7.15), curvilinear forms analogous to those obtained above. It follows that:

$$
\begin{aligned}
u_q^* &= u_c^* \cos \phi_q + v_c^* \sin \phi_q \\
&= \mathrm{Re}[\,F\{e^{i\phi_q}\bar{z}/z - T_3 e^{-i\phi_q} \log z\bar{z}\}\,] \\
&= F_1\{\cos(\phi_q - 2\theta) - 2T_3 \log r \cos \phi_q\} \\
&\quad + F_2\{-\sin(\phi_q - 2\theta) - 2T_3 \log r \sin \phi_q\},
\end{aligned} \tag{8.1}
$$

$$
\begin{aligned}
v_q^* &= \mathrm{Re}[\,iF\{e^{i\phi_q}\bar{z}/z + T_3 e^{-i\phi_q} \log z\bar{z}\}\,] \\
&= F_1\{-\sin(\phi_q - 2\theta) + 2T_3 \log r \sin \phi_q\} \\
&\quad + F_2\{-\cos(\phi_q - 2\theta) - 2T_3 \log r \cos \phi_q\}.
\end{aligned} \tag{8.2}
$$

Using the operators (7.22), together with above complex expressions for u_q^* and v_q^*, the complex form for ζ_q^* in equation (7.21) can be obtained, leading to

$$r\zeta_q^*/\mu = F_1\{\sin(2\phi_q-3\theta)+T_3\sin(2\phi_q-\theta)\}$$
$$+F_2\{\cos(2\phi_q-3\theta)-T_3\cos(2\phi_q-\theta)\}, \tag{8.3}$$

and similarly

$$r\sigma_q^* = F_1[2\mu\{\cos(2\phi_q-3\theta)+T_3\cos(2\phi_q-\theta)\}+2(\lambda+\mu)T_4\cos\theta]$$
$$+F_2[2\mu\{-\sin(2\phi_q-3\theta)+T_3\sin(2\phi_q-\theta)\}+2(\lambda+\mu)T_4\sin\theta],$$

$$T_4 = 1-T_3. \tag{8.4}$$

It is seen that the displacements and stresses are infinite at $r=0$, and consequently this point must be excluded from the region of the solution. The point $r=0$ is excluded by a cylindrical cavity of radius r. On identifying the cavity with C in Figure 4, it follows that $\phi_q = (\pi/2)+\theta$, and stresses (8.3) and (8.4) become

$$r\zeta_q^* = \mu_1(F_1\sin\theta-F_2\cos\theta)$$
$$r\sigma_q^* = \mu_2(F_1\cos\theta+F_2\sin\theta)$$
$$\mu_1 = 2\mu(1-T_3) \tag{8.5}$$
$$\mu_2 = 2(\lambda+\mu)(1-T_3)-2\mu(1+T_3).$$

We note that

$$\mu_1+\mu_2 = -r\mu(1+T_3) \tag{8.6}$$

Since y_q' is inwards into the cavity, the action of the cavity on the outer material causes stresses ζ_q^* and σ_q^* as shown. The resultant forces along the x and y-axes follow readily as

$$F_x = \int_0^{2\pi}(-\sigma_q^*\cos\theta-\zeta_q^*\sin\theta)\,rd\theta$$
$$= -F_1(\mu_1+\mu_2)\pi$$

$$F_y = \int_0^{2\pi}(-\sigma_q^*\sin\theta+\zeta_q^*\cos\theta)\,rd\theta$$
$$= -F_2(\mu_1+\mu_2)\pi,$$

or, on using result (8.6)

$$F_1 = F_x/4\pi\mu(1+T_3); \quad F_2 = F_y/4\pi\mu(1+T_3). \tag{8.7}$$

Accordingly, the action of the cavity on the outer material is equivalent to that of a line load at $r=0$ with components F_x and F_y, the corresponding coefficients F_1 and F_2 being as given above.

IX. SINGULAR LOADINGS ON BOUNDARY

The multivalued solutions (7.10), (7.11) and (7.12), together with the conjugate form involving $\log \bar{z}$

$$\bar{u} = \mathrm{Re}\,(H \log \bar{z})$$
$$\bar{v} = \mathrm{Re}\,(-iH \log \bar{z}) \tag{9.1}$$

can be combined together to produce a solution that gives zero stresses on the radial lines $\theta = \alpha$ and $\theta = \beta$ in Figure 5, with $r = 0$ at the vertex P.

The work is facilitated by noting that superposition of displacements (7.10) and (7.11) can be expressed in the complex variable form

$$u = \mathrm{Re}\left[F\left(\frac{\bar{z}}{z} + T_3 \log \frac{\bar{z}}{z}\right)\right]$$
$$v = \mathrm{Re}\left[iF\left(\frac{\bar{z}}{z} - T_3 \log \frac{\bar{z}}{z}\right)\right], \tag{9.2}$$

from which

$$\mu_q^* = \mathrm{Re}\left[F\left\{e^{i\phi_q}\frac{\bar{z}}{z} + T_3 e^{-i\phi_q}\log\frac{\bar{z}}{z}\right\}\right]$$
$$v_q^* = \mathrm{Re}\left[iF\left\{e^{i\phi_q}\frac{\bar{z}}{z} - T_3 e^{-i\phi_q}\log\frac{\bar{z}}{z}\right\}\right] \tag{9.3}$$

and the stress components on radial lines $\phi_q = 0$ give

$$r\zeta_q^*/2\mu = -F_1(1+T_3)\sin\theta + F_2(1+T_3)\cos\theta$$
$$r\sigma_q^*/2\mu = -F_1(1+T_3)\cos\theta - F_2(1+T_3)\sin\theta. \tag{9.4}$$

Likewise, displacements (7.12) and (9.1) yield

$$u_q^{**} = \mathrm{Re}\,[De^{i\phi_q}\log z + He^{i\phi_q}\log \bar{z}]$$
$$v_q^{**} = \mathrm{Re}\,[iDe^{i\phi_q}\log z - He^{i\phi_q}\log z] \tag{9.5}$$

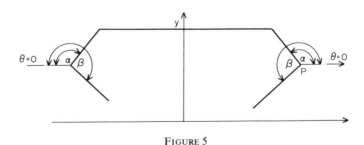

FIGURE 5

from which the corresponding stress components on $\phi_q = \theta$ are

$$r\zeta_q^{**}/2\mu = -(D_1 + H_1)\sin\theta - (D_2 - H_2)\cos\theta$$
$$r\sigma_q^{**}/2\mu = -(D_1 + H_1)\cos\theta + (D_2 - H_2)\sin\theta. \tag{9.6}$$

On superposing systems (9.4) and (9.6) to make

$$\zeta_q^* + \zeta_q^{**} = 0,$$

we require

$$F_1(1 + T_3) + D_1 + H_1 = 0$$
$$F_2(1 + T_3) - D_2 + H_2 = 0, \tag{9.7}$$

which gives

$$G_1 = D_1 = -F_1 T_5; \quad G_2 = -D_2 = -F_2 T_5;$$
$$T_5 = (1 + T_3)/2 \tag{9.8}$$

The resulting curvilinear displacements and stresses are

$$u_q^* = F_1\{\cos(\phi_q - 2\theta) + T_4\theta\sin\phi_q + 2T_3\log r \cos\phi_q\}$$
$$\quad + F_2\{-\sin(\phi_q - 2\theta) - T_4\theta\cos\phi_q - 2T_5\log r \sin\phi_q\}$$
$$v_q^* = F_1\{-\sin(\phi_q - 2\theta) + T_4\theta\cos\phi_q + 2T_5\log r \sin\phi_q\}$$
$$\quad + F_2\{-\cos(\phi_q - 2\theta) + T_4\theta\sin\phi_q - 2T_5\log r \cos\phi_q\}$$
$$r\zeta_q^*/2\mu = F_1\{\sin(2\phi_q - 3\theta) + \sin(2\phi_q - \theta)\}$$
$$\quad + F_2\{\cos(2\phi_q - 3\theta) - \cos(2\phi_q - \theta)\}$$
$$r\sigma_q^*/2\mu = F_1\{\cos(2\phi_q - 3\theta) + \cos(2\phi_q - \theta) - 2\cos\theta\}$$
$$\quad + F_2\{-\sin(2\phi_q - 3\theta) + \sin(2\phi_q - \theta) - 2\sin\theta\}. \tag{9.9}$$

On putting $\phi_q = (\pi/2) + \theta$ the resultant stresses on the cavity C, in Figure 5, used to exclude the singular point P follow as integrals (8.7) for F_x and F_y, except that the integration is now from α to β. The resulting forces applied at P to the material outside the cavity are:

$$F_x/2\mu = F_1\{2(\beta - \alpha) + \sin 2\beta - \sin 2\alpha\} + F_2\{\cos 2\alpha - \cos 2\beta\}$$
$$F_y/2\mu = F_1\{\cos 2\alpha - \cos 2\beta\} + F_2\{2(\beta - \alpha) - \sin 2\beta + \sin 2\alpha\}, \tag{9.10}$$

from which

$$F_1 = (F_x z_6 - F_y z_5)/\Delta_0$$
$$F_2 = -(F_x z_5 - F_y z_4)/\Delta_0, \tag{9.11}$$

where

$$z_4 = 2(\beta - \alpha) + \sin 2\beta - \sin 2\alpha$$
$$z_5 = \cos 2\alpha - \cos 2\beta$$
$$z_6 = 2(\beta - \alpha) - \sin 2\beta + \sin 2\alpha$$
$$\Delta_0 = 2\mu(z_4 z_6 - z_5^2). \tag{9.12}$$

When point P is on a straight line boundary then $\beta = \alpha + \pi$, and accordingly equation (9.10) give

$$F_1 = F_x/4\pi\mu; \quad F_2 = F_y/4\pi\mu. \tag{9.13}$$

Note, displacement components u_q^*, v_q^* in (9.9) contain θ explicitly and consequently a suitable cut — which cannot be in the material body — must be used to make θ single valued. Consequently, both cuts to the right and left may be required in the same problem, depending on the location of the line loads on the boundary.

X. SYMMETRICAL FUNCTIONS

Consider a doubly symmetric region with j' sides in each quadrant in Figure 6. The edge-axes for the corresponding jth sides are as shown and corresponding to solution (2.7) we now take similar edge-functions from the other quadrants.

Physical consideration of the deformation component $u(x, y)$ shows that, in a problem with symmetry around Oy, u must be an odd function of x and, if it has further symmetry around Ox, u must be an even function of y. Similarly, v must be an even function of x and an odd function of y.

On denoting the corresponding unknowns* by $A_{M,k}^j$, $B_{M,k}^j$; $k = 2, 3, 4$, where k is the quadrant indicator, the symmetry properties of u require that

(i) if symmetrical around Oy

$$A_{M,2}^j = -A_M^j; \quad B_{M,2}^j = -B_M^j \tag{10.1}$$

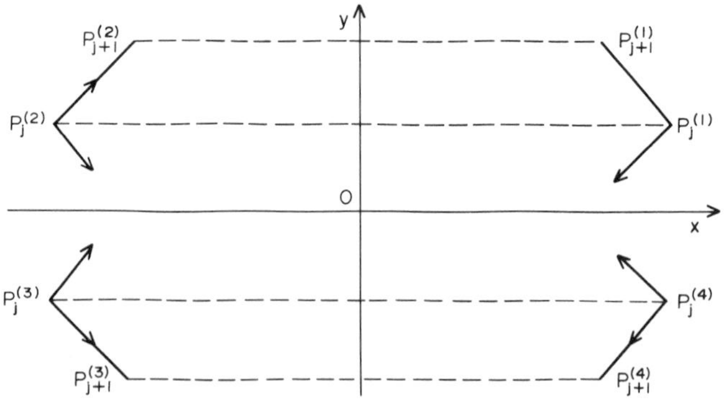

FIGURE 6

---

* We will retain A_M^j and B_M^j for quantities in the first quadrant.

(ii) if symmetrical around both Ox and Oy

$$-A^j_{M,2} = -A^j_{M,3} = A^j_{M,4} = A_M{}^j$$

$$-B^j_{M,2} = -B^j_{M,3} = B^j_{M,4} = B_M{}^j. \tag{10.2}$$

Consequently, on introducing the additional subscript k to indicate the appropriate quadrant, e.g., $x'_{j,k}$, $y'_{j,k}$, we obtain from edge-functions (2.5) the symmetrical edge-functions:

$$u'_j = \sum_{k=1}^{k'} \mu_k(A_M{}^j + m_j y'_{j,k} B_M{}^j)e^{-m_j y' j,k} \sin (m_j x'_{j,k} + \alpha_M) \tag{10.3}$$

where $k' = 2$ or 4, according as u has single or double symmetry, and the set μ_k is defined as:

$$\mu_k = 1, -1, -1, 1. \tag{10.4}$$

The derived symmetrical edge-functions follow similarly from formula (3.4) on inserting the factor μ_k, and are denoted by the symbol

$$Q_t S_M{}^j. \tag{10.5}$$

The subroutine for edge-functions can be extended quite simply to incorporate symmetrical edge-functions, as in Quinlan (1968) merely by the insertion of an additional k loop. The subroutine obviously requires the necessary parameters to evaluate the required edge-function, viz:

(1) x_k, y_j and ϕ_j to designate edge-axes;
(2) function indicator t, and symmetry number k';
(3) α_M and m_j;
(4) evaluation point (x, y) and associated function axes ϕ_q, and then a symmetrical edge-function is as simple to use in a problem as any of the usual functions, e.g., $\sin x$, $\log x$.

XI. SYMMETRICAL SINGULAR LOADINGS

It remains to develop symmetrical forms for the singular loadings in Sections VIII and IX. The cuts for $\theta = 0$ introduced in Figure 5 require a corresponding symmetrical system of cuts and axes.

Care is required in applying an additional k-loop in a computer subroutine for Sections VIII and IX to ensure that, since the curvilinear quantities (8.1) to (8.4) for interior line-loads, or (9.3) and (9.4) for boundary line-loads at any point P are obtained with respect to their corresponding axes x'_q and y'_q, their signs must be adjusted to refer to the first quadrant axes x'_q and y'_q as shown for P.

XII. APPLICATIONS

To illustrate the power and scope of the edge-function method in elasto-statics, two examples are presented in Appendices B and C for beams (a) and (b) as in Figure 7 involving line-loads on beams of given cross-sections.

Each problem is doubly symmetric. Consequently, both symmetrical edge-functions and singular functions are required.

Beam (a) is of square cross-section with four symmetrically placed unit line-loads acting parallel to OX on the boundary. The boundary residuals are of order $1/10^5$ after eight harmonics, involving but 45 equations.

Displacements and stresses are given for 11 equidistant points on EF for 4, 6 and 8 harmonics, and the stability of the results is poetry to any numerical analyst. In fact, 4 harmonics give the numerical value for each quantity to at least the order of $1/10^3$, and this involves but 21 equations.

Beam (b) is presented as a final example involving an irregular polygon with two sets of four symmetrically placed line loads acting in both the interior and on the boundary. Naturally, due to the irregularity of the figure, the results are not as spectacular as in (a), though an accuracy of at least two per cent is obtained with 10 harmonics, involving 89 equations.

In all cases the outstanding feature is that the stresses—involving derivatives of the displacement components—are as accurate as the displacements. No other method purporting to give reliable numerical results has this remarkable feature.

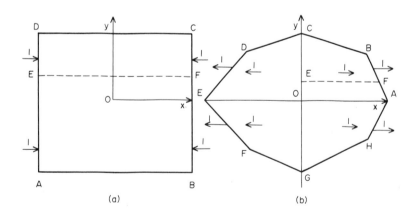

(a)   (b)

ACKNOWLEDGMENTS

This research was grant-aided (Grants AF EOAR 68-0018 and 69-0049) by The Air Force Office of Scientific Research, OAR, through the European Office of Aerospace Research, United States Air Force, for which the author is deeply appreciative. The author is indebted to Dr. D. J. Evans, The University, Sheffield, and Dr. P. O'Regan, University College, Cork, for their assistance in programming, and to University College, Cork, for its computer facilities.

REFERENCES

1. I. Fredholm, Solution d'un problème fondamental de la théorie de l'élasticité, *Ark. Mat. Astr. Fys.* **2**, 3–8 (1905).
2. G. Lauricella, Sur l'integration de l'equation relative a l'equilibre des plaques elastiques encastrees, *Acta. Math.* **32**, 201–256 (1909).
3. D. I. Sherman, *Dokl. Akad. Nauk SSSR* 27–28 (1940); 32 (1941).
4. S. G. Mikhlin, *Integral equations*, Pergamon Press, London, (1957).
5. N. I. Muskhelishvili, *Some basic problems of the mathematical theory of elasticity*, Noordhoff, Groningen (1953).
6. M. A. Jaswon, Integral equation methods in potential theory, I. *Proc. Roy. Soc.* **A275**, 23–32 (1963).
7. P. M. Quinlan, The torsion of an irregular polygon, *Proc. Roy. Soc.*, **A282**, 208–227 (1964).
8. F. J. Rizzo, An integral equation approach to boundary value problems of classical elastostatics, *Quart. Appl. Math.* **25**, 83–895 (1967).
9. P. M. Quinlan, The λ-method for rectangular plates, *Proc. Roy. Soc.* **A228**, 371–394 (1965).
10. P. M. Quinlan, Polygonal and swept-back plates with cut-outs and column supports, OAR Research Applications Conference, (1968).
11. G. V. Kelly *The λ-method for linear boundary value problems*. Ph. D. thesis, National University of Ireland, (1967).
12. S. Timoshenko and J. N. Goodier, *Theory of Elasticity*, McGraw-Hill, New York, (1951).

APPENDIX A

Quadrilateral Beam under Specified Surface Displacements

Table III
Root Mean Square of Boundary Residuals

Side	Function	Truncation Level		
		$L = 5$	$L = 6$	$L = 7$
AB	1	0·3305E-05	0·1555E-05	0·1767E-05
	2	0·1279E-05	0·6753E-06	0·6367E-06
BC	1	0·2571E-05	0·3097E-05	0·1518E-05
	2	0·2260E-05	0·1381E-05	0·1280E-05
CD	1	0·1914E-05	0·4206E-05	0·6024E-06
	2	0·6460E-06	0·3699E-05	0·1051E-06
DA	1	0·9167E-06	0·9660E-06	0·4753E-06
	2	0·4355E-07	0·3191E-06	0·1728E-07

Table IV
u-Displacements at points on EF

Point	$L = 5$	$L = 6$	$L = 7$
1	−0·1249E-03	−0·1248E-03	−0·1250E-03
2	−0·8724E-04	−0·8744E-04	−0·8742E-04
3	−0·5553E-04	−0·5560E-04	−0·5559E-04
4	−0·2901E-04	−0·2905E-04	−0·2904E-04
5	−0·6064E-05	−0·6091E-05	−0·6068E-05
6	0·1492E-04	0·1490E-04	0·1492E-04
7	0·3464E-04	0·3459E-04	0·3464E-04
8	0·5245E-04	0·5240E-04	0·5244E-04
9	0·6692E-04	0·6691E-04	0·6693E-04
10	0·7721E-04	0·7750E-04	0·7739E-04
11	0·8406E-04	0·8746E-04	0·8657E-04

Table V

σ_q-Normal stresses at points on EF

Point	$L = 5$	$L = 6$	$L = 7$
1	0·2414E 01	0·2638E 01	0·2572E 01
2	0·2603E 01	0·2607E 01	0·2606E 01
3	0·2816E 01	0·2814E 01	0·2814E 01
4	0·3011E 01	0·3010E 01	0·3010E 01
5	0·3150E 01	0·3149E 01	0·3150E 01
6	0·3262E 01	0·3262E 01	0·3262E 01
7	0·3415E 01	0·3415E 01	0·3416E 01
8	0·3669E 01	0·3668E 01	0·3668E 01
9	0·3997E 01	0·3994E 01	0·3998E 01
10	0·4245E 01	0·4214E 01	0·4216E 01
11	0·4226E 01	0·3797E 01	0·3910E 01

APPENDIX B

Square Beam under Four Symmetrically placed Line-Loads on Boundary

Table VI

Displacement component u

Point	$L = 4$	$L = 6$	$L = 8$
1	−0·7834E-04	−0·7848E-04	−0·7847E-04
2	−0·7650E-04	−0·7656E-04	−0·7656E-04
3	−0·7415E-04	−0·7418E-04	−0·7418E-04
4	−0·6743E-04	−0·6746E-04	−0·6746E-04
5	−0·5793E-04	−0·5795E-04	−0·5796E-04
6	−0·4717E-04	−0·4719E-04	−0·4719E-04
7	−0·3587E-04	−0·3588E-04	−0·3589E-04
8	−0·2434E-04	−0·2435E-04	−0·2435E-04
9	−0·1271E-04	−0·1272E-04	−0·1272E-04
10	−0·1060E-05	−0·1060E-05	−0·1060E-05
11	0·1060E-04	0·1060E-04	0·1060E-04

Table VII

Shear stress ζ_q

Point	$L = 4$	$L = 6$	$L = 8$
1	$-0 \cdot 9587$E-03	$-0 \cdot 1960$E-02	$0 \cdot 6746$E-03
2	$-0 \cdot 3562$E 00	$-0 \cdot 3559$E 00	$-0 \cdot 3546$E 00
3	$-0 \cdot 6106$E 00	$-0 \cdot 6104$E 00	$-0 \cdot 6102$E 00
4	$-0 \cdot 5322$E 00	$-0 \cdot 5322$E 00	$-0 \cdot 5322$E 00
5	$-0 \cdot 3715$E 00	$-0 \cdot 31715$E 00	$-0 \cdot 3716$E 00
6	$-0 \cdot 2343$E 00	$-0 \cdot 2343$E 00	$-0 \cdot 2343$E 00
7	$-0 \cdot 1376$E 00	$-0 \cdot 1375$E 00	$-0 \cdot 1376$E 00
8	$-0 \cdot 7428$E-01	$-0 \cdot 7420$E-01	$-0 \cdot 7423$E-01
9	$-0 \cdot 3290$E-01	$-0 \cdot 3285$E-01	$-0 \cdot 3287$E-01
10	$-0 \cdot 2564$E-02	$-0 \cdot 2560$E-02	$-0 \cdot 2561$E-02
11	$0 \cdot 2687$E-01	$0 \cdot 2683$E-01	$0 \cdot 2684$E-01

Table VIII

Normal stress σ_q

Point	$L = 4$	$L = 6$	$L = 8$
1	$-0 \cdot 8408$E 00	$-0 \cdot 8330$E 00	$-0 \cdot 8331$E 00
2	$0 \cdot 1787$E 00	$0 \cdot 1784$E 00	$0 \cdot 1788$E 00
3	$0 \cdot 2798$E 00	$0 \cdot 2798$E 00	$0 \cdot 2801$E 00
4	$0 \cdot 1350$E 00	$0 \cdot 1350$E 00	$0 \cdot 1352$E 00
5	$0 \cdot 2886$E-01	$0 \cdot 2863$E-01	$0 \cdot 2880$E-01
6	$-0 \cdot 3091$E-01	$-0 \cdot 3139$E-01	$-0 \cdot 3126$E-01
7	$-0 \cdot 6409$E-01	$-0 \cdot 6476$E-01	$-0 \cdot 6465$E-01
8	$-0 \cdot 8231$E-01	$-0 \cdot 8309$E-01	$-0 \cdot 8300$E-01
9	$-0 \cdot 9146$E-01	$-0 \cdot 9230$E-01	$-0 \cdot 9223$E-01
10	$-0 \cdot 9449$E-01	$-0 \cdot 9535$E-01	$-0 \cdot 9528$E-01
11	$-0 \cdot 9242$E-01	$-0 \cdot 9326$E-01	$-0 \cdot 9319$E-01

APPENDIX C

Trapezoidal Beams under Internal and Boundary Line Loads

Table IX
u-Displacement component at points on EF

Point	$L = 6$	$L = 7$	$L = 8$
1	$-0\cdot1362$E-07	$-0\cdot1387$E-07	$-0\cdot1399$E-07
2	$0\cdot2301$E-04	$0\cdot2299$E-04	$0\cdot2298$E-04
3	$0\cdot4617$E-04	$0\cdot4613$E-04	$0\cdot4611$E-04
4	$0\cdot6894$E-04	$0\cdot6889$E-04	$0\cdot6885$E-04
5	$0\cdot8949$E-04	$0\cdot8944$E-04	$0\cdot8939$E-04
6	$0\cdot1079$E-03	$0\cdot1079$E-03	$0\cdot1078$E-03
7	$0\cdot1284$E-03	$0\cdot1284$E-03	$0\cdot1283$E-03
8	$0\cdot1471$E-03	$0\cdot1471$E-03	$0\cdot1470$E-03
9	$0\cdot1645$E-03	$0\cdot1646$E-03	$0\cdot1646$E-03
10	$0\cdot1758$E-03	$0\cdot1762$E-03	$0\cdot1763$E-03
11	$0\cdot1755$E-03	$0\cdot1771$E-03	$0\cdot1769$E-03

Table X
ζ_q Shear stress at points on EF

Point	$L = 6$	$L = 7$	$L = 8$
1	$0\cdot2296$E-04	$0\cdot2345$E-04	$0\cdot2353$E-04
2	$0\cdot2047$E 00	$0\cdot2042$E 00	$0\cdot2047$E 00
3	$0\cdot4404$E 00	$0\cdot4392$E 00	$0\cdot4402$E 00
4	$0\cdot7114$E 00	$0\cdot7089$E 00	$0\cdot7098$E 00
5	$0\cdot8300$E 00	$0\cdot8260$E 00	$0\cdot8265$E 00
6	$0\cdot7812$E 00	$0\cdot7751$E 00	$0\cdot7745$E 00
7	$0\cdot1301$E 01	$0\cdot1293$E 01	$0\cdot1293$E 01
8	$0\cdot1642$E 01	$0\cdot1636$E 01	$0\cdot1635$E 01
9	$0\cdot2126$E 01	$0\cdot2119$E 01	$0\cdot2120$E 01
10	$0\cdot2260$E 01	$0\cdot2253$E 01	$0\cdot2251$E 01
11	$0\cdot1008$E 01	$0\cdot9622$E 00	$0\cdot1046$E 01

Generalised Lobatto Quadrature Formulas
for Contour Integrals

S. J. Maskell† and R. A. Sack

University of Salford, Salford, England

I. INTRODUCTION

Functionals of the form

$$I_{ab} \equiv I_1 = \int_a^b w(z)f(z)\,dz = \int_a^b (z-a)^\alpha(b-z)^\beta h(z)f(z)\,dz \qquad (1.1a)$$

where a, b, α and β are complex constants and $h(z)$ and $f(z)$ are functions of the complex variable z, analytic in a connected region D containing a and b, can be defined as line integrals provided the whole of the path of integration lies within D and

$$\mathrm{Re}(\alpha) > -1, \quad \mathrm{Re}(\beta) > -1. \qquad (1.1b)$$

In many applications, such functionals arise as contour integrals, the possible paths being shown in Figure 1:

$$I_{ab} \equiv I_2 = \frac{\oint wf\,dz}{1-e^{-2\pi\beta i}}, \quad \text{valid for } \mathrm{Re}(\alpha) > -1, \quad \beta \in \mathbb{Z}, \qquad (1.2)$$

$$I_{ab} \equiv I_3 = \frac{\oint wf\,dz}{1-e^{2\pi\alpha i}}, \quad \text{valid for } \mathrm{Re}(\beta) > -1, \quad \alpha \notin \mathbb{Z}, \qquad (1.3)$$

$$I_{ab} \equiv I_4 \text{ as in } I_2 \text{ or } I_3, \text{ valid for } \alpha \notin \mathbb{Z}, (\alpha+\beta) \in \mathbb{Z}, \qquad (1.4)$$

$$I_{ab} \equiv I_5 = \frac{\oint wf\,dz}{(1-e^{2\pi\alpha i})(1-e^{-2\pi\beta i})}, \text{ valid for } \alpha \notin \mathbb{Z}, \beta \in \mathbb{Z}. \,(1.5)$$

---

† Present address: Department of Mathematics, University of Exeter, Exeter, Devon, England.

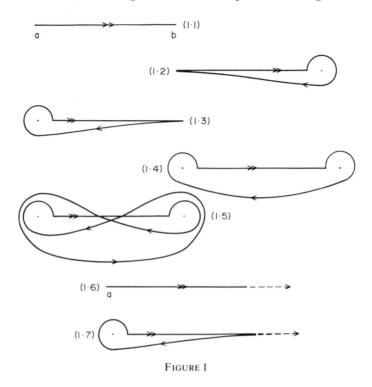

FIGURE 1

Here $\mathbb{Z}$ denotes the set of real integers and the contour of integration for each of the above formulas is shown in the correspondingly numbered diagram. Whenever α and β satisfy more than one of the set of conditions appropriate to (1.1)–(1.5) the same value of I_{ab} is obtained, whichever formula is applied. Between them the conditions (1.1)–(1.5) cover all possible values of α and β, except real negative integers.

Similarly the improper integral

$$I_{a\infty} = \int_a^\infty (z-a)^\alpha h(z)f(z)\,dz \equiv \int_a^\infty w(z)f(z)\,dz \qquad (1.6)$$

valid for $\mathrm{Re}(\alpha) > -1$ and suitable convergence conditions near $z = \infty$ is equivalent to

$$I_{a\infty} = (1-e^{2i\pi\alpha})^{-1} \oint wf\,dz, \quad \alpha \notin \mathbb{Z}, \qquad (1.7)$$

When there is no risk of confusion these integrals will be abbreviated to $I(w;f)$ or simply $I(f)$. One important property of these functionals is that they have real value if a, b, α and β are real constants (we always assume $b > a$) and $h(z)$ and $f(z)$ are real functions for real z in the intervals $[a,b]$ or $[a,\infty)$ respectively. This real-valuedness of $I(w;f)$ applies even when it cannot be expressed as a line integral (1.1) or (1.6); it implies for all the graphs in Fig. 1 that the branch of the weight function (kernel) $w(z)$ corresponding to the straight (real) section of the path marked by a double arrow is real-valued.

There exists an extensive literature regarding the numerical approximation (quadrature) of the real line-integrals (1.1) and (1.6) by finite sums

$$I(f) \simeq Q_N(f) = \sum_1^N C_\nu f(x_\nu). \qquad (1.8)$$

In particular, for every weight function $w(x)$ which is non-negative in the interval $[a,b]$ or $[a,\infty)$, there exists a Gauss quadrature formula which optimizes the abscissas $\{x_j\}$ so that the product integral (1.8) is rigorously rendered by an N-point summation for any polynomial $f(x)$ up to degree $2N - 1$.

Such Gauss formulas have the following properties: (Kopal 1961, Lanczos 1957)

(i) the set $\{x_j, C_j\}$ (the Christoffel numbers) is unique for any given $a, b, w(x)$;
(ii) the abscissas $\{x_j\}$ are given by the zeros of the N-th orthogonal polynomial $T_N(x)$;
(iii) the x_j are all real, distinct and contained in the interval $[a,b]$;
(iv) the weights C_j are all real and positive.

To the writers' knowledge no corresponding numerical formulas are known if (1.1b) is not satisfied and the integral is defined by (1·2) – (1·5) or (1·7). Even if it is known from other considerations that the result must have real value, one usually has to resort to complex quadrature points and weights, which are mainly selected empirically or on an *ad hoc* basis with all the inefficiency inherent in the use of complex arithmetic. It is clear that in the general case it is impossible to find completely analogous formulas to (1·8) for which the conditions (i)–(iv) are all satisfied. As a counter-example one need only consider $\alpha + \beta = -2$ (or $-3, -4, \dots$), $h \equiv f \equiv 1$; in this case the contour (1·4) can be expanded into an infinite circle, the integral vanishes and in view of (1·8) the C_j must add up to zero. For the same weight function any attempt to construct a set of orthogonal polynomials $T_n(x)$ is bound to lead to inconsistencies as from the above consideration the (constant) polynomial $T_0(x)$ would have to be orthogonal to itself.

One way of restricting the quadrature points to real arguments would be to transform the integrand analytically into a form satisfying (1.1b) using integration by parts, and then to apply (1.8). Even if only one of the exponents α and β is fractional, this procedure would involve the evaluation of derivatives of $f(x)$ (and possibly of $h(x)$ as well) at all the quadrature points; if both exponents are fractional there is the added difficulty that the repeated integrals of $(x-a)^\alpha(b-x)^\beta$ are themselves complicated functions and the new kernel $w(x)$ may be of a more complicated nature than indicated in the form (1.1). It is, therefore, advisable to develop quadrature formulas which involve explicitly the values of $f(x)$ for real values of x in the interval $[a, b]$, and derivatives of $f(x)$ at a few selected points only. The obvious choice for these special points are the end-points a and b of the range, so that $I(f)$ is approximated by

$$I(f) \simeq Q_{NLM}(f) = \sum_1^N C_\nu f(x_\nu) + \sum_0^{L-1} A_\lambda f^{(\lambda)}(a) + \sum_0^{M-1} B_\mu f^{(\mu)}(b). \quad (1.9)$$

As the arguments x_ν in the first sum are not prescribed a priori, an optimized formula of this type should correctly render $I(f)$ if $f(x)$ is any polynomial of degree $2N+L+M-1$. This type of quadrature can be employed even if the condition (1.1b) applies. Thus for $L = M = 1$ and $w(x) \equiv 1$ the corresponding procedure is known as Lobatto quadrature[1], for $w(x) \equiv 1$ and $L = M = 2$ formulas have been derived by Stancu and Stroud (1963), though the special case $N = 1$ had been discussed previously, especially by Lanczos (1957) who calls it the corrected Simpson's rule (cf. also Sack 1962).

The purpose of the present paper is to give a consistent treatment of generalized Lobatto quadrature formulas (1.9) for real values of a, b, α and β, not necessarily satisfying (1.1b), but any one of the conditions appropriate to (1.1)–(1.5) as well as

$$\alpha + L > -1, \quad \beta + M > -1. \quad (1.10)$$

In addition $h(x)$, though not necessarily $f(x)$, should be real and non-negative in the interval (a,b). The general theory is developed in Section II; the special case that the kernel in (1.1)–(1.5) is given by

$$w(x) = (x-a)^\alpha(b-x)^\beta; \quad h(x) \equiv 1 \quad (1.11)$$

is treated in Section III. A number of numerical tests are described in Section IV, and the results discussed in Section V.

II. GENERAL FORMULAS FOR THE COEFFICIENTS

It is clear that under the conditions stated just before and in (1.10) the kernel

$$v(x) = (x-a)^L(b-x)^M w(x) \quad (2.1)$$

satisfies (1.1b), and hence there exist for $v(x)$ quadrature formulas of the form (1.8)

$$I(v;g) \simeq \sum_1^N c_\nu g(x_\nu) \qquad (2.2)$$

with the properties (i)–(iv) listed after (1.8). Now if we take the special case for the function $f(x)$

$$f(x) = (x-a)^L (b-x)^M g(x) \qquad (2.3)$$

then on substitution into (1.9) all the terms in the 2nd and 3rd sum vanish. But since

$$I(v;g) \equiv I(w;f) \qquad (2.4)$$

comparison with (2.2) shows that the set $\{x_\nu\}$ is the same in (2.2) and (1.9), whereas the weights are related by

$$C_\nu = c_\nu (x_\nu - a)^{-L} (b - x_\nu)^{-M}. \qquad (2.5)$$

Hence once the quadrature formula (2.2) has been established by any of the numerous numerical or analytical methods available[2,4,5,6,11], it remains to find the coefficients A_λ and B_μ in (1.9). It is assumed that integrals

$$I_{lm} \equiv I(w; (x-a)^l (b-x)^m), \quad l,m \in \mathbb{Z}, \quad l,m \geq 0, \qquad (2.6)$$

can be evaluated exactly. With this choice of $f(x)$

$$\left(\frac{d}{dx}\right)^\lambda f(x)\bigg|_{x=a} = (-)^{l-\lambda} \frac{\lambda!}{(\lambda-l)!} \frac{m!}{(m+l-\lambda)!} c^{m+l-\lambda} \qquad (2.7a)$$

$$\left(\frac{d}{dx}\right)^\mu f(x)\bigg|_{x=b} = (-)^m \frac{\mu!}{(\mu-m)!} \frac{l!}{(m+l-\mu)!} c^{m+l-\mu} \qquad (2.7b)$$

valid for $m+l \geq \lambda \geq l$ and $m+l \geq \mu \geq m$ respectively (otherwise the derivatives vanish) where we abbreviate

$$c = b - a. \qquad (2.8)$$

Substitution of (2.6), (2.7) into (1.10) yields

$$I^{lM} = \sum_1^N C_\nu (x_\nu - a)^l (b-x_\nu)^M + \sum_l^{L-1} (-)^{l-\lambda} A_\lambda \frac{\lambda! M! c^{M+l-\lambda}}{(\lambda-l)!(M+l-\lambda)!} \qquad (2.9a)$$

$$I^{Lm} = \sum_1^N C_\nu (x_\nu - a)^L (b-x_\nu)^m + \sum_m^{M-1} (-)^m B_\mu \frac{\mu! L! c^{L+m-\mu}}{(\mu-m)!(L+m-\mu)!} \qquad (2.9b)$$

If in these expressions l is reduced from $L-1$ to 0, or m from $M-1$ to 0, each A_l or B_m can be expressed in terms of the integrals, the C_ν and the previously calculated $A_\lambda(\lambda > l)$ or $B_\mu(\mu > m)$. For $M = 0$, of which (1.7) is a special case, (2.9a) simplifies to

$$I^l = \sum_1^N C_\nu(x_\nu - a)^l + B_l l!$$ (2.10)

The sum involving the internal quadrature points x_ν can be avoided with the use of the more general integral

$$I^{lm}(P_N) = I(w; (x - a)^l(b - x)^m P_N(x))$$ (2.11)

where $P_N(x)$ is the N-th orthogonal polynomial for the kernel $v(x)$ in (2.1). As the $\{x_\nu\}$ in (2.2) are given by the zeros of this polynomial, we obtain, by using the generalizations of (2.7) for the appropriate derivatives, in analogy to (2.7)–(2.9)

$$I^{lM}(P_N) = \sum_{\lambda=l}^{L-1} \lambda! A_\lambda U_{MN}^{\lambda-l}(a),$$ (2.12a)

$$I^{Lm}(P_N) = \sum_{\mu=m}^{M-1} (-)^\mu \mu! B_\mu U_{LN}^{\mu-m}(b),$$ (2.12b)

where

$$U_{MN}^{\lambda-l}(a) = \sum_\tau \frac{(-)^\tau M! c^{M-\tau}}{\tau!(\lambda-l-\tau)!(M-\tau)!} \left(\frac{d}{da}\right)^{\lambda-l-\tau} P_N(a),$$ (2.13a)

$$U_{LN}^{\mu-m}(b) = \sum_\sigma \frac{L! c^{L-\sigma}}{\sigma!(\mu-m-\sigma)!(L-\sigma)!} \left(\frac{d}{db}\right)^{\mu-m-\sigma} P_N(b).$$ (2.13b)

Here the sum over τ in (2.13a) is to be taken from 0 to the smaller of M and $\lambda - l$, and that over σ in (2.13b) from 0 to the smaller of L and $\mu - m$. Again for $M = 0$ the summation simplifies, with the result

$$I^l(P_N) = \sum_{\lambda=l}^{L-1} \frac{\lambda! A_\lambda}{(\lambda-l)!} \left(\frac{d}{da}\right)^{\lambda-l} P_N(a).$$ (2.14)

As with (2.9), the coefficients A_l (and B_m) can be calculated successively by letting l run from $L-1$ to 0 in (2.12a) or (2.14) (and m from $M-1$ to 0 in (2.12b)).

III. GENERALIZED LOBATTO-MEHLER QUADRATURE

More specific formulas can be derived, if the kernel $w(z)$ is of the form (1.11). If the conditions (1.1b) apply, the corresponding quadrature using (1.8) is known as Gauss-Mehler quadrature. In particular, if $a = -1$, $b = 1$ the corresponding orthogonal polynomials are the Jacobi polynomials

$$P_N^{(\beta,\alpha)}(x) = (\alpha+1)_N \, {}_2F_1(-N, \alpha+\beta+N+1; \alpha+1; \tfrac{1}{2}+\tfrac{1}{2}x)/N! \qquad (3.1a)$$

$$= (-)^N(\beta+1)_N \, {}_2F_1(-N, \alpha+\beta+N+1; \beta+1; \tfrac{1}{2}-\tfrac{1}{2}x)/N! \qquad (3.1b)$$

(cf. Abramowitz and Stegun 1965, Ch. 15 and 22; further references to this work will be abbreviated by A.S.). Here ${}_2F_1$ is the hypergeometric function

$$\,{}_2F_1(a, b; c; x) = \sum (a)_n(b)_n x^n/[(c)_n n!] \qquad (3.2)$$

and

$$(a)_n = a(a+1)(a+2)\cdots(a+n-1) = \Gamma(a+n)/\Gamma(a) \qquad (3.3)$$

In the literature the exponents α and β are usually associated with the upper and lower limits respectively; to conform with the standard notation the superscripts in the definition of P_N in (3.1) appear reversed. For arbitrary values of a and b the last arguments in (3.1a) and (3.1b) are to be replaced by $(x-a)/c$ and $(b-x)/c$ respectively. When applied to (1.9), the indices in (3.1) are $\alpha+L$ and $\beta+M$, in view of (2.1).

For given values of α, β, L and M the integrals (2.6) simplify to beta functions

$$I^{lm} = \int_a^b (x-a)^{\alpha+l}(b-x)^{\beta+m}dx = \frac{\Gamma(\alpha+l+1)\Gamma(\beta+m+1)c^{\alpha+\beta+l+m+1}}{\Gamma(\alpha+\beta+l+m+2)} \qquad (3.4)$$

(cf. A.S. 6.2.1, 2). This formula is usually derived on the assumption that all the arguments of the gamma functions in (3.4) have positive real parts; but with the definitions (1.2)–(1.5) it holds equally under the appropriate conditions. However, the more complicated integrals $I^{lm}(P_N)$ in (2.11) can also be expressed in closed form, so that the coefficients A_λ and B_μ can be determined by means of (2.12) rather than by (2.9). Substitution of (3.1b) and (3.4) into (2.11) yields

$$I^{lM}(P_N) = (-)^N \frac{(\beta+1+M)_N}{N!} c^{\alpha+\beta+l+M+1}$$

$$\times \sum \frac{(-N)_s(\alpha+\beta+L+M+N+1)_s\,\Gamma(\beta+M+1+s)\Gamma(\alpha+l+1)}{(\beta+M+1)_s\Gamma(\alpha+\beta+l+M+2+s)s!}$$

$$= (-)^N \frac{\Gamma(\beta+1+M+N)\Gamma(\alpha+l+1)c^{\alpha+\beta+l+M+1}}{N!\,\Gamma(\alpha+\beta+l+M+2)}$$

$$\times \, {}_2F_1\left(\begin{matrix} -N, \alpha+\beta+N+L+M+1; \\ \alpha+\beta+l+M+2; \end{matrix} 1\right)$$

$$= (-)^N \frac{\Gamma(\beta+1+M+N)\Gamma(\alpha+l+1)c^{\alpha+\beta+l+M+1}(1+l-L-N)_N}{N!\,\Gamma(\alpha+\beta+l+M+N+2)}. \qquad (3.5)$$

In the last step use has been made of the formula for terminating hypergeometric series of unit argument (A.S. 15.1, 20)

$$F(-n, \beta; \gamma; 1) = (\gamma - \beta)_n / (\gamma)_n. \tag{3.6}$$

The last factor in the numerator of (3.5) can also be written as $(-)^N(L-l)_N$. Similarly

$$I^{Lm}(P_N) = (-)^N \frac{\Gamma(\alpha+1+L+N)\Gamma(\beta+m+1)\,(M-m)_N\,c^{\alpha+\beta+L+m+1}}{N!\Gamma(\alpha+\beta+L+m+N+2)}. \tag{3.7}$$

With the formula for the derivatives of the hypergeometric function (A.S. 15.2.2.) (2.13) and (3.1) yield

$$U^\lambda_{MN}(a) = M!c^{M-\lambda} \sum_\tau \frac{(-)^{\lambda-\tau}(\alpha+1)_N(-N)_\tau(\alpha+\beta+N+1)_\tau}{(M-\lambda+\tau)!N!\tau!(\alpha+1)_\tau(\lambda-\tau)!} \tag{3.8a}$$

$$U^\mu_{LN}(b) = L!c^{L-\mu} \sum_\sigma \frac{(-)^\sigma(\beta+1)_N(-N)_\sigma(\alpha+\beta+N+1)_\sigma}{(L-\mu+\sigma)!N!\sigma!(\beta+1)_\sigma(\mu-\sigma)!}. \tag{3.8b}$$

These sums can be expressed in terms of the generalized hypergeometric function $_3F_2$ of unit argument, but as this function cannot, in general, be expressed in closed form, (3.8) is best left as it stands. The summation indices τ and σ are complementary to the correspondingly denoted indices in (2.13) and are to be taken over all integers for which the arguments of all the factorials in the denominators of (3.8a) or (3.8b) are non-negative.

For $l = L-1$ and $m = M-1$ respectively (2.12) and (3.8) simplify to

$$A_{L-1} = \frac{(\beta+M+1)_N}{(\alpha+L+1)_N} \frac{N!}{(L-1)!} \frac{\Gamma(\alpha+L)\Gamma(\beta+M+1)}{\Gamma(\alpha+\beta+L+M+N+1)} c^{\alpha+\beta+l}, \tag{3.9a}$$

$$B_{M-1} = (-)^{M-1} \frac{(\alpha+L+1)_N}{(\beta+M+1)_N} \frac{N!}{(M-1)!} \frac{\Gamma(\beta+M)\Gamma(\alpha+L+1)}{\Gamma(\alpha+\beta+L+M+N+1)} c^{\alpha+\beta+M}. \tag{3.9b}$$

For $l = L-2$ and $m = M-2$ the corresponding results are obtained after elementary, but tedious, manipulations

$$A_{L-2} = \frac{A_{L-1}(L-1)}{(\alpha+L-1)c} \left[\frac{2N(\alpha+\beta+L+M+N+1)(\alpha+L)}{(\alpha+L+1)} \right.$$
$$\left. + (\alpha+L)(M+1) + \beta \right], \tag{3.10a}$$

$$B_{M-2} = -\frac{B_{M-1}(M-1)}{(\beta+M-1)c} \left[\frac{2N(\alpha+\beta+L+M+N+1)(\beta+M)}{(\beta+M+1)} \right.$$
$$\left. + (\beta+M)(L+1) + \alpha \right]. \tag{3.10b}$$

The calculation of the higher coefficients becomes progressively more complicated, but since the cases $L, M = 1, 2$ are likely to be the most important to arise, all the coefficients required in (1.9) are given in (3.9) and (3.10).

For integrals of the form

$$\int_0^\infty x^\alpha e^{-kx} f(x) dx \tag{3.11}$$

the numerical quadrature corresponding to the kernel (2.1) with (1.10) and $M = 0$ is known as Gauss-Laguerre quadrature (A.S. Ch. 13, 22, Kopal 1961). The abscissas $\{x_\nu\}$ and weights $\{c_\nu\}$ are determined by the zeros of generalized Laguerre polynomials. The coefficients A_λ associated with the lower limit $a = 0$ are found most easily from (3.9) and (3.10) by a limiting process, using the definition

$$e^{-kx} = \lim [(c-x)/c]^\beta; c \to \infty, \beta \to \infty, \beta/c = k.$$

Since for large values of β, $(\beta)_n = \beta^n$, substitution of this limit into (3.9a) and (3.10a) yields with $M = 0$

$$A_{L-1} = \frac{N! \Gamma(\alpha+L)}{(L-1)!(\alpha+L+1)_N k^{\alpha+L}}, \tag{3.12}$$

$$A_{L-2} = \frac{N! \Gamma(\alpha+L-1)}{(L-2)!(\alpha+L+1)_N k^{\alpha+L-1}} \left(2N \frac{\alpha+L}{\alpha+L+1}+1\right). \tag{3.13}$$

The same arguments as follow (3.10) are applicable in this case though the higher coefficients, if required, can be found with less effort than in the case $b \neq \infty$.

IV. NUMERICAL TESTS

A number of numerical tests were carried out on the ICL KDF9 computer of the University of Salford; the single-length precision of the machine for floating point arithmetic is 39 bits ($11\frac{1}{2}$ decimal digits).

The abscissas $\{x_\nu\}$ were determined, in all cases, by diagonalizing the "terminal matrix", i.e. the tri-diagonal matrix containing the coefficients in the recurrence relations between the (normalized) orthogonal polynomials $P_i(x)$ for the weight function (2.1).

$$\alpha_i P_{i+1}(x) + (\beta_i - x) P_i(x) + \alpha_{i-1} P_{i-1}(x) = 0 \tag{4.1}$$

(Note that the α_i, β_i are coefficients, not the exponents in (1.1)). The following methods were used to find the eigenvalues of these matrices:

(a) Givens' method of Sturm sequences, accelerated by the Newton–Raphson procedure once the eigenvalues are separated[2].

(b) The QR-algorithm in its standard irrational form.

(c) The recent stable rational modifications of the QR-algorithm (Reinsch 1971, Sack 1972).

(d) For the case of Gauss–Chebyshev quadrature only, $\alpha + L = \beta + M = \frac{1}{2}$ or $-\frac{1}{2}$, the abscissas are simply the cosines of equidistant arguments (cf. e.g. A.S. 22.16.4,5).

Similarly the weights c_ν in (2.2) have been shown by Golub and Welsch (1969) to be

$$c_\nu = I(v; 1) q_{1\nu}^2 \qquad (4.2)$$

where $q_{1\nu}$ is the leading component of the normalized eigenvector of the terminal matrix belonging to the eigenvalue x_ν (in Sack and Donovan 1972 this was erroneously attributed to Wilf 1962). These components $q_{1\nu}$ or their squares, can again be determined by a variety of methods (the term *a posteriori* means here that the weights are found after the abscissas have been determined, by whatever method):

(i) All the eigenvectors are calculated *a posteriori* by Wilkinson's (1958) method and only the leading components retained.

(ii) The weights are found as the ratios of two determinants (Sack and Donovan 1972, eqn. 4.4b) either *a posteriori* or simultaneously with method (a).

(iii) The leading components only of the eigenvectors are determined by the irrational QR-algorithm, either *a posteriori* or simultaneously with method (b)[4]; this method was not used in the current work.

(iv) The squares of the leading components are found *a posteriori* by the rational QR-algorithm[11]; this method was used only in conjunction with method (c)[10].

(v) For Gauss-Chebyshev quadrature, the weights are again known analytically as in (d).

For a given kernel $w(x)$, once the abscissas $\{x_\nu\}$ and weights $\{c_\nu\}$ had been found for a range of values of N by any of the above methods, the coefficients A_λ and B_μ were calculated (for values of L and M not exceeding 2) by means of (2.9) or minor modifications thereof. If $w(x)$ was of the form (1.11) or (3.11) these computed values were compared with the theoretical results (3.9)–(3.13). After that a family of functions $f(x)$ was chosen so that the integrals could be expressed in an analytic form, and the numerical results obtained by means of the generalized Lobatto formula (1.9) compared with the theoretical values.

The most extensive tests were carried out for the kernel of the form (3.11)

$$w = e^{-x} x^{-5/2} \pi^{-1/2}, \quad a = 0, L = 2, \qquad (4.3)$$

for which (3.12) and (3.13) simplify to

$$A_1 = -2N! / (\tfrac{1}{2})_N, \quad A_0 = -(\tfrac{8}{3}) N! / (\tfrac{1}{2})_{N-1}. \qquad (4.4)$$

The elements of the terminal matrix for $v = e^{-x}x^{-1/2}$ are given by (cf. AS 22.7.12)

$$\beta_i = 2i + \tfrac{1}{2}, \qquad \alpha_i^2 = (i+1)(i+\tfrac{1}{2}). \tag{4.5}$$

To determine the $\{x_\nu\}$ and $\{c_\nu\}$ from this matrix, all the methods listed above, except (iii), were employed in turn for values of $N = 1(1)15$; method (c) was also repeated in double precision, but the results subsequently used in single precision only. No significant discrepancies were found between the various single precision computations for the abscissas; the errors were confined to the last (12th) significant digits, except for the root nearest to 0, which dropped to near 0.04 and in consequence lost two significant digits (in the algorithm combining (c) and (iv) a further digit was lost). Similar results were found for the weights except that with increasing N the weights associated with the highest abscissas became very small, down to the order of 10^{-20}, and while these small values were correctly rendered by methods (i) and (iv), and presumably (iii), the absolute error in method (ii) remained of the order of the machine accuracy, and negative weights occurred from $N = 10$ onward.

On comparing the numerically computed coefficients A_0 and A_1 with those known from (4.4), discrepancies occurred in the 12th significant digit only, if the $\{x_\nu\}$ were calculated in double precision by method (c). For the other methods the errors were of similar magnitude; for methods (c) and (i) they are listed, regardless of sign and to single-digit accuracy, for even N in the first rows of Table I. Again 1 or 2 additional digits were lost with the subroutine combining (c) with (iv), but only for $N > 10$.

Table I

Approximate values and errors of the computed values of A_0 and A_1 for the kernel $x^{-5/2}e^{-x}$ and of the integrals $G_0(k)$ and $G_1(k)$ of (4.6).

k		N 2	4	6	8	10	12	G_0 / $-G_1$
	$-A_1$	5.333	7.314	8.866	10.18	11.35	12.41	
	δA_1	0	$4_{10}-11$	$2_{10}-10$	$1_{10}-9$	$3_{10}-8$	$2_{10}-8$	
	$-A_0$	10.67	34.13	65.02	101.8	143.8	190.3	
	δA_0	0	$6_{10}-10$	$2_{10}-9$	$2_{10}-8$	$5_{10}-7$	$8_{10}-7$	
$-.5$	δG_0	$8_{10}-5$	$3_{10}-7$	$2_{10}-9$	$2_{10}-10$	$6_{10}-11$	$5_{10}-10$	.4714
$-.5$	δG_1	$1_{10}-3$	$7_{10}-6$	$6_{10}-8$	$5_{10}-10$	$6_{10}-14$	$6_{10}-12$	1.414
$+.5$	δG_0	$2_{10}-5$	$9_{10}-9$	$2_{10}-10$	$3_{10}-10$	$4_{10}-10$	$4_{10}-10$	2.449
$+.5$	δG_1	$2_{10}-4$	$2_{10}-7$	$2_{10}-10$	$2_{10}-11$	$9_{10}-12$	$2_{10}-11$	2.449
1.5	δG_0	$4_{10}-3$	$6_{10}-5$	$1_{10}-6$	$2_{10}-8$	$1_{10}-10$	$6_{10}-10$	5.270
1.5	δG_1	$1_{10}-2$	$3_{10}-4$	$6_{10}-6$	$2_{10}-7$	$5_{10}-9$	$7_{10}-11$	3.162
2.5	δG_0	$4_{10}-2$	$2_{10}-3$	$8_{10}-5$	$5_{10}-6$	$3_{10}-7$	$2_{10}-8$	8.731
2.5	δG_1	$7_{10}-2$	$4_{10}-3$	$3_{10}-4$	$2_{10}-5$	$2_{10}-6$	$1_{10}-7$	3.742

Finally the integrals

$$G_0(k) = \pi^{-1/2} \int x^{-5/2} e^{-x} e^{-kx} dx = (1+k)^{3/2}(\tfrac{4}{3})$$

$$G_1(k) = \pi^{-1/2} \int x^{-3/2} e^{-x} e^{-kx} dx = -2(1+k)^{1/2} \qquad (4.6)$$

were approximated by (1.9) for the same range of N and k ranging from -0.9 to $+2.9$; for the same diagonalization procedure as above the errors are listed for a representative selection of k in the bottom eight rows of Table I; the approximate values of G_0 and G_1 are given at the end of each row. Where the discrepancies represented genuine differences, the same results were obtained by all methods; where they represented rounding errors all methods led to fluctuations of the same order of magnitude. Only if the weights had been determined by method (ii) catastrophic errors could build up for negative k and large N; this was due to enhancement of the wrong coefficients c_ν by the positive exponential and had nothing to do with the coefficients at $x = 0$; the same effect occurred if the singularity remained integrable, i.e. with the standard generalized Laguerre quadrature. The results obtained on using (c) and (iv) were no worse than those listed in Table I in spite of the less accurate coefficients, and those obtained by the double-precision use of (c) were no better.

Less extensive tests were carried out for the kernel

$$w = (1-x^2)^{5/2}/\pi, \quad b = -a = 1, \quad L = M = 2 \qquad (4.7)$$

for which (3.9) and (3.10) simplify to

$$A_0 = B_0 = -N(N^2+2)/6; \ B_1 = -A_1 = N/4. \qquad (4.8)$$

Here the accuracy of the quadrature points and weights for $v = (1-x^2)^{-1/2}$ can be checked analytically by methods (d) and (v); the errors were confined to the 12th decimal place for all methods, except for the sub-routine incorporating (c) and (iv) where for large N one additional figure was lost. The errors in the computed approximations to (4.8) using (c) and (i) are tabulated for even N in the upper part of Table II. Only the larger errors of each pair $(|\delta A_0|, |\delta B_0|)$ and $(|\delta A_1|, |\delta B_1|)$ are listed; they increase marginally less fast with N than in Table I.

The actual quadrature tests were performed for the integrals

$$H_j(k) = \pi^{-1} \int_{-1}^{1} (1-x^2)^{-5/2+j} e^{kx} dx, \quad j = 0, 1, 2 \qquad (4.9)$$

the theoretical values of which are

$$H_2(k) = I_0(k), \quad H_1(k) = -kI_1(k), \quad H_0(k) = (k^2/3)I_2(k) \qquad (4.10)$$

Table II

Approximate values and errors of the computed values of A_0, B_0 and A_1, B_1 for the kernel $(1-x^2)^{-5/2}$ and of the integrals $H_0(k)$ and $H_1(k)$ of (4.9).

N		2	4	6	8	10	12	
k								H_0
	$-A_0$	2	12	38	88	170	292	$-H_1$
	δA_0	$1_{10}{-}11$	$1_{10}{-}10$	$4_{10}{-}9$	$2_{10}{-}8$	$1_{10}{-}7$	$5_{10}{-}8$	
	δA_1	$4_{10}{-}12$	$4_{10}{-}12$	$6_{10}{-}11$	$2_{10}{-}10$	$8_{10}{-}10$	$2_{10}{-}10$	
.5	δH_0	$1_{10}{-}8$	$7_{10}{-}11$	$2_{10}{-}9$	$4_{10}{-}9$	$4_{10}{-}8$	$1_{10}{-}7$	.00266
.5	δH_1	$3_{10}{-}6$	$1_{10}{-}11$	$7_{10}{-}11$	$6_{10}{-}11$	$5_{10}{-}10$	$9_{10}{-}10$	.1289
1.5	δH_0	$9_{10}{-}5$	$2_{10}{-}9$	$3_{10}{-}9$	$6_{10}{-}9$	$6_{10}{-}8$	$1_{10}{-}7$	.2534
1.5	δH_1	$2_{10}{-}3$	$1_{10}{-}7$	$1_{10}{-}10$	$2_{10}{-}10$	$8_{10}{-}10$	$9_{10}{-}10$	1.4725
2.5	δH_0	$6_{10}{-}3$	$1_{10}{-}6$	$8_{10}{-}9$	$1_{10}{-}8$	$1_{10}{-}7$	$2_{10}{-}7$	2.6593
2.5	δH_1	$6_{10}{-}2$	$2_{10}{-}5$	$3_{10}{-}9$	$4_{10}{-}10$	$2_{10}{-}9$	$2_{10}{-}9$	6.2918
3.5	δH_0	$1_{10}{-}1$	$7_{10}{-}5$	$1_{10}{-}8$	$4_{10}{-}8$	$3_{10}{-}7$	$5_{10}{-}7$	15.647
3.5	δH_1	$5_{10}{-}1$	$8_{10}{-}4$	$3_{10}{-}7$	$1_{10}{-}9$	$5_{10}{-}9$	$5_{10}{-}9$	21.720

where $I_l(k)$ denotes the modified Bessel function of order l; the derivation of H_2 is a standard definition, the other results are most easily found by integration by parts. The differences between the true values of H_0 and H_1 according to (4.10) and those computed by means of (1.9) are listed in the lower part of Table II for a representative range of k. It is plain that once the systematic discrepancies have been reduced, rounding-off errors can become larger with further increase in N. That this effect is not primarily due to the computational errors in the coefficients is shown by the fact that it occurs almost to the same degree when all the abscissas and coefficients are taken as their analytic values from (d), (v) and (4.10). By contrast, the approximations to H_2, which do not involve the function at $x = \pm1$, do not show any enhancement of δH_2 with increasing N.

A further test was concerned with the kernel

$$w = \pi^{-1}(1-x^2)^{-5/2}e^{qx}, \quad b = -a = 1, L = M = 2. \qquad (4.11)$$

In this case the terminal matrix for $v = \pi^{-1}(1-x^2)^{-1/2}e^{qx}$ is not known *a priori* (unless $q = 0$), but can be found from the modified moments

$$\nu_l = \pi^{-1}\int_{-1}^{1} T_l(x)e^{qx}\mathrm{d}x = I_l(q) \qquad (4.12)$$

(I_l denoting again the modified Bessel functions and $T_l(x)$ the Chebyshev polynomials) by means of the long quotient-modified difference algorithm[11,3]. As the coefficients A_λ and B_μ are not known analytically, no comparison with the computed values was possible. The actual tests referred to the computation of the functions $H_0(q+k)$ and $H_1(q+k)$ of (4.9)–(4.10), for which the part e^{qx} was drawn into the kernel and e^{kx} was

Table III

Errors in the computed values of $H_0(0)$, $H_1(0)$, $H_0(1)$, $H_1(1)$ of (4.9) and (4.13) on drawing the factor e^{qx} into the kernel.

k	q		N 2	4	6	8	10	12
3	-3	δH_0	$1_{10}-2$	$7_{10}-6$	$9_{10}-9$	$2_{10}-8$	$2_{10}-7$	$2_{10}-7$
3	-3	δH_1	$8_{10}-2$	$1_{10}-4$	$2_{10}-8$	$5_{10}-10$	$4_{10}-10$	$9_{10}-10$
2	-2	δH_0	$7_{10}-4$	$6_{10}-8$	$3_{10}-9$	$5_{10}-9$	$9_{10}-9$	$9_{10}-9$
2	-2	δH_1	$9_{10}-3$	$2_{10}-6$	$2_{10}-10$	$2_{10}-10$	$2_{10}-10$	$6_{10}-10$
1	-1	δH_0	$3_{10}-6$	$1_{10}-10$	$4_{10}-11$	$4_{10}-10$	$1_{10}-9$	$9_{10}-10$
1	-1	δH_1	$2_{10}-4$	$2_{10}-9$	0	$7_{10}-11$	$6_{10}-11$	$6_{10}-11$
3	-2	δH_0	$2_{10}-2$	$9_{10}-6$	$6_{10}-9$	$1_{10}-8$	$2_{10}-8$	$3_{10}-8$
3	-2	δH_1	$1_{10}-1$	$1_{10}-4$	$3_{10}-8$	$2_{10}-10$	$6_{10}-10$	$1_{10}-9$
2	-1	δH_0	$1_{10}-3$	$1_{10}-7$	$1_{10}-9$	$4_{10}-10$	$5_{10}-9$	$1_{10}-9$
2	-1	δH_1	$1_{10}-2$	$2_{10}-6$	$1_{10}-10$	$2_{10}-10$	$6_{10}-11$	$3_{10}-10$
1	0	δH_0	$3_{10}-6$	$1_{10}-10$	$4_{10}-10$	$3_{10}-10$	$4_{10}-10$	$3_{10}-9$
1	0	δH_1	$2_{10}-4$	$2_{10}-9$	$8_{10}-11$	$2_{10}-10$	$3_{10}-10$	$4_{10}-10$
-1	2	δH_0	$3_{10}-6$	$4_{10}-10$	$3_{10}-10$	$1_{10}-9$	$4_{10}-9$	$4_{10}-9$
-1	2	δH_1	$1_{10}-4$	$2_{10}-9$	$1_{10}-10$	$2_{10}-10$	$1_{10}-10$	$8_{10}-10$
-2	3	δH_0	$6_{10}-4$	$6_{10}-8$	$4_{10}-9$	$1_{10}-8$	$6_{10}-8$	$6_{10}-8$
-2	3	δH_1	$7_{10}-3$	$2_{10}-6$	$2_{10}-10$	$1_{10}-10$	$2_{10}-10$	$1_{10}-9$

taken as $f(x)$. In Table III are listed the differences between the analytic and computed values for

$$H_0(0) = H_1(0) = 0;$$
$$H_0(1) = .0452492232556, \quad H_1(1) = -0.565159103991. \tag{4.13}$$

The results show again the increase of scatter as N exceeds an optimal value, more markedly so for H_0 than for H_1.

V. CONCLUSIONS

It has been shown in this paper that generalized Lobatto quadrature formulas with multiple terminal nodes (1.9) can serve to approximate real-valued integrals of all types (1.1)–(1.7), provided only the condition (1.10) is satisfied. For branch-points compatible with (1.1b) and $h(x) \equiv 1$ special cases have long been known as alternatives to the plain Gauss formulas. Thus the standard Lobatto quadrature corresponds to

$$\alpha = \beta = 0, \quad L = M = 1, \quad A_0 = B_0 = c/[(N+1)(N+2)] \tag{5.1}$$

and the method of Stancu and Stroud (1963) to

$$\alpha = \beta = 0, \quad L = M = 2, \quad A_1 = -B_1 = 2c^2/(N+1)_4$$
$$A_0 = B_0 = 4c[\tfrac{2}{3}N(N+5)+3]/(N+1)_4. \tag{5.2}$$

For:

$$\alpha = \beta = -\tfrac{1}{2}, \quad L = M = 1, \quad A_0 = B_0 = \pi/(2N+2) \qquad (5.3)$$

one obtains the shifted Chebyshev quadrature where the internal abscissas are given by the zeros of the Chebyshev polynomials of the second kind, and the terminal weights are equal to half the internal weights.

The advantage of the new approach is that it is equally applicable to integrals not described by (1.1). However in such cases the positive-definiteness of the coefficients is lost, for it is clear that the integrals, taken as functions of the exponents, have simple poles whenever α or β are equal to a negative integer. Since the internal weights C_ν vary continuously with α and β this singularity must be reflected in at least one of the coefficients A_λ or B_μ, with a consequent change of sign on passing through the pole. Another new feature is that the magnitude of the coefficients will increase as the number N of internal quadrature points is raised. Application of Stirling's formula to (3.9) and (3.10) and to the recurrence relations (2.12), (2.13) shows that

$$A_\lambda \sim N^{-2(\alpha+\lambda+1)}, B_\mu \sim N^{-2(\beta+\mu+1)}; \qquad (5.4)$$

to the extent that $\alpha + \lambda$ or $\beta + \mu$ are smaller than -1, the coefficients will increase more rapidly with N, with a corresponding reduction in the numerical accuracy. This is borne out by the results shown in Tables II and III, where the errors in H_0 increase more markedly than those in H_1; the effective exponents in H_1 are $-3/2$, compared with $-5/2$ for H_0.

Applications of the new approach to the evaluation of multi-centre integrals occurring in molecular quantum mechanics will be published elsewhere.

ACKNOWLEDGEMENT

One of us (S. J. M.) wishes to thank the Science Research Council for a grant (B/SR 7480).

REFERENCES

1. Abramowitz, M. and Stegun, I. A. (editors) 1965. *Handbook of Mathematical Functions*. New York: Dover.
2. Gautschi, W. 1968. Construction of Gauss-Christoffel quadrature formulas. *Math. Comp.* **22**, 251–270.
3. Gautschi, W. 1970. On the construction of Gaussian quadrature rules from modified moments. *Math. Comp.* **24**, 245–260.

4. Golub, G. H., and Welsch, J. H. 1969. Calculation of Gauss quadrature rules. *Math. Comp.* **23**, 221–230.
5. Kopal, Z. 1961. *Numerical Analysis.* 2nd edition. London: Chapman and Hall.
6. Lanczos, C. 1957. *Applied Analysis.* London: Pitman.
7. Reinsch, C. H. 1971. A stable rational QR algorithm for the computation of the eigenvalues of an Hermitian tridiagonal matrix. *Math. Comp.* **25**, 591–597.
8. Sack, R. A. 1962. Newton-Cotes type quadrature formulas with terminal corrections. *Comp. J.* **5**, 230–237.
9. Sack, R. A. 1972. A fully stable rational version of the QR algorithm for tridiagonal matrices. *Numer. Math.* **18**, 432–441.
10. Sack, R. A. 1974. *Com. A.C.M.* To be published.
11. Sack, R. A., and Donovan, A. F. 1972. An algorithm for Gaussian quadrature given modified moments. *Numer. Math.* **18**, 465–478.
12. Stancu, D. D. and Stroud, A. H. 1963. Quadrature formulas with simple Gaussian nodes and multiple fixed nodes. *Math. Comp.* **17**, 384–394.
13. Wilf, H. S. 1962. *Mathematics for the Physical Sciences.* New York: Wiley.
14. Wilkinson, J. H. 1958. The calculation of the eigenvectors of codiagonal matrices. *Comp. J.* **1**, 90–96.

Spline Functions and Differential Equations* — First Order Equations

I. J. SCHOENBERG

Mathematics Research Center
University of Wisconsin—Madison, Wisconsin, U.S.A.

I. A REMARK DUE TO C. LANCZOS

One of the usual ways of solving numerically a differential equation (D.E.) is the so-called collocation method. It consists in requiring functions of a certain family to satisfy the D.E. at prescribed isolated points. Lanczos shows the inadequacy of its uncritical application by the following remark (see [3]). Let

$$y' = f(x, y) \tag{1.1}$$

be a D.E. and let us try to approximate its solution

$$y = y(x), \ (0 \leqslant x \leqslant l), \text{ such that } y(0) = y_0 \text{ is prescribed.} \tag{1.2}$$

Let

$$x_\nu = \nu h, \ (\nu = 0, \cdots, n), \quad \text{where } h = l/n. \tag{1.3}$$

We first recall that the D.E. (1.1) determines a field of directions, i.e. it attaches to every point (x, y) a line-element

$$E_1(x, y) = (x, y; f(x, y)). \tag{1.4}$$

Let

$$y_1, y_2, \cdots, y_n \text{ be arbitrarily fixed quantities,} \tag{1.5}$$

and let us construct a function $S(x)$, $(0 \leqslant x \leqslant l)$, satisfying the following two conditions:

$$S(x_\nu) = y_\nu, \qquad (\nu = 0, \cdots, n), \tag{1.6}$$

$$S'(x_\nu) = f(x_\nu, S(x_\nu)), \qquad (\nu = 0, \cdots, n). \tag{1.7}$$

A function $S(x)$ satisfying these conditions is easily constructed as

*Sponsored by the United States Army under Contract No.: DA-31-124-ARO-D-462.

follows. The points (x_ν, y_ν) determine the line-elements

$$E_1(x_\nu, y_\nu), \qquad (\nu = 0, \cdots, n). \tag{1.8}$$

Any two consecutive among them, $E_1(x_{\nu-1}, y_{\nu-1})$ and $E_1(x_\nu, y_\nu)$ say, may be interpolated, by 2-point Hermite interpolation, producing a cubic polynomial defined in the interval $[x_{\nu-1}, x_\nu]$. The sequence of n cubic polynomials defines a composite function $S(x)$ that satisfies, by construction, the conditions (1.6) and (1.7). $S(x)$ may also be described as a cubic spline of the class $C^1[0, l]$ having knots at $x_1, \cdots, x_{n-1}$. In view of (1.5) $S(x)$ can in no way be regarded as an approximation to the solution (1.2). This is in substance Lanczos' remark, which he also extends to qth-order differential elements ($q \geqslant 1$) (see Section II below).

The cubic spline $S(x)$ just constructed depends evidently on the n arbitrary parameters (1.5), and certain n further conditions are needed to turn $S(x)$ into a good approximation of (1.2). We point out that this will be achieved if we require that

$$S(x) \text{ be a } quadratic \text{ spline, rather than cubic.} \tag{1.9}$$

Indeed, writing $y'_\nu = f(x_\nu, y_\nu)$, the cubic arc in $[x_{\nu-<}, x_\nu]$ will be quadratic iff *the two tangents at its endpoints,*

$$y = y_{\nu-1} + y'_{\nu-1}(x - x_{\nu-1}) \text{ and } y = y_\nu + y'_\nu(x - x_\nu), \tag{1.10}$$

intersect on the vertical line $x = (x_{\nu-1} + x_\nu)/2$. This will be the case provided that the right sides in (1.10) become equal if we substitute $x - x_{\nu-1} = h/2$ and $x - x_\nu = -h/2$. This leads to the equations

$$y_\nu - y_{\nu-1} = \frac{h}{2}(y'_{\nu-1} + y'_\nu),$$

or

$$y_\nu - y_{\nu-1} = \frac{h}{2}(f(x_{\nu-1}, y_{\nu-1}) + f(x_\nu, y_\nu)), \qquad (\nu = 0, \cdots, n), \tag{1.11}$$

which furnish successively the values of $y_1, \cdots, y_n$. This is the classical *trapezoidal method* for the numerical solution of differential equations.

The last paragraph shows the advantage of the spline function approach over the usual discrete variable methods. They furnish smooth approximations (the splines) which are valuable whenever these approximations are needed for subsequent analytical processes requiring smooth functions (see the Introduction of my 1946 paper[9]). The simultaneous approximation of the derivatives (Theorem 2 below) should be particularly stressed.

The purpose of this paper is to extend the result concerning the trapezoidal rule so as to obtain spline approximations of degree $2q$ and class $C^q[0, l]$. Just as the case when $q = 1$ reduced to the trapezoidal rule (1.11), so will the results for the case of a general $q > 1$ reduce to a class of methods due to Milne[7]. For this reference I am indebted to Ben Noble. (See also [8, Chap. 5]). The new element is the spline approximation $S(x)$ that we obtain. The main results below were obtained in the Fall of 1966 and are to be found in Loscalzo's thesis[5] which is now out of print (see also[4] and[6], also for further references).

In a second paper to follow I shall try to show how the spline function approach systematizes a number of multistep methods that have been used for the integration of second and higher order differential equations. It will also lead in a natural way to new ones.

In [5,67–75] there is a result for analytic differential equations. It is shown that $S(x)$ converges to the solution $y(x)$ for fixed h, as $q \to \infty$. A more adequate setting for this should be the complex plane and this may be carried out on a future occasion.

II. THE SPLINE APPROXIMATIONS OF HIGHER DEGREE

Lanczos generalizes his remark in a natural way. Let q be a natural number. The D.E. (1.1) determines at every point (x,y) a differential element of order q

$$E_q(x,y) = (x,y;y',y'',\cdots,y^{(q)}), \tag{2.1}$$

where the $y^{(j)}$ are computed from (1.1) by successive total differentiations of its right side. Thus $y' = f(x,y) \equiv g_1(x,y)$, while

$$y'' = f_x + f_y y' = f_x(x,y) + f_y(x,y)f(x,y) \equiv g_2(x,y)$$

and generally

$$y^{(j)} = \frac{d}{dx} g_{j-1}(x;y) = g_{j-1,x} + g_{j-1,y}f \equiv g_j(x,y) \qquad (j = 2,\cdots,q). \tag{2.2}$$

In terms of the functions $g_j(x,y)$ we may write (2.1) explicitly as

$$E_q(x,y) = (x,y; g_1(x,y),\cdots,g_q(x,y)). \tag{2.3}$$

It is useful to use the following terminology. We say that the function $S(x)$ satisfies (1.1) q times at the point $q = \xi$, provided that

$$S^{(j)}(\xi) = g_j(\xi, S(\xi)) \quad \text{for} \quad j = 1,\cdots,q. \tag{2.4}$$

Clearly $S(x)$ satisfies (1.1) q times at $x = \xi$ if

$$S'(x) - f(x, S(x)) = O(x-\xi)^q) \quad \text{as} \quad x \to \xi. \tag{2.5}$$

Again we select the y_ν as in (1.5) and consider the differential elements

$$E_q(x_\nu, y_\nu) = (x_\nu, y_\nu; g_1(x_\nu, y_\nu), \cdots, g_q(x_\nu, y_\nu)), \quad (\nu = 0, \cdots, n). \quad (2.6)$$

Any two consecutive elements,

$$E_q(x_{\nu-1}, y_{\nu-1}) \quad \text{and} \quad E_q(x_\nu, y_\nu), \quad (2.7)$$

can be interpolated within the interval $[x_{\nu-1}, x_\nu]$ by a polynomial of degree $\leq 2q+1$ by 2-point Hermite interpolation. The composite function $S(x)$ so obtained is a spline function of degree $2q+1$. By construction $S(x)$ enjoys the two properties:

$$S(x) \in C^q[0,l], \quad (2.8)$$

$$S(x) \text{ satisfies } (1.1) \, q \text{ times at } x = 0, h, \cdots, nh = l. \quad (2.9)$$

As the y_ν are arbitrary, it is clear that $S(x)$ is as yet no useful approximation of (1.2). We will show that $S(x)$ turns into a useful approximation if we require that

$$S(x) \text{ be a spline function of degree } 2q \text{ (rather than } 2q+1). \quad (2.10)$$

III. THE CONSTRUCTION OF THE SPLINE APPROXIMATION $S(x)$

In the case when $q = 1$ we could invoke the tangential property of the parabola (due to the Greeks) to obtain the trapezoidal rule. The case $q > 1$ requires results obtained by Hermite nearly a hundred years ago. He found [1] for the divided difference of order $2q + 1$

$$\overbrace{}^{q+1} \quad \overbrace{}^{q+1}$$
$$F(\alpha,\alpha,\cdots, \alpha, \; \beta,\cdots, \beta)$$

based on the points α, β, each of multiplicity $q+1$, the explicit expression

$$F(\alpha, \cdots, \alpha, \beta, \cdots, \beta) = (-1)^q \binom{2q}{q} h^{-2q-1} H_q(F; \alpha, \beta), \quad (h = \beta - \alpha), \quad (3.1)$$

where

$$H_q(F; \alpha, \beta) = F(\beta) - F(\alpha) - c_{1q}h(F'(\alpha) + F'(\beta)) - c_{2q}h^2(F''(\alpha)$$
$$- F''(\beta)) - \cdots - c_{qq}h^q(F^{(q)}(\alpha) + (-1)^{q-1}F^{(q)}(\beta)). \quad (3.2)$$

The coefficients $c_{\nu q}$ are the rational numbers

$$c_{\nu q} = \frac{1}{\nu!} \frac{q(q-1) \cdots (q-\nu+1)}{2q(2q-1) \cdots (2q-\nu+1)}, \quad (\nu = 1, \cdots, q). \tag{3.3}$$

Notice that $c_{1q} = \frac{1}{2}$.

These formulae will be applied only to functions $F(x)$ that satisfy (1.1) q times at $x = \alpha$ and $x = \beta$, when

$$F^{(\nu)}(\alpha) = g_\nu(\alpha; F(\alpha)), \quad F^{(\nu)}(\beta) = g_\nu(\beta, F(\beta)).$$

For convenience we introduce the two functions

$$\Phi(x, y) = y + \tfrac{1}{2}hg_1(x, y) + c_{2q}h^2 g_2(x, y) + \cdots + c_{qq}h^q g_q(x, y), \tag{3.4}$$

$$\Psi(x, y) = y - \tfrac{1}{2}hg_1(x, y) + c_{2q}h^2 g_2(x, y) - \cdots (-1)^q c_{qq}h^q g_q(x, y), \tag{3.5}$$

and (3.2) is seen to become

$$H_q(F; \alpha, \beta) = \Psi(\beta, F(\beta)) - \Phi(\alpha, F(\alpha)), \tag{3.6}$$

where $h = \beta - \alpha$.

We apply these formulae to the spline function $S(x)$ for $\alpha = x_{\nu-1}$, $\beta = x_\nu$, which is legitimate in view of (2.9). From the requirement (2.10), the polynomial component of $S(x)$ that interpolates the elements (2.7) is to be of degree $< 2q + 1$ and this is expressed by the vanishing of the divided difference

$$S(x_{\nu-1}, \cdots, x_{\nu-1}, x_\nu \cdots, x_\nu). \tag{3.7}$$

Writing

$$\varphi_\nu(y) = \Phi(x_\nu, y), \quad \psi_\nu(y) = \Psi(x_\nu, y) \tag{3.8}$$

and

$$s_\nu = S(x_\nu) \tag{3.9}$$

(formerly we wrote $y_\nu = S(x_\nu)$), we may express the vanishing of (3.7) by the relation

$$\psi_\nu(s_\nu) - \varphi_{\nu-1}(s_{\nu-1}) = 0 \quad (\nu = 1, \cdots, n). \tag{3.10}$$

These are the relations that allow us to compute the s_ν successively, because the first relation $\psi_1(s_1) - \varphi_0(y_0) = 0$ will give s_1, then the second relation $\psi_2(s_2) - \varphi_1(s_1) = 0$ will produce s_2, a.s.f.

A few pertinent questions arise:

Does the solution $s_\nu = S(x_\nu)$ of the system (3.10) exist, and is it uniquely defined by (3.10)? (3.11)

If it does, and it produces the spline $S(x) = S_n(x)$, how does $S_n(x)$ behave as $n \to \infty$? (3.12)

These and other questions will be answered below under appropriate assumptions.

IV. THE UNIQUE EXISTENCE OF $S_n(x)$
AND AN ERROR ESTIMATE

At this point we need some assumptions. In the first place we assume that the equation

$$y' = f(x, y) \qquad (4.1)$$

has a continuous solution

$$y(x) \ (0 \leqslant x \leqslant l), \text{ such that } y(0) = y_0. \qquad (4.2)$$

Secondly, we assume that

$$f(x, y) \in C^{2q+1} \text{ if } (x, y) \in D_H, \qquad (4.3)$$

where

$$D_H = \{(x, y); 0 \leqslant x \leqslant l, |y - y(x)| \leqslant H\}. \qquad (4.4)$$

Our first result is

Theorem 1.
There are positive constants h_0 and K_2 such that if

$$0 < h \leqslant h_0 \text{ and } n = l/h \text{ is an integer,} \qquad (4.5)$$

then the spline approximation $S(x)$ of Section III exists uniquely and satisfies the inequality

$$|y(x) - S(x)| < K_2 h^{2q} \leqslant H \text{ if } 0 \leqslant x \leqslant l. \qquad (4.6)$$

We add the obvious remark that the error function

$$E(x) = S'(x) - f(x, S(x))$$

is of the class $C^{q-1}[0, l]$ and has q-fold zeros at the grid-points x_ν.

Proof: We begin by inspecting the functions $g_\nu(x, y)$ defined by (2.2) for $\nu = 1, \cdots, q$. These are seen to be polynomials in the partial derivatives of f of orders not exceeding $q - 1$. By (4.3) the $g_\nu(x, y)$ have continuous partial derivatives. Let

$$L_\nu = \max \left| \frac{\partial g_\nu}{\partial y} \right| \text{ for } (x, y) \in D_H \qquad (4.7)$$

and therefore

$$|g_\nu(x, \eta_2) - g_\nu(x, \eta_1)| \leqslant L_\nu(\eta_2 - \eta_1)$$

$$\text{if } y(x) - H \leqslant \eta_1 < \eta_2 \leqslant y(x) + H. \qquad (4.8)$$

LEMMA 1. *If we select h^* such that*

$$0 < h \leqslant h^* \qquad (4.9)$$

this implies the two inequalities

$$1 + \tfrac{1}{2} hL_1 + c_{2q}h^2L_2 + \cdots + c_{qq}h^qL_q \leqslant 1 + hL_1 = \gamma, \tag{4.10}$$

$$1 - \tfrac{1}{2} hL_1 - c_{2q}h^2L_2 - \cdots - c_{qq}h^qL_q \geqslant 1/(1 + hL_1) = \gamma^{-1}, \tag{4.11}$$

then (4.9) *also implies that*

$$\gamma^{-1} \leqslant \{\Phi(x, \eta_2) - \Phi(x, \eta_1)\}/(\eta_2 - \eta_1) \leqslant \gamma, \tag{4.12}$$

and

$$\gamma^{-1} \leqslant \{\Psi(x, \eta_2) - \Psi(x, \eta_1)\}/(\eta_2 - \eta_1) \leqslant \gamma, \tag{4.13}$$

whenever

$$0 \leqslant x \leqslant l, y(x) - H \leqslant \eta_1 < \eta_2 \leqslant y(x) + H. \tag{4.14}$$

PROOF OF LEMMA 1: From (3.4), (4.8) and (4.11) we obtain

$$\Phi(x, \eta_2) - \Phi(x, \eta_1) \geqslant \eta_2 - \eta_1 - \sum h^\nu c_{\nu q} |g_\nu(x, \eta_2) - g_\nu(x, \eta_1)|$$
$$\geqslant \left(1 - \sum h^\nu c_{\nu q} L_\nu\right)(\eta_2 - \eta_1) \geqslant \gamma^{-1}(\eta_2 - \eta_1),$$

and using (4.10), also that

$$\Phi(x, \eta_2) - \Phi(x, \eta_1) \leqslant \left(1 + \sum h^\nu c_{\nu q} L_\nu\right)(\eta_2 - \eta_1) \leqslant \gamma(\eta_2 - \eta_1).$$

Evidently, the same proof holds also for $\Psi(x, y)$.

The lemma implies that the functions (3.8), in particular $\psi_\nu(y)$, are strictly increasing. Therefore the equation (3.10) has at most a single solution s_ν, if one exists at all. There remains the question of existence.

For this purpose we use the fact that also the exact solution $y(x)$ of (4.1) also satisfies the equations (3.10) if only approximately. Writing

$$y_\nu = y(x_\nu), \ (\nu = 0, \ldots, n), \tag{4.15}$$

we again apply the relation (3.6), this time for $F(x) = y(x)$, $\alpha = x_{\nu-1}$, $\beta = x_\nu$. For the divided difference (3.1) we use the mean-value theorem

$$y(x_{\nu-1}, \ldots, x_{\nu-1}, x_\nu, \ldots, x_\nu) = \frac{1}{(2q+1)!} y^{(2q+1)}(\xi_\nu), \ (x_{\nu-1} < \xi_\nu < x_\nu). \tag{4.16}$$

Now (3.6), in view of our notations (3.8), shows that

$$\psi_\nu(y_\nu) - \varphi_{\nu-1}(y_{\nu-1}) = R_\nu \tag{4.17}$$

where

$$R_\nu = (-1)^q \frac{(q!)^2}{(2q)!(2q+1)!} h^{2q+1} y^{(2q+1)}(\xi_\nu). \tag{4.18}$$

Writing

$$M_{2q+1} = \max_x |y^{(2q+1)}(x)| \tag{4.19}$$

and

$$\sigma = \frac{(q!)^2}{(2q)!(2q+1)!} M_{2q+1} \tag{4.20}$$

we conclude from (4.18) that

$$|R_\nu| \leq \sigma h^{2q+1}, \qquad (\nu = 1, \ldots, n). \tag{4.21}$$

We are now able to estimate the errors

$$e_\nu = y_\nu - s_\nu = y(x_\nu) - S(x_\nu), \qquad (\nu = 1, \ldots, n). \tag{4.22}$$

Indeed, from (3.10) and (4.17) we obtain

$$\psi_\nu(y_\nu) - \psi_\nu(s_\nu) = \varphi_{\nu-1}(y_{\nu-1}) - \varphi_{\nu-1}(s_{\nu-1}) + R_\nu, \tag{4.23}$$

hence

$$|\psi_\nu(y_\nu) - \psi_\nu(s_\nu)| \leq |\varphi_{\nu-1}(y_{\nu-1}) - \varphi_{\nu-1}(s_{\nu-1})| + |R_\nu|,$$

and using the properties (4.12), (4.13) of the function (3.8), we conclude that $\gamma^{-1}|e_\nu| \leq \gamma|e_{\nu-1}| + |R_\nu|$ and finally that

$$|e_\nu| \leq \gamma^2|e_{\nu-1}| + \gamma|R_\nu| \tag{4.24}$$

Explicitly, this means that

$$|e_1| \leq \gamma|R_1|,$$
$$|e_2| \leq \gamma^2|e_1| + \gamma|R_2|,$$
$$\vdots$$
$$|e_n| \leq \gamma^2|e_{n-1}| + \gamma|R_n|, \tag{4.25}$$

where we assume that

$$x_n = nh = l. \tag{4.26}$$

Multiplying the inequalities (4.25) in reverse order by $1, \gamma^2, \gamma^4, \ldots, \gamma^{2n-2}$ respectively, and adding them, we obtain

$$|e_n| \leq \gamma\{|R_n| + \gamma^2|R_{n-1}| + \cdots + \gamma^{2n-2}|R_1|\}. \tag{4.27}$$

Using (4.21) we can estimate the right hand side as follows

$$|e_n| \leq \gamma(1 + \gamma^2 + \cdots + \gamma^{2n-2})\sigma h^{2q+1} \leq \gamma \frac{\gamma^{2n}-1}{\gamma^2-1} \sigma h^{2q+1} < \frac{\gamma^{2n+1}}{\gamma^2-1} \sigma h^{2q+1}, \tag{4.28}$$

and in view of $\gamma = 1 + hL_1$ and (4.26)

$$\frac{\gamma^{2n+1}}{\gamma^2-1} \sigma < \frac{(1+hL_1)^{2n+1}}{2hL_1} \sigma = \frac{\sigma}{2hL_1}\left(1+\frac{lL_1}{n}\right)^{2n+1} < \frac{\sigma}{2hL_1}(e^{lL_1/n})^{2n+1}$$

$$= \frac{\sigma}{2hL_1} e^{2lL_1+(1/n)lL_1} \leq \frac{\sigma}{2hL_1} e^{3lL_1}.$$

From (4.28) we now obtain

$$|e_n| < \frac{\sigma}{2L_1} e^{3lL_1} h^{2q} = Kh^{2q}. \tag{4.29}$$

We further restrict h such that

$$Kh^{2q} \leqslant H \tag{4.30}$$

and rewrite (4.29) as

$$|y(x_\nu) - S(x_\nu)| < Kh^{2q} \quad (\nu = 0, \cdots, n). \tag{4.31}$$

These estimates also allow to show the existence of the solutions s_ν of the equations (3.10). However, we prefer to discuss the equivalent equations (4.23). Suppose that $s_0 = y_0, s_1, \cdots, s_{\nu-1}$ have already been determined ($\nu \leqslant n$) and we wish to show that (4.23) has a solution s_ν. As s_ν increases from $y_\nu - H$ to $y_\nu + H$, we know that the left side of (4.23) increases through a range of values that will certainly contain the interval $[-\gamma^{-1}H, \gamma^{-1}H]$, by Lemma 1. However, we already know that the right side does not exceed in absolute value the quantity

$$\gamma |e_{\nu-1}| + |R_\nu| \leqslant \gamma Kh^{2q} + \sigma h^{2q+1}.$$

Therefore s_ν, satisfying (4.23), must exist if $\gamma Kh^{2q} + \sigma h^{2q+1} \leqslant \gamma^{-1}H$, or

$$K(1 + hL_1)^2 h^{2q} + (1 + hL_1)\sigma h^{2q+1} \leqslant H.$$

This is clearly satisfied by a further appropriate adjustment of the constant K in (4.31).

We have yet to investigate the size of

$$\varphi(x) = y(x) - S(x) \tag{4.32}$$

for values of x between the points x_ν. For this purpose we apply Hermite's 2-point interpolation formula

$$F(t) = \sum_0^q \{F^{(j)}(0)l_j(t) + (-1)^j F^{(j)}(1)l_j(1-t)\}$$
$$+ t^{q+1}(t-1)^{q+1}F^{(2q+2)}(\eta)/(2q+2)!, \quad (0 < \eta < 1). \tag{4.33}$$

Let $x_{\nu-1} \leqslant x \leqslant x_\nu$ and let us apply (4.33) to

$$F(t) = \varphi(x_{\nu-1} + ht), \quad \text{where} \quad x = x_{\nu-1} + ht. \tag{4.34}$$

We obtain that

$$\varphi(x) = \sum_0^q h^j\{\varphi^{(j)}(x_{\nu-1})l_j(t) + (-1)^j\varphi^{(j)}(x_\nu)l_j(1-t)\} + R(x), \tag{4.35}$$

where

$$|R(x)| \leqslant \frac{h^{2q+2}}{2^{2q+2}(2q+2)!} \max_{[x_{\nu-1}, x_\nu]} |y^{(2q+2)}(x)|. \tag{4.36}$$

As both $y(x)$ and $S(x)$ satisfy our D.E. q times at x_ν we have

$$|\varphi^{(j)}(x_\nu)| = |y^{(j)}(x_\nu) - S^{(j)}(x_\nu)| = |g_j(x_\nu,y_\nu) - g_j(x_\nu,s_\nu)|$$
$$\leq L_j|y_\nu - s_\nu| \leq L_j Kh^{2q}, \tag{4.37}$$

by (4.8) and (4.30). Setting

$$Q_j = \max_{[0,1]} (|l_j(t)| + |l_j(1-t)|), \tag{4.38}$$

(4.35) and (4.37) yield

$$|y(x) - S(x)| < \left(\sum_0^q h^j Q_j L_j \right) Kh^{2q} + K_1 M_{2q+2} h^{2q+2}, \tag{4.39}$$

where

$$K_1 = \frac{1}{2^{2q+2}(2q+2)!}, \quad M_{2q+2} = \max_{[0,l]} |y^{(2q+2)}(x)|. \tag{4.40}$$

Clearly (4.39) implies the estimate (4.6) for an appropriate K_2. $\square$

V. ALL DERIVATIVES OF $S(x)$ APPROXIMATE THOSE OF $y(x)$

We keep the assumptions, notations and results obtained and wish to establish

Theorem 2.
There are positive constants A_r such that

$$|y^{(r)}(x) - S^{(r)}(x)| < A_r h^{2q-r} \quad in [0,1], \quad for \quad r = 0,1,\cdots, \ 2q. \tag{5.1}$$

If $q < r \leq 2q$, we mean by $S^{(r)}(x_\nu)$ either one of the two values $S^{(r)}(x_\nu \pm 0)$ $(0 < \nu < n)$.

PROOF: We write the interpolation formula (4.33) in the more explicit form

$$F(t) = \sum_0^q \{F^{(j)}(0)l_j(t) + (-1)^j F^{(j)}(1)l_j(1-t)\}$$
$$+ \int_0^1 K(t,v) F^{(2q+2)}(v) \, dv, \tag{5.2}$$

where, by Peano's theorem

$$K(t,v) \in C^{2q} \quad if \ (t,v) \in [0,1] \times [0,1].$$

Using the notation (4.32), we apply (5.2) to the function

$$F(t) = \varphi(x_{\nu-1} + ht),$$

so that

$$F^{(j)}(0) = h^j \varphi^{(j)}(x_{\nu-1}), \quad F^{(j)}(1) = h^j \varphi^{(j)}(x_\nu),$$

and finally

$$F^{(2q+2)}(v) = h^{2q+2}y^{(2q+2)}(x_{\nu-1}+hv),$$

because $S^{(2q+2)}(x_{\nu-1}+hv) = 0$ if $v \in (0,1)$.

Writing $x = x_{\nu-1} + ht$, and differentiating the resulting equation r times with respect to t, we obtain

$$h^r\varphi^{(r)}(x) = \sum_0^q h^j\{\varphi^{(j)}(x_{\nu-1})l_j^{(r)}(t) + (-1)^{j+r}\varphi^{(j)}(x_\nu)l_j^{(r)}(1-t))$$
$$+ h^{2q+2}\int_0^1 K_t^{(r)}(t,v)y^{(2q+2)}(x_{\nu-1}+hv)\,dv. \qquad (5.3)$$

Writing

$$Q_{j,r} = \max |l_j^{(r)}(t)| \text{ in } [0,1],$$

and using the estimates (4.29), we obtain from (5.3), that

$$|\varphi^{(r)}(x)| \leqslant \sum_{j=0}^q h^{2q+j-r}2KL_jQ_{j,r} + K_3h^{2q+2-r}.$$

The dominant term on the right side is obtained for $j = 0$ and the inequalities (5.1) are established. $\qquad\square$

Theorem 2 is decidedly an improvement over [5, Theorem 4.2 on page 60].

VI. EXAMPLES

Many examples are discussed by Loscalzo in [5, Chap. 5] as applications of his computer program SPLINDIF described in MRC Tech. Sum. Report No 842 of 1968. Here we discuss the simplest linear cases only.

1. Our first example is

$$y' = f(x), \quad y(0) = 0, \qquad (6.1)$$

hence the problem of approximating the integral of $f(x)$. The spline $S(x)$ must satisfy (6.1) q times at the knots $x_\nu = \nu h$ and this means that

$$S^{(j)}(x_\nu) = f^{(j-1)}(x_\nu), \qquad (j = 1, \cdots, q; \nu = 0,1, \cdots). \qquad (6.2)$$

Evidently

$$s(x) = S'(x) \qquad (6.3)$$

is a spline of degree $2q-1$, of class C^{q-1}, which is the unique solution of the spline interpolation problem

$$s^{(i)}(x_\nu) = f^{(i)}(x_\nu), \qquad (i = 0, \cdots, q-1), \text{ for all } \nu. \qquad (6.4)$$

This problem (its nodes are q-fold!) falls apart into successive 2-point

Hermite interpolation problems. The integral

$$S(x) = \int_0^x s(t)\,dt$$

is the solution of our original problem. This is an excellent method if a very accurate table of $S(x)$ is wanted and the requisite derivatives of $f(x)$ are available.

   2. We choose the equation

$$y' = y, \, y(0) = 1, \tag{6.5}$$

the corresponding spline approximation $S(x)$ being perhaps worthy of independent interest. Firstly, $S(x)$ must satisfy (6.5) q times at its knots, and this means that

$$S(\nu h) = S'(\nu h) = \cdots = S^{(q)}(\nu h) \text{ for all } \nu. \tag{6.6}$$

Secondly, on writing

$$\rho_q(h) = \frac{1 + \tfrac{1}{2}h + c_{2q}h^2 + \cdots + c_{qq}h^a}{1 - \tfrac{1}{2}h + c_{2q}h^2 - \cdots + (-1)^q c_{qq}h^a}, \tag{6.7}$$

we find that the fundamental relations (3.10) become

$$S(\nu h) = \rho_q(h)S((\nu - 1)h), \qquad (\nu = 1, 2, \ldots). \tag{6.8}$$

From (6.6) and (6.8) we find that each polynomial component of the spline $S(x)$ is obtained from its left neighboring component by shifting the latter to the right by the amount h and then multiplying it by the constant factor $\rho_q(h)$. It follows that $S(x)$ satisfies the functional equation

$$S(x + h) = \rho_q(h)S(x) \tag{6.9}$$

for all positive x. However, the method works as well for negative steps h, and (6.9) is seen to hold for all real x. Since $S(0) = 1$,

$$S(nh) = (\rho_q(h))^n. \tag{6.10}$$

If $x \neq 0$, then (6.10) will produce $S(x)$ if we choose $h = x/n$. This gives the rational function $(\rho_q(x/n))^n$ as an approximation to e^x. The relation

$$\lim_{n \to \infty} \left(\rho_q \left(\frac{x}{n} \right) \right)^n = e^x \tag{6.11}$$

actually holds uniformly in every bounded region of the complex plane, e.g. if we choose $q = 3$, $x = i$ and $n = 4$ we obtain by direct calculation

$$(\rho_3(i/4))^4 = 0.54030\,23081 + 0.84147\,09828\,i$$

which agrees with $e^i = \cos 1 + i \sin 1$ to eight decimal places in the real and the imaginary parts (See Lanczos' book [2, 425–426]).

As expected, there are connections with continued fractions. Since $S(h) = \rho_q(h)$, the first relation (4.25) and (4.21) show that

$$e^h = \rho_q(h) + \mathrm{O}(h^{2q+1}) \text{ as } h \to 0. \tag{6.12}$$

This shows that the rational function $\rho_q(x)$ must be connected in a simple way with Lambert's continued fraction

$$\frac{e^x - e^{-x}}{e^x + e^{-x}} = \frac{x|}{|1} + \frac{x^2|}{|3} + \frac{x^2|}{|5} + \frac{x^2|}{|7} + \cdots. \tag{6.13}$$

Indeed, if

$$\frac{e^{2x} - 1}{e^{2x} + 1} = \frac{A_q(x)}{B_q(x)} + \mathrm{O}(x^{2q+1}) \tag{6.14}$$

exhibits the qth convergent (or approximant) A_q/B_q of (6.13), then (6.12) shows that we must have the relation

$$\rho_q(x) = \frac{B_q(x/2) + A_q(x/2)}{B_q(x/2) - A_q(x/2)}.$$

REFERENCES

1. Hermite, Ch., Sur la formule d'interpolation de Lagrange, *J. für Reine u. Angew. Math.*, 84 (1878), 70–79; *Oeuvres*, 3, 432–443.
2. Lanczos, C., *Applied Analysis*, Englewood Cliffs, N.J., 1956.
3. ———, Solution of ordinary differential equations by trigonometric interpolation, *Proceedings of the PICC Symposium, Rome 1960* (Birkhäuser Verlag), 22–32.
4. Loscalzo, F. R. and Schoenberg, I. J., *On the use of spline functions for the approximation of solutions of ordinary differential equations*, MRC Tech. Sum. Report No. 723, Madison, January 1967, 8 pages.
5. Loscalzo, F. R., *On the use of spline functions for the numerical solution of ordinary differential equations*, MRC Tech. Sum. Report No. 869, Madison, May 1968.
6. ———, An introduction to the application of spline functions to initial value problems, 37–64 in "*Theory and Applications of Spline Functions*", edited by T. N. E. Greville, Academic Press, New York-London, 1969.
7. Milne, W. E., A note on the numerical integration of differential equations. *J. Res. of the Nat. Bur. of Standards*, 43 (1949), 537–542.
8. A. Ralston, *A first course in Numerical Analysis*, McGraw-Hill Book Co., New York, 1965.
9. Schoenberg, I. J., Contributions to the problem of approximation of equidistant data by analytic functions, *Quart. Appl. Math.* **44**, (1946), 45–99 and 112–141.

Added in proof. July 1973:

1. I am indebted to J. Barkley Rosier for his constructive criticism which helped to tighten the proof of Theorem 1.
2. Dr. Loscalzo informs me that he made extra copies of his paper [5] which are available on request by writing to Dr. F. R. Loscalzo, 2 Ivy Lane, Montvale, New Jersey 07645., U.S.A.

author index

Subject Index